AF844764

Mechanical Engineering Series

Frederick F. Ling
Editor-in-Chief

The Mechanical Engineering Series features graduate texts and research monographs to address the need for information in contemporary mechanical engineering, including areas of concentration of applied mechanics, biomechanics, computational mechanics, dynamical systems and control, energetics, mechanics of materials, processing, production systems, thermal science, and tribology.

Advisory Board/Series Editors

Applied Mechanics	F.A. Leckie University of California, Santa Barbara
	D. Gross Technical University of Darmstadt
Biomechanics	V.C. Mow Columbia University
Computational Mechanics	H.T. Yang University of California, Santa Barbara
Dynamic Systems and Control/ Mechatronics	D. Bryant University of Texas at Austin
Energetics	J.R.Welty University of Oregon, Eugene
Mechanics of Materials	I. Finnie University of California, Berkeley
Processing	K.K. Wang Cornell University
Production Systems	G.-A. Klutke Texas A&M University
Thermal Science	A.E. Bergles Rensselaer Polytechnic Institute
Tribology	W.O. Winer Georgia Institute of Technology

For further volumes:
http://www.springer.com/series/1161

Vincenzo Vullo • Francesco Vivio

Rotors: Stress Analysis and Design

Springer

Vincenzo Vullo
Dip. di Ingegneria Industriale
University of Rome Tor Vergata
Rome
Italy

Francesco Vivio
Dip. di Ingegneria Industriale
University of Rome Tor Vergata
Rome
Italy

ISSN 0941-5122 ISSN 2192-063X (electronic)
ISBN 978-88-470-2561-5 ISBN 978-88-470-2562-2 (eBook)
DOI 10.1007/978-88-470-2562-2
Springer Milan Dordrecht Heidelberg London New York

Library of Congress Control Number: 2012939161

© Springer-Verlag Italia 2013
This work is subject to copyright. All rights are reserved by the Publisher, whether the whole or part of the material is concerned, specifically the rights of translation, reprinting, reuse of illustrations, recitation, broadcasting, reproduction on microfilms or in any other physical way, and transmission or information storage and retrieval, electronic adaptation, computer software, or by similar or dissimilar methodology now known or hereafter developed. Exempted from this legal reservation are brief excerpts in connection with reviews or scholarly analysis or material supplied specifically for the purpose of being entered and executed on a computer system, for exclusive use by the purchaser of the work. Duplication of this publication or parts thereof is permitted only under the provisions of the Copyright Law of the Publisher's location, in its current version, and permission for use must always be obtained from Springer. Permissions for use may be obtained through RightsLink at the Copyright Clearance Center. Violations are liable to prosecution under the respective Copyright Law.
The use of general descriptive names, registered names, trademarks, service marks, etc. in this publication does not imply, even in the absence of a specific statement, that such names are exempt from the relevant protective laws and regulations and therefore free for general use.
While the advice and information in this book are believed to be true and accurate at the date of publication, neither the authors nor the editors nor the publisher can accept any legal responsibility for any errors or omissions that may be made. The publisher makes no warranty, express or implied, with respect to the material contained herein.

Printed on acid-free paper

Springer is part of Springer Science+Business Media (www.springer.com)

Foreword

It is with great pleasure and lively interest that I introduce this new and unusual text.

Great pleasure, first because of my esteem and affection for Vincenzo Vullo, who was my first coworker at the Politecnico di Torino, where he began his academic career: a career I followed with considerable satisfaction as he advanced to become full professor, first at the Università di Napoli Federico II and later at the Università di Roma Tor Vergata. I have also had the pleasure of establishing a close professional relationship with Francesco Vivio who, in the two short years that I have known him and been able to observe his work with Vincenzo Vullo, which has been as synergistic as it has been profitable, has demonstrated all the qualities that I regard as worthy of the tradition of our school and my best coworkers.

Lively interest, first because the topic dealt with in this book is one that has always been dear to our school of *Progettazione Meccanica e Costruzione di Macchine* (Machine Design Theory and Methodology) in Torino, and thus carries on its aims and thinking; second, because mindful of another of Torino's great schools, that of mathematics, the book uses highly sophisticated analytical methods but never loses sight of their ultimate practical purpose in engineering applications and design. In this connection, it is with satisfaction that I can testify that the book has not been plucked out of thin air, but treads a methodological route based on analytical solutions that began in Torino with Giovannozzi in the 1950s, continued with my own contribution, in particular as regards the tabular computation methods fashionable at that time, and, immediately thereafter, with the contribution of my first coworker, who prepared the course notes for my classes. This route was to some extent interrupted in the 1980s (though in reality, Vincenzo Vullo amplified his course notes in that period for the classes he taught at University of Tor Vergata) when the finite element method burst onto the scene, sweeping everything and everyone before it, but was resumed at the beginning of this century as part of the renewed interest in the analytical methods that had briefly fallen out of favor.

It is not my intention here to ignore the importance, either of the book's opening chapter, which, on the basis of the concepts clearly set forth in the introduction, presents the governing equations and the advantages, as well as the limitations, of the mono-dimensional theory for a rotating disk subjected to a temperature gradient along the radius, or of the chapters dealing with the three rotating disk profiles whose differential equations can be solved in closed form. For these topics, which are those normally addressed in the standard texts, the authors provide a comprehensive coverage, enlivened by concrete, fully developed application examples that clarify the design process for these machine parts.

That said, however, I would like to draw the reader's attention to the authors' theoretical treatment of disks whose thickness varies according to functions that lead to governing differential equations that could not previously be integrated in closed form. Regarding the approach used for these questions, which is as complete as it is new and original (and, once again, is masterful from the analytical standpoint, but always firmly rooted in conceptual design, whose acknowledged virtues are raised to an even higher level, in a sort of Pindaric flight, by the authors' unusual mathematical approach), I would like to permit myself a brief historical survey spanning nearly a century of theoretical studies, with major contributions from the German and Italian schools.

The first theoretical analysis of solid or annular converging conical disks subjected to centrifugal loading and thus featuring a profile whose thickness varies with the radius according to a linear function was presented by Emil Von Honegger in 1927. Honegger's approach is based on the formulation developed by Ernst Meissner in 1913, which transformed the homogeneous second-order differential equation governing the stress and strain states of conical shells of revolution under axisymmetric loads, and whose thickness increases or decreases from the axis to the periphery, into a Gauss hypergeometric differential equation. Honegger returned to the problem of the conical shell with linearly varying thickness a few years later, and in 1919 presented the exact analytical solution of the Gauss-type equation. In 1927, Honegger extended this approach to rotating converging conical disks.

Both the theoretical work by Meissner and Honegger on conical shells and Honegger's work on rotating conical disks were based on Felix Klein's fundamental contributions to solving hypergeometric differential equations. In the course notes written for a class he taught at the university of Göttingen in 1894, whose content was then incorporated in the treatise on hypergeometric differential equations he published in 1933, Klein brought about a turning point in the treatment used for this type of equations, solving them via a linear combination of independent hypergeometric functions.

Later, in 1950, Renato Giovannozzi broadened Honegger's analysis, extending it to solid or annular converging conical disks subjected to thermal loads. With two further studies in 1956, Giovannozzi again increased the scope of his analysis to include solid or annular converging and diverging conical disks subjected to centrifugal and thermal load and featuring a fictitious variation of the density along the radius. Although it is not among the solutions involving the use of hypergeometric series, the contributions of the Torino school include one of my own studies in 1960 regarding the analysis of the centrifugal and thermal stresses in hyperbolic disks whose temperature and density vary along the radius according to entirely general polynomial functions. This study also outlined the tabular computation method, in line with that developed by Grammel, for disks divided into adjacent partial annular disks, and thus completed Giovannozzi's work on conical disks. Mention should also be made of the work done by Filippo Manna in 1968 on nonconventional disk profiles in simple rotation, which also continued to pursue the goals of the Torino school.

To reduce the complexity of the calculation procedure employed for the areas adjacent to the singular points in the interval of interest and thus adapt it to the computational tools available in their day, both Honegger and Giovannozzi split the interval of the independent variable into two contiguous partial intervals and, to express the general solution for the rotating conical disk in terms of radial displacement, use four hypergeometric functions combined in pairs to ensure the continuity of the displacement function and its first derivative at the point where the two partial intervals are interconnected. As the new computation method presented in this book demonstrates, the resulting method is extremely elaborate, with no advantage in terms of the accuracy of the results it yields.

After Honegger and Giovannozzi, noteworthy developments in the theory of disks subjected to centrifugal and thermal load and having profiles leading to differential equations that cannot be solved in closed form were published by Eraslan et al. starting in 2000. In addition to other considerations, which are in any case covered in this book, Eraslan arrives at analytical relations that are solved numerically. Unlike the purely theoretical and analytical methods discussed in this textbook, this is thus a hybrid method in which the governing differential equations must necessarily be solved by numerical means, and need not detain us further.

What I have highlighted in this historical survey is the core and foundation on which the authors base their original developments, which apply to:

- Conical disks subjected in the elastic field to centrifugal and thermal load and whose density is constant or varies along the radius
- Disks having thickness varying nonlinearly along the radius according to a power of a linear function, subjected in the elastic field to centrifugal and thermal load and whose density is constant or varies along the radius
- Disks having thickness varying nonlinearly along the radius, subjected in the elastic field to centrifugal and thermal load, whose density is constant or varies along the radius and which are constructed of strain-hardening material.

It should also be mentioned, however, that the purely theoretical treatment presented by the authors owes a considerable debt to the contributions made by Francesco Giacomo Tricomi in 1967 to the solution of hypergeometric differential equations or Gauss equations, which should be considered as important as those made by Klein. This work, as the authors themselves acknowledge, was decisive for their method of handling disks whose thickness varies linearly or nonlinearly.

For conical disks stressed in the elastic field, the solution furnished by Vullo and Vivio is simpler and less elaborate than that proposed by Honegger and Giovannozzi and consequently involves a much more straightforward approach to the calculation procedure. In addition, it generalizes Giovannozzi's solution both for thermal load and for the effect of the fictitious density variation along the radius.

The treatment for non-linearly variable profile disks generalizes that for conical disks, and is a particular case thereof. This treatment is especially valuable for specifying the optimized disk profile at the conceptual design stage (note that this is the noblest phase of the design), as it permits immediate comparison of the structural behavior of a wide range of profiles, including converging or diverging concave disks, solid and annular converging or diverging convex disks, and the particular cases of hyperbolic disks, conical disks, and uniform thickness disks. It should also be noted that the analytical solution can be further generalized, as the passage from the nonlinear profile defined by a power of a linear function to that defined by a power of a nonlinear function is immediate. Without introducing changes from the standpoint of the formalism expressing the relations used for solution, this further generalization broadens the horizons of design, as it is possible to describe profiles with an inflection point, such as uniform strength disks.

For rotating disks stressed beyond yielding, the authors start from the well-known simplified treatment for constant thickness disks consisting of elastic-perfectly plastic, i.e., non-hardening, material, discussing this treatment's limitations as well as its conceptual importance in understanding the physical phenomena associated with prestressing by overspeeding and the residual stress field that can thus be obtained. This simplified treatment is the basis for another of the bold flights that are such a feature of this book: a method for dealing with disks whose thickness varies nonlinearly according to a power of a linear function, consisting of a strain-hardening material loaded beyond yielding. This original analytical formulation employs the Von Mises yield criterion and the associated flow rule, and introduces a polynomial correlation that links equivalent plastic strain to equivalent stress and can approximate any experimental true stress–true strain curve with the desired margin of confidence.

This formulation, which yields solutions based on analytical relations, overcomes the limitations of the elastic-perfectly plastic model and permits a structural characterization of the material which is entirely consistent with the actual mechanical behavior of the generic material used for the disk. In this context, I have no doubt that the expert reader will appreciate the ample space devoted to two methods of enormous value in the history of rotor theory. Two entire sections of the book, in fact, are devoted respectively to Manson's finite difference method for determining the stress state in the elastic field, and to Millenson-Manson's finite difference method, in its more general formulation that takes plastic flow, creep, and the effects of prestressing by overspeeding into account.

Equally noteworthy is the chapter devoted to an analysis of rotating cylinders subjected to heat flow and stressed in the elastic field. This chapter also clarifies the effect of such cylinders' triaxial stress state as opposed to that of the biaxial state exhibited by a constant thickness disk with the same radial dimensions, and demonstrates that the differences between the results obtained using a biaxial model and a triaxial model are insignificant in the conceptual design of a turbine disk, whose axial dimension is small compared to its outside diameter. This aspect, which is extremely important from the design standpoint, is confirmed by the many calculation examples, where three-dimensional finite element models are also employed.

The calculation examples are a particularly valuable feature of this book. Many concern disks whose profile vary with the radius, while all show each step of the calculation using the analytical procedures preferred by these authors, verifying the results by comparison both with those in the literature, where available, and with those obtained with finite element models.

To conclude this foreword, I would like to credit the authors for having produced a book which embodies all of the rigor and concrete attention to structural analysis that distinguishes the Torino school, and which in a certain sense strips away the years and carries me back in time to a topic that, though challenging indeed, figured so prominently in the verdant spring of my academic career. I congratulate the authors for having been so successful in combining a rigorous analytical treatment with a lucid attention to practical utility in engineering, stimulating interest in a closer alliance between mathematics and structural design methods: an alliance in which neither is subservient to the other; both are on an equal footing, in the awareness that together they can make the designer's role even more valuable, enabling him to arrive at structures that are to a large extent optimized at the time of product conception and thus significantly reducing the amount of work that will be involved in subsequent detail design with finite element methods.

In this light, the book's merits are twofold. First, it puts high mathematical concepts at the foundation of rotor structural design, but does so in order that the use of higher mathematics in mechanical design can help us achieve an immediate grasp of the physical phenomena linked to strain mechanics that often remain elusive when numerical techniques such as the finite element method are used. Second, it steers the helm of mathematics toward solving complex problems of practical interest in engineering, and thus aims in a sense to instill mathematics with the engineering spirit. In all of this, the *two* authors – who are credible because they have long used the finite element method, and continue to use it daily – demonstrate that mathematics, the higher mathematics which can often be used to advantage in solving challenging structural problems, still has its place, even though finite element techniques are now so widely used and well established that they must be regarded as indispensable.

I am convinced that the book, which is a credit to the our school and is addressed to engineers and, more generally, to advanced scholars who are comfortable with the use of mathematics, and to Ph.D. and other postgraduate degree programs, will be welcomed and appreciated by a broad spectrum of academics and engineering practitioners.

Politecnico di Torino

Pasquale Mario Calderale
Professor of Mechanical
and Biomechanical Engineering

Preface

Stress and strain analysis in rotating disks and cylinders subjected to surface and body loads, as well as to thermal loads deriving from temperature variation along the radius, is a classic subject of machine design. It is a fascinating subject indeed, but also an extremely complex one, as the basic background of the elasticity theory applied to two- or three-dimensional structures is not sufficient for thorough analysis. In fact, it is often necessary to broaden the design spectrum in order to investigate what happens in the case of centrifugal loads that cause stresses beyond the material's yield point, or strong thermal fields that give rise to creep.

The elastic analysis of stresses and strains in rotors is usually the subject of several chapters in standard machine design textbooks. As a rule, these chapters focus on elastic problems for which an exact analytical solution in the form of algebraic formulas for the quantities of interest can be obtained, but seldom deal with inelastic or creep problems. Moreover, attention is limited to disks for which governing equations are solvable in closed form (constant-thickness, hyperbolic-profile, and uniform-strength disks). Only in a few texts is the conical disk taken into consideration, and always under elastic load.

However, very few actual disks have a geometry that is sufficiently simple for such solutions to be directly applicable. Normally, the geometry of rotors is far from that of an axisymmetrically loaded, rotating solid: this is dictated by design requirements. In such cases, numerical solution methods are clearly the only practicable ways for accurately defining the main quantities that are crucial to designers. The most widespread numerical technique for stress and strain analysis of such structures is the finite element method (FEM). Nevertheless, there are some cases where the finite difference method still has several advantages. The boundary integral equation or boundary element method (BEM) can also be used very effectively in a few specific cases.

Designers of disks and rotors are well aware that very few practical engineering issues involve structures for which stress and strain analysis in the linear elastic field and, to an even greater extent, under nonlinear conditions (i.e., plastic or viscoelastic conditions) produces equations to be solved in closed form in order to express relevant parameters through algebraic formulas.

However, formulating the problem in strict analytical terms, and in compliance with the physical phenomenon to be modeled, is essential for arriving at a full understanding of the problem from a continuum mechanics standpoint. Moreover, when a product is still in the design stage, such a formulation is simpler and more versatile than FEM or BEM numerical methods, and makes it possible to quickly define a general configuration which can then be fine-tuned using these numerical methods. Indeed, working with an analytical equation at this stage is much easier than working with a numerical model, even when the latter is expressed in a parametric form. Thus, a more rational design layout is obtained, which makes it possible to save on calculation resources and to streamline the process that leads to the final project.

In a setting where numerical methods, and FEM in particular, are used almost exclusively, a text that returns to an emphasis on theoretical/analytical methods cannot fail to come as something of a surprise. There are several reasons for choosing such an emphasis for the topic covered here, including the following:

- There can be no doubt that these methods are the foundation of numerical techniques. A better knowledge of the former can thus open up new prospects for the latter, while the coordinated use of both can contribute significantly to optimizing the structural calculation procedures used for rotors, bringing major computational advantages.
- Though FEM methods have many advantages during detail design, theoretical/analytical models are much more flexible and versatile during the conceptual development of a new rotor: not only are they entirely general, but they require at most that a differential equation be integrated using simple numerical analysis procedures in the cases where it cannot be solved in closed form.
- When theoretical/analytical models are used, rotor optimization (in terms of profile shape, mass and hence inertia, strength, etc.) is simpler than with FEM modeling which, despite the introduction of new and increasingly sophisticated procedures, still hinges on the initial discretization of the structure in question.
- The theoretical/analytical models presented here are entirely general, as they make it possible to address and solve any singularity problem, including that arising at the rotational axis. During conceptual design, this is another aspect of theoretical/analytical models that makes them highly advantageous from the computational standpoint.
- Lastly, a familiar fact should be borne in mind: the results obtained with FE models are verified experimentally, by creating a simple FEM model whose theoretical solution is known, or by direct comparison with the results that can be obtained with theoretical/analytical model.

It should be borne in mind, however, that the theoretical treatment based on the biaxial stress state leads to results that are more approximate than those of the finite element method, which can account for the effects of triaxial stress. Experimental evidences as well as direct comparison between bidimensional theoretical analysis and FE numerical results demonstrate that the differences are insignificant in the conceptual design of a turbine disk, whose axial dimension is small compared to its outside diameter.

Accordingly, the focus is maintained on analytical-methodological aspects in dealing with each subject. To ensure that analytical models can be even more readily used at the product design stage – thus allowing for extensive operational flexibility – a function is introduced whereby it is possible to define a fourfold infinity of disk profiles, solid or annular, concave or convex, converging or diverging. With this function, even constant-thickness disks or conical disks can be examined as particular cases. Most solutions are expressed in an analytical form, by using linear combinations of hypergeometric functions and particular integrals in closed form. However, actual engineering applications are considered in the analytical formulation of all problems.

To make this text as comprehensive as possible, numerical methods – especially those based on finite differences – are introduced and utilized. The finite element method is used only for verifying and/or validating the analytical approach in the various examples shown, which make the theoretical approach easier to understand and more practical for use in design. With the exception of a few instances, these example cases are developed in full, in order to provide designers with a detailed description of all successive subsequent steps of the analysis process, which is the basis for calculating and verifying disk profiles and configurations of actual practical interest.

This layout mirrors the course in Machine Design offered to final year students of Mechanical Engineering at the University of Rome Tor Vergata. However, this book covers a far broader range of topics than were dealt with in the course. From the outset, in fact, the intention was to provide a textbook for researchers and industry specialists facing actual issues pertaining to rotor conceptual design, as well as for advanced university courses, such as those for Ph.D.s and other postgraduate degrees. This work, which addresses the various subjects in increasing levels of complexity, also covers the basics, including application examples to provide a better understanding of each problem, and is thus also intended for university teaching at master's degree level.

This textbook consists of an introduction which describes and discusses general hypotheses and the associated assumptions underlying theoretical approaches and solution methods for disks, which apply both to elastic conditions and beyond yielding, plus 13 chapters.

The first nine chapters deal with stress analysis in rotating disks in the linear elastic field. More specifically, after outlining a few general considerations, these chapters address the following aspects: mono-dimensional theory of thin disks; equilibrium and compatibility equations; general differential equation for a rotating disk subjected to thermal load; solid and annular disks of constant thickness subjected to various loads – including thermal load – and featuring a fictitious density variation along the radius; hyperbolic-profile annular disk subjected to various loads including thermal load; uniform strength disk; solid and annular conical disks with either variable or constant density along the radius, subjected to various loads including thermal load; disk whose thickness varies according to a power of a linear function and having either variable or constant density along its radius, subjected to various loads including thermal load; disk of arbitrary profile:

Timoshenko-Grammel's method and Manson's method; disk of constant thickness and disk of variable thickness according to a power of a linear function, subjected to angular acceleration; and yield criteria for rotating disks and stress concentration.

Chapter 10 concerns stress analysis in rotating cylinders in the linear elastic field. Specifically, after outlining a few general notions, this chapter addresses the following subjects: fundamental governing equations; cylindrical solid, either with clamped ends or indefinitely extended in the direction of its axis, subjected to centrifugal and thermal loads; cylindrical solid of finite length with free ends, subjected to centrifugal and thermal loads (with edge effect stress on the free ends); and cylindrical solid rod of finite length with free ends, subjected to thermal transient load.

Chapters 11 and 12 deal with stress analysis of disks subjected to centrifugal loads that cause stresses beyond the material's yield point. More specifically, after outlining few necessary general ideas, these chapters analyze the following aspects: approximated basic hypotheses and their limitations regarding disks in non-hardening materials: Millenson-Manson's method as an extension of Manson's method from elastic load to elasto-plastic load; general analytical method; design-related use of plasticization and limit-design factor.

The last chapter contains three appendixes dealing respectively with: calculation of one-dimensional elements subjected to centrifugal load, such as rotating bars, paddles, and blades; in-depth analysis of the solution of the hypergeometric differential equation; and the finite element method for elasto-plastic problems.

To the authors' knowledge, a textbook such as this, which focuses specifically on the analysis and design of rotors (disks and cylinders), does not exist in the international technical literature. It should be pointed out that it not only covers "classic" subjects but also includes advanced scientific contributions by the authors. All these contributions are aimed at designing rotors at the concept stage, i.e., prior to detailed design. Special mention should be made of the following: linear elastic analysis of a conical disk subjected to thermal load and showing fictitious variable density along its radius, by implementing a procedure that simplifies and generalizes the solution put forth by Honegger and Giovannozzi; linear elastic analysis of a disk whose profile varies along the radius according to a power of a linear function, subjected to thermal load and with variable density, which makes the previous calculation method for conical disks more general; analysis of a variable-profile disk subjected to centrifugal load beyond the material's yield point, according to Von Mises' yield criterion, and introducing the completely general law $\sigma = \sigma(\varepsilon)$, expressed by an n-grade polynomial; linear elastic analysis of Stodola's hyperbolic disk, subjected to thermal load along its radius; and linear elastic analysis of a disk whose thickness varies according to a power of a linear function, subjected to angular acceleration.

Rome
March 2012

Vincenzo Vullo
Francesco Vivio

Contents

Introduction and Basic Hypotheses

Rotating members known as rotors are used in turbomachines and many power-generating or power-consuming thermal and hydraulic machines (turbines, turbopumps, turbochargers, centrifugal pumps, centrifugal compressors, fans, molecular pumps, centrifuges, ultracentrifuges, etc.), turbogenerators, turbo-alternators, turbo-dynamos and many synchronous and asynchronous electrical machines; certain types of aircraft (helicopters, autogyros or gyroplanes, gyrogliders, rotoplanes, etc.); propulsion units for turboprop, turbofan, and turbojet airplanes and those of rotor ships and turbine-driven tankers; the drives of various ground vehicles featuring mechanical energy storage (flywheels) and of wind generators; many machine tools (spindles, flywheels, etc.); as well as other power-consuming machines such as turbo extractors and turboexpanders.

The calculation and design problems presented by these rotating members are in general highly specific and are in many cases linked with the rotation of the structural element. A rotary element is subjected to a wide range of dissimilar load conditions. Some are independent of rotation (e.g., torque in drive mechanisms, axial thrust on turbine rotors, and so forth), while others are directly correlated with angular velocity and/or angular acceleration, e.g., stresses arising from centrifugal loading, angular acceleration and gyroscopic moments. For rotors such as turbine disks, brake disks, and the like, uneven temperature distributions often make operating conditions even more severe.

Technological advances have lead to higher and higher rotational speeds, whether in order to lower the mass of the rotating members, which can generally be achieved by reducing torque for any given amount of power transmitted (as is the case for drive mechanisms, but also for electric motors and generators, gas turbines used in aeronautical applications and turbopumps for spacecraft), to increase machine productivity (as for machine tools, textile machinery, and manufacturing machinery in general), or to reach the high peripheral velocities required by machines such as molecular pumps and ultracentrifuges for enriching uranium. On the other hand, the availability of heat-resistant materials and the need to improve the efficiency of thermal machines have resulted in increasingly demanding operating conditions, with elevated temperatures and high thermal gradients.

Together, these two trends toward higher speeds and higher temperatures make it essential that the stress and strain states of machine rotors be analyzed as rigorously as possible to ensure that parts can be correctly sized. They also require that materials be carefully selected, that production be subject to rigorous process controls, that the finished component be tested to failure in the prestressing equipment, that stringent maintenance requirements be scrupulously observed, and that appropriate nondestructive tests be carried out regularly during service. All of these results from the fact that these rotating members can be extremely hazardous: when they fail during service, they generally do so catastrophically, bursting apart to liberate ballistic projectiles whose mass is often one-third to one-half of that of the rotor. These projectiles travel at very high speeds, frequently reaching peripheral velocities in the order of several 100 m/s.

It should also be borne in mind that rotating machine parts should always be protected by appropriate guards, as they are hazardous not only in the event of catastrophic failure – though this should never occur during operation – but also in actual service conditions, when they could be contacted by machine operators or other personnel. In this connection, the designer must abide by all regulatory requirements in order to guarantee the necessary operating safety at all times and under all circumstances.

In the following pages, we will address the issues involved in structural calculation and conceptual design of rotating machine members subjected to loads arising from rotation (centrifugal loads and angular acceleration), as well as to nonuniform temperature distributions. The designer is not always in a position to choose design parameters freely, as they often depend on other functional or performance-related considerations (examples include pump impellers, where geometrical dimensions and angular velocity are basic design data, as is the form of many surfaces, and energy storage flywheels, where maximum and minimum velocities, the quantity of energy to be stored, and, in some cases, the package dimensions are all specified). Aside from simple cases where true design calculation is possible, it is generally necessary to proceed through successive approximations, verifying each step through calculations and gradually refining the design.

Discussion will be limited here to rotors, in steady state conditions (fatigue phenomena have not considered), in which biaxial and/or triaxial stress and strain states are generated by surface loads to the outer radius and inner radius, centrifugal loading, thermal loading, and that associated with angular acceleration. Calculations for rotors involving uniaxial structures will be covered in Appendix A, where procedures are presented for determining stresses and strains in bars, paddles, and blades subjected only to centrifugal load.

The stress state in most rotating members is three-dimensional, both because of their shape, which almost always is not of simple geometry, and because complex states of stress overlap. Consequently, it is necessary to use versatile and generally valid methods such as finite element analysis, which, though in no way difficult to apply from a conceptual standpoint, involves certain cost problems. At least in the early stages, i.e., those involved in so-called conceptual design, the structural engineer thus uses analytical models which, though of general validity, require that a certain number of simplifying assumptions be made in order to arrive at results that are meaningful and useable in actual applications. These assumptions reduce the problem's complexity, leading at times to closed form solutions.

The first simplifying assumption that is usually introduced is that of axial symmetry in the broad sense (generalized axisymmetry), i.e., referring not only to the rotor's geometrical shape but also to the distribution of body and surface forces, the thermal loads acting on the rotor, the boundary conditions, and the material's elastic properties. From the geometrical standpoint, however, this assumption is not always entirely justified. Take, e.g., the case of rotors featuring holes, blade attachments, and other disuniformities, all of which make axisymmetry a rather rough approximation. If the irregularities are minor compared to the problem's significant magnitudes, the designer first assumes that the influence on the rotor's stress state is negligible, and introduces it later as a localized effect in the areas affected by the irregularities. This is the procedure followed for holes, for which the notch effect is considered. Blades, on the other hand, and especially those that are closely spaced, are simulated either by means of an axisymmetric distribution of surface forces or by a fictitious increase in material density in the areas adjacent to those to which the blades are attached. As indicated above, the assumption of generalized axisymmetry also includes that of the material's mechanical properties, which must thus be axisymmetrically orthotropic or isotropic.

Introducing only the assumption of axisymmetry, through simplifying the problem, is not in itself sufficient to reach closed form solutions. Numerical methods are still required, though the procedures involved are far more straightforward and thus less costly. Further simplifying assumptions must be introduced: plane stress state ($\sigma_z = 0$) for solids of small axial thickness, plane strain state ($\varepsilon_z = 0$) for cylindrical solids constrained at the ends or infinitely extended along the axis, and strain state with $\varepsilon_z = const$ (the so-called generalized plane strain state) for cylindrical solids of finite length and free at the ends.

As illustrated in Fig. I.1, if the axial dimensions of the rotor are not too large (as for a thin rotor, thin disk, or simple disk) and the loads acting on it are contained within plane xy (it is thus further assumed that the disk's geometry, including the boundary conditions, the mechanical properties of the disk material, and the stresses on it are symmetrical with respect to the disk's mid-plane, at axial coordinate $z = 0$), it can be assumed that $\sigma_z = \tau_{xz} = \tau_{yz} = 0$ at every point of the rotor, meaning that the stress state is plane. Strictly speaking, this is true only where thickness h is constant and infinitesimal, although in many cases of interest in actual applications, the errors that result from using this assumption are sufficiently small that they can be considered negligible. Given the major simplifications it permits, this assumption is also frequently used for fairly thick rotors, with significant gradients in relation $h = h\ (r)$, and even when there are discontinuities in this relation. With this assumption, all stress characteristics are independent of the axial coordinate z, and two coordinates (r and ϑ) are sufficient to describe the phenomenon completely. In other words, we are dealing with a plane, i.e., two-dimensional, problem.

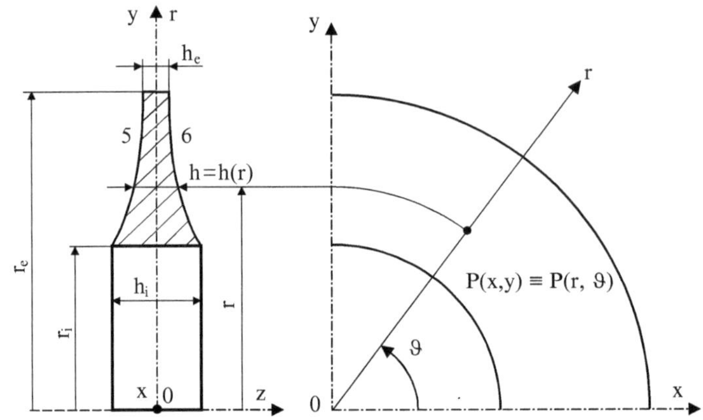

Fig. I.1 Cross section with an axial plane and side view of an axisymmetric rotor; geometry and coordinate axes are shown in a Cartesian reference system O(x, y, z) and in a superimposed polar reference system O(r, ϑ, z)

Where the assumptions of a plane stress state and generalized axisymmetry coexist, the problem becomes one-dimensional, and the stress and strain characteristics can be described by means of coordinate r alone.

In the following pages, it will be assumed that the material is homogeneous and isotropic unless otherwise noted. In some cases, discussion will be extended to axisymmetric orthotropic materials for the sake of greater generality. The cases in which the material works in the linear elastic field and in which it is loaded beyond yielding will be handled separately. In the latter case, and again unless otherwise noted, it will be assumed that the material shows non-hardening elastic-plastic behavior, or in other words, that behavior is perfectly elastic up to the yield stress (i.e., in this event, the limit of proportionality, the elastic limit, and the yield point coincide), and perfectly plastic, henceforth, up to failure. In specific cases, however, rather than using this restrictive assumption of a perfectly elastic-plastic material, we will introduce the general assumption of a strain-hardening material, i.e., a material whose $\sigma = \sigma(\varepsilon)$ curve (NB: true stress – true strain) is that actually plotted from the results of a tensile test on a standard specimen, and which can be modeled from one of the many strain-hardening laws found in the literature on the subject (Ludwik, Ramberg-Osgood, Swift, etc.).

A rotor to which the plane stress assumption can be applied is called a *disk* or *thin disk*. Obviously, the stress state calculated through the thin disk theory is an approximation of the real stress state, and the greater the thickness of the disk and the higher the thickness gradient along the radius, the rougher this approximation will be.

In a disk stressed symmetrically around its axis and loaded uniformly along the latter (here, assuming plane stress, we will have $\sigma_z = 0$), the radial, tangential, and axial directions will be the principal directions of both stresses and strains. From the design standpoint, most interest focuses on cases where strains are infinitesimal (the small strain assumption) and those where strains have a nonnegligible finite value (the large strain assumption).

Below, it will be assumed that the strains are small, and that the associated field is thus described by conventional (or engineering) strains ε_r, ε_t, and ε_z. In cases involving large strains, it will clearly be necessary to introduce the true (or natural) strains $\bar{\varepsilon}_r = \ln(1+\varepsilon_r)$, $\bar{\varepsilon}_t = \ln(1+\varepsilon_t)$, and $\bar{\varepsilon}_z = \ln(1+\varepsilon_z)$. The assumption that strains are small is fully verified for materials that show well-defined yielding in the $\sigma = \sigma(\varepsilon)$curve. In such cases, there is a gradual passage from an elastic strain state to the plastic strain state, with a plastic/elastic interface that, from the permanent residual deformation that first appears, shifts to mark the boundary between two zones, one plastic and one elastic. Because of the two zones' continuity and contiguity, strains in the plastic region must be of comparable magnitude with those in the elastic region, or in other words must be small because those of the elastic zone are small.

Below the limit of proportionality, there are only a few cases in which the stress state in a non-axisymmetric thin disk can be determined in closed form, namely, non-axisymmetric disks consisting of orthotropic material, but with axisymmetric geometry and loading. In all other cases, numerical solutions must be used, though they will be simpler because of the bidimensional nature of the problem. If the disk can be considered axisymmetric, the resulting mono-dimensional problem can be solved in closed form in a larger number of cases. Where this is not possible, simple numerical solutions have been developed which are mostly based on the finite difference method and, more recently, on numerical integration of the differential equations governing the stress and strain state. Beyond the yield point, the stress state in a thin disk can be solved in closed form by means of the mono-dimensional theory only by assuming axisymmetry, plane stress state, constant thickness, and elastic-plastic behavior with no strain hardening on the part of the disk material.

It should also be recognized that the stress state in a rotor having large axial dimensions (a thick disk or simple cylinder) but with constant thickness and subjected to centrifugal and/or thermal load is not plane. In each elementary disk of axial thickness dz, in fact, centrifugal loading generates a stress state which can in the first approximation be considered as plane. However, because of the material's lateral contraction, as unit axial strain ε_z arise which in general varies along radius r, the sections tend to undergo distortion. This distortion is free to occur in disks of infinitesimal thickness, but is prevented in the case of large thicknesses, where a state of coaction is consequently produced.

In analyzing rotors of large axial thickness, it is thus more realistic to assume a plane strain state ($\varepsilon_z = 0$) if the rotor is constrained at its ends or infinitely extended along the axis, or uniform axial translation of a generic rotor cross section ($\varepsilon_z = const$) if the rotor has finite axial length and is free at the ends. Both assumptions of a plane strain state ($\varepsilon_z = 0$) and of uniform axial translation of a generic cross section ($\varepsilon_z = const$) coincide with the plane stress state assumption ($\sigma_z = 0$) only where axial strain ε_z is independent of variable r. This is the case of cylindrical solids satisfying the above constraint and generalized axisymmetry conditions, consisting of homogenous isotropic material and subjected only to uniform pressures, internal and external if the solid is hollow and only internal if it is non-hollow, all providing that Saint-Venant's principle is respected. In other words, thick-walled tubes, cylindrical bars, and thin disks all feature the same stress state if they are subjected only to surface force distributions. If body forces or thermal gradients are also at work, the assumptions of plane stress state, plane strain state, and uniform translation of a generic cross section will lead to different results.

The general treatment of rotating cylinders with large axial dimensions will also assume that the material is homogeneous and isotropic. However, only the case in which the material works in the linear elastic field will be considered.

Chapter 1
Mono-Dimensional Elastic Theory of Thin Disk

Take (Fig. I.1) a thin disk having a geometry of revolution, featuring a plane of symmetry perpendicular to its axis (plane of axial coordinate $z = 0$, in which the other two axes x, y that complete the three-dimension rectangular Cartesian reference system O(x, y, z) lie), rotating at angular velocity ω and subject to angular acceleration $\dot{\omega}$. Let it also be assumed that no surface force is applied to disk side faces 5 and 6. In addressing this type of problem, it is advantageous to use a polar reference system O(r, ϑ, z), where the position of a generic point P (x, y) $\equiv$ P (r, ϑ), in the midplane of the disk is defined by coordinate r, which establishes its distance from pole O, and by angle ϑ (angular coordinate) between axis O-r and abscissa O-x taken as reference.

The disk's mass elements will be subjected to centrifugal forces and to elementary tangential forces. These forces generate the stress and strain states that will be analysed here using the mono-dimensional theory, whose range of validity was illustrated in the introduction. Within the limits of this theory, radial stress σ_r, hoop stress σ_t and radial displacement u are functions of coordinate r alone, or in other words are constant above a cylinder of radius r whose axis is the axis of the disk.

To analyse thin disk stress and strain states, the equilibrium equations and the compatibility equations must be considered simultaneously.

1.1 Equilibrium Equations

Take an isolated mixtilinear volume element of the disk located between side surfaces 5 and 6 shown in Fig. I.1 and delimited by two diametral planes with angular coordinates ϑ and $\vartheta + d\vartheta$, which consequently form the angle $d\vartheta$ between them, and by two cylinders having radii r and $r + dr$ (Fig. 1.1). This element thus has finite dimensions in the direction of the z axis and infinitesimal dimensions along the other two polar coordinates. Consider the equilibrium of all forces acting on the element, both along line OA perpendicular to the disk axis and passing through the centre of gravity G, and along line GB perpendicular in G to OA.

V. Vullo and F. Vivio, *Rotors: Stress Analysis and Design*,
Mechanical Engineering Series, DOI 10.1007/978-88-470-2562-2_1,
© Springer-Verlag Italia 2013

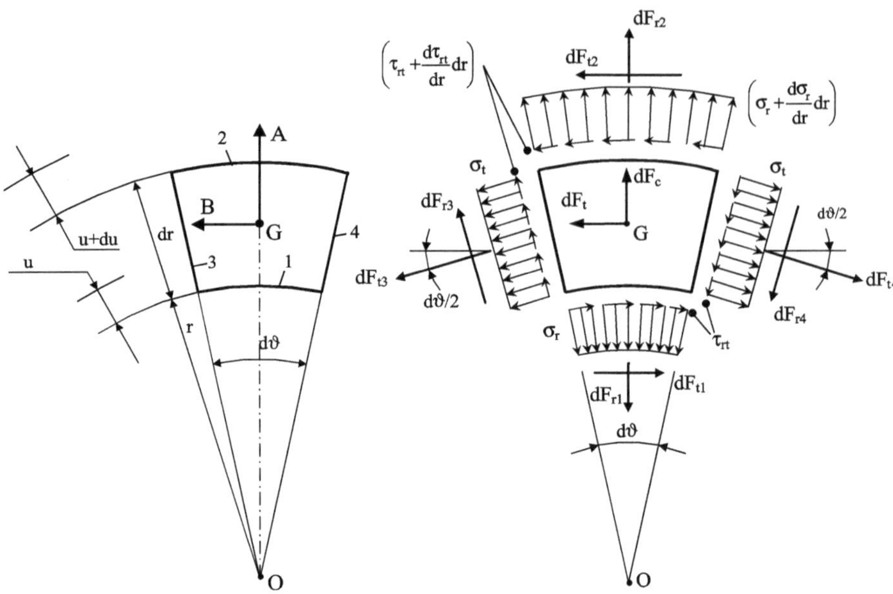

Fig. 1.1 Volume element, distribution of normal and shear stress components on its faces assuming axisymmetry, and elementary surface and body forces acting on the element

The volume element is subjected to two elementary body forces applied at the centre of gravity: force dF_c acting in a radial centrifugal direction and due to the centrifugal load (this is generally the most severe load condition), and force dF_t acting in the tangential direction, i.e., perpendicular to the radius and due to angular acceleration $\dot{\omega}$. These forces are given by the following relations:

$$\begin{aligned} dF_c &= r \cdot \omega^2 \cdot dm = r \cdot \omega^2 \cdot \gamma \cdot dV = \gamma \cdot \omega^2 \cdot r^2 \cdot h \cdot d\vartheta \cdot dr, \\ dF_t &= r \cdot \omega \cdot dm = r \cdot \omega \cdot \gamma \cdot dV = \gamma \cdot \omega \cdot r^2 \cdot h \cdot d\vartheta \cdot dr. \end{aligned} \tag{1.1}$$

where dm and dV denote the elementary mass and volume of the isolated element, and γ is the mass per unit volume, or density, of the material.

The other elementary forces acting on the volume element are:

- The radial force on the element's cylindrical face 1, at radius r, directed radially and facing the interior:

$$dF_{r1} = \sigma_r \cdot h \cdot r \cdot d\vartheta; \tag{1.2}$$

- The tangential force on the element's cylindrical face 1, at radius r, directed tangentially:

$$dF_{t1} = t_{rt} \cdot h \cdot r \cdot d\vartheta, \tag{1.3}$$

where τ_{rt} is the shear stress component;

- The radial force on the element's cylindrical face 2, at radius $r + dr$, directed radially and facing the exterior:

$$dF_{r2} = \left(\sigma_r + \frac{d\sigma_r}{dr} \cdot dr\right) \cdot \left(h + \frac{dh}{dr} \cdot dr\right) \cdot (r + dr) \cdot d\vartheta = dF_{r1} + \frac{d}{dr}(dF_{r1}) \cdot dr = \\ = \left[\sigma_r \cdot h \cdot r + \frac{d}{dr}(\sigma_r \cdot h \cdot r) \cdot dr\right] \cdot d\vartheta; \tag{1.4}$$

- Tangential force on the element's cylindrical face 2, at radius $r + dr$, directed tangentially:

$$dF_{t2} = \left(\tau_{rt} + \frac{d\tau_{rt}}{dr} \cdot dr\right) \cdot \left(h + \frac{dh}{dr} \cdot dr\right) \cdot (r + dr) \cdot d\vartheta = dF_{t1} + \frac{d}{dr}(dF_{t1}) \cdot dr = \\ = \left[\tau_{rt} \cdot h \cdot r + \frac{d}{dr}(\tau_{rt} \cdot h \cdot r) \cdot dr\right] \cdot d\vartheta\ ; \tag{1.5}$$

- Two equal radial forces acting on the element's plane side faces 3 and 4 and facing in the centrifugal direction on face 3 and in the centripetal direction on face 4 (these forces are equal in modulus, neglecting higher-order infinitesimals):

$$|dF_{r3}| = |dF_{r4}| = \tau_{rt} \cdot h \cdot dr; \tag{1.6}$$

- Two equal tangential forces acting on the element's plane side faces 3 and 4, perpendicular to them and facing the exterior of the element:

$$|dF_{t3}| = |dF_{t4}| = \sigma_t \cdot h \cdot dr. \tag{1.7}$$

In formulating these relations, which express the elementary forces acting on the mixtilinear volume element, terms introducing higher-order infinitesimals in the equilibrium equations are omitted. In accordance with the initial assumptions, in any case, no forces act on the volume element's side faces 5 and 6, which are those delimiting the disk laterally (Fig. I.1).

Accordingly, the dynamic equilibrium equations along radial direction *GA* and tangential direction *GB* are:

$$\begin{cases} -dF_{r1} + dF_{r2} + dF_{r3} \cdot cos\dfrac{d\vartheta}{2} - dF_{r4} \cdot cos\dfrac{d\vartheta}{2} - dF_{t3} \cdot \sin\dfrac{d\vartheta}{2} \\ \quad - dF_{t4} \cdot \sin\dfrac{d\vartheta}{2} + dF_c = 0 \\ -dF_{t1} + dF_{t2} + dF_{r3} \cdot \sin\dfrac{d\vartheta}{2} + dF_{r4} \cdot \sin\dfrac{d\vartheta}{2} + dF_{t3} \cdot cos\dfrac{d\vartheta}{2} \\ \quad - dF_{t4} \cdot cos\dfrac{d\vartheta}{2} + dF_t = 0. \end{cases} \tag{1.8}$$

By introducing relations (1.1, 1.2, 1.3, 1.4, 1.5, 1.6, and 1.7) derived earlier, which express the elementary forces acting on the element, in the above equilibrium equations, noting that $cos(d\vartheta/2) = 1$ and $sin(d\vartheta/2) = d\vartheta/2$ for infinitesimal angle $d\vartheta/2$, developing calculations and taking care to omit higher-order infinitesimals, we arrive at the following dynamic equilibrium equations for the volume element:

$$\begin{cases} \sigma_r \cdot h + \dfrac{d\sigma_r}{dr} \cdot h \cdot r + \dfrac{dh}{dr} \cdot \sigma_r \cdot r - \sigma_t \cdot h + \gamma \cdot \omega^2 \cdot r^2 \cdot h = 0 \\ \tau_{rt} \cdot h + \dfrac{d\tau_{rt}}{dr} \cdot h \cdot r + \dfrac{dh}{dr} \cdot \tau_{rt} \cdot r + \tau_{rt} \cdot h + \gamma \cdot \dot{\omega} \cdot r^2 \cdot h = 0. \end{cases} \tag{1.9}$$

These equations can also be rewritten in the following more compact form:

$$\begin{cases} \dfrac{d}{dr}(\sigma_r \cdot h \cdot r) - \sigma_t \cdot h + \gamma \cdot \omega^2 \cdot r^2 \cdot h = 0 \\ \dfrac{d}{dr}(\tau_{rt} \cdot h \cdot r) + \tau_{rt} \cdot h + \gamma \cdot \dot{\omega} \cdot r^2 \cdot h = 0. \end{cases} \tag{1.10}$$

Note that the shear stress components do not appear in the first of (1.10), which expresses dynamic equilibrium in the radial direction, while the normal stress components do not appear in the second of (1.10), which expresses dynamic equilibrium in the direction perpendicular to the radius. This decoupling of normal and shear stress components is because axisymmetry was assumed. Had this assumption not been made, the equilibrium equations would have been:

$$\begin{cases} \dfrac{\partial}{\partial r}(\sigma_r \cdot h \cdot r) + \dfrac{\partial}{\partial \vartheta}(\tau_{rt} \cdot h) - \sigma_t \cdot h + \gamma \cdot \omega^2 \cdot r^2 \cdot h = 0 \\ \dfrac{\partial}{\partial r}(\tau_{rt} \cdot h \cdot r) + \dfrac{\partial}{\partial \vartheta}(\sigma_t \cdot h) + \tau_{rt} \cdot h + \gamma \cdot \dot{\omega} \cdot r^2 \cdot h = 0. \end{cases} \tag{1.11}$$

These are more general than (1.10), which are a specific case for axisymmetry.

1.2 Compatibility Equations

In the general case in which there is no axisymmetry, normal strain components (radial strain ε_r and tangential strain ε_t) and shear strain component γ_{rt} are linked to components of the displacement in the radial and tangential directions u and v by the following geometric relations:

$$\begin{cases} \varepsilon_r = \dfrac{\partial u}{\partial r} \\ \varepsilon_t = \dfrac{u}{r} + \dfrac{1}{r} \cdot \dfrac{\partial v}{\partial \vartheta} \\ \gamma_{rt} = \dfrac{\partial v}{\partial r} - \dfrac{v}{r} + \dfrac{1}{r} \cdot \dfrac{\partial u}{\partial \vartheta}. \end{cases} \tag{1.12}$$

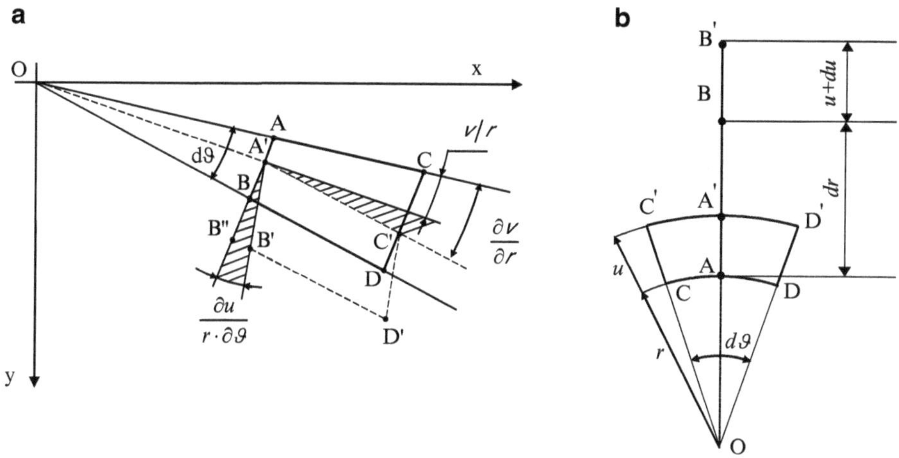

Fig. 1.2 (**a**) Displacements and rotations of a mixtilinear element, without axisymmetry; (**b**) elementary radial and tangential fibers and displacements of their ends, with axisymmetry

The first of relations (1.12) is obtained by considering (Fig. 1.2a) the undeformed mixtilinear element *ABCD* located between two cylinders of radius r and $r + dr$ and between two diametral planes with angular coordinates ϑ and $\vartheta + d\vartheta$ (given that $d\vartheta$ is an elementary angle, arcs *AB* and *CD*, which are not shown in the figure in order to avoid over-complication, can be replaced by their chords) and bearing in mind that the radial displacements of sides $\overline{AB}$ and $\overline{CD}$ are given by u and by $u + (\partial u/\partial r) \cdot dr$ respectively. It follows that the unit elongation of this element in the radial direction is given by relation $\varepsilon_r = \partial u/\partial r$, or in other words by the first of the two expressions (1.12).

Tangential strain component ε_t, on the other hand, depends on both radial displacement u and tangential displacement v. For side $\overline{AB}$ of the element in question, the contribution of radial displacement u is evaluated by considering that this displacement causes initial arc $AB = r \cdot d\vartheta$ to acquire a length $(r + u) \cdot d\vartheta$. Consequently (Fig. 1.2b), the tangential strain related to displacement u is given by relation:

$$\frac{(r + u) \cdot d\vartheta - r \cdot d\vartheta}{r \cdot d\vartheta} = \frac{u}{r}. \tag{1.13}$$

The contribution of tangential displacement v is determined by considering that it moves point A to A', with a tangential displacement v, and point B moves to point B'', with a tangential displacement $v + (\partial v/\partial\vartheta) \cdot d\vartheta$. It follows that there is a difference of tangential displacement equal to $(\partial v/\partial\vartheta) \cdot d\vartheta$, and the related tangential strain will be $(1/r) \cdot \partial v/\partial\vartheta$. Adding the two contributions gives the total tangential strain expressed by the second of relations (1.12).

Shear strain component γ_{rt} represents the total variation of angle *BAC*, initially a right angle. It is determined by comparing the element's position $A'B'C'D'$ after strain-induced deformation, and its initial position *ABCD*. The angle between directions *AB* and $A'B'$ is due to radial displacement u and is $(1/r) \cdot \partial u/\partial\vartheta$.

Likewise, the angle between directions AC and $A'C'$ is due to tangential displacement v and is $\partial v/\partial r$. However, only a part of the latter contributes to shear strain component (the part represented with a dashed area), as the remaining part, v/r represents the angular displacement due to rotation of the element $ABCD$ as a rigid body about the axis through O. It can be concluded from these considerations that shear strain component γ_{rt} is given by the third relation (1.12).

Where the assumption of axisymmetry applies, relations (1.12) are reduced to the simpler form:

$$\varepsilon_r = \frac{du}{dr}; \quad \varepsilon_t = \frac{u}{r}; \quad \gamma_{rt} = \frac{dv}{dr} - \frac{v}{r}. \tag{1.14}$$

In the case of axisymmetry, as can thus be seen, normal strain components ε_r and ε_t are linked only to radial displacement u, while shear strain component γ_{rt} is linked only to tangential displacement v. Assuming axisymmetry, expressions ε_r and ε_t can be derived from even simpler geometrical considerations. Take, for instance, a radial fiber, considering (Fig. 1.2b) a portion thereof of infinitesimal length dr prior to deformation, located between points A and B at distances r and $r + dr$ from the centre respectively: after deformation, end A moves to A', displaced by u, while end B moves to B', displaced by $u + du$. Similarly, if we take a tangential fiber, considering (again with reference to Fig. 1.2b) an infinitesimal arc thereof of length $r{\cdot}d\vartheta$ prior to displacement, located between points C and D both at distance r from the centre, after displacement end C moves to C', displaced by u, while end D moves to D', also displaced by u. We thus have:

$$\varepsilon_r = \frac{\overline{A'B'} - \overline{AB}}{\overline{AB}} = \frac{du}{dr}, \tag{1.15}$$

$$\varepsilon_t = \frac{(r+u)\cdot d\vartheta - r\cdot d\vartheta}{r\cdot d\vartheta} = \frac{u}{r}. \tag{1.16}$$

It is also clear that, where loading is entirely centrifugal ($\omega \neq 0$ and $\dot{\omega} = 0$) and for an axisymmetric stress field, tangential displacement v will be zero, and $\gamma_{rt} = 0$. Essentially, then, if axisymmetry is assumed, when the mixtilinear element shown in Fig. 1.1 is subjected to axisymmetric centrifugal load, its movement will be entirely radial, with no tangential displacement.

The first two geometric relations (1.14), which apply when axisymmetry is assumed, give the following compatibility equation in terms of strain:

$$\varepsilon_r = \frac{d}{dr}(r\cdot\varepsilon_t). \tag{1.17}$$

Were the assumption of axisymmetry not to apply, the compatibility equation would be expressed as follows:

$$r\cdot\frac{\partial^2\gamma_{rt}}{\partial r\cdot\partial\vartheta} + \frac{\partial\gamma_{rt}}{\partial\vartheta} - r^2\cdot\frac{\partial^2\varepsilon_t}{\partial r^2} - 2r\cdot\frac{\partial\varepsilon_t}{\partial r} + r\cdot\frac{\partial\varepsilon_r}{\partial r} - \frac{\partial^2\varepsilon_r}{\partial\vartheta^2} = 0, \tag{1.18}$$

which is more general than (1.17) and subsumes it.

For an axisymmetric orthotropic material, the constitutive equations which relate the stress components to strain components (these equations, in the linear elastic range, represent the generalised Hooke's law) can be written in the following form:

$$\begin{Bmatrix} \varepsilon_r \\ \varepsilon_t \\ \gamma_{rt} \end{Bmatrix} = [\mathbf{S}] \cdot \begin{Bmatrix} \sigma_r \\ \sigma_t \\ \tau_{rt} \end{Bmatrix} = \begin{bmatrix} \dfrac{1}{E_r} & \dfrac{-\nu_{tr}}{E_t} & 0 \\ \dfrac{-\nu_{rt}}{E_r} & \dfrac{1}{E_t} & 0 \\ 0 & 0 & \dfrac{1}{G_{rt}} \end{bmatrix} \cdot \begin{Bmatrix} \sigma_r \\ \sigma_t \\ \tau_{rt} \end{Bmatrix}. \tag{1.19}$$

As will be recalled, an orthotropic material is a non-isotropic material which has three mutually orthogonal planes of symmetry at every point. From the standpoint of its elastic behaviour, complete characterization requires nine parameters, viz., the three Young's moduli E, the three moduli of rigidity G and the three Poisson's ratios ν in the directions normal to the planes of symmetry. For complete elastic characterization of an isotropic material, on the other hand, only two of the three parameters E, G, ν, are required, as they are linked by a linear dependence. If the structure of the orthotropic material is such that it has three mutually orthogonal planes of symmetry at every point, but also an overall axis of symmetry, the material is *axisymmetric orthotropic*. A composite material produced by winding a fiber embedded in a matrix around a cylinder fulfils these conditions if the fiber is wound tangentially: in this case, the radial, tangential and axial directions are normal to the planes of symmetry at every point.

In formulating the constitutive equations given above, the tangential and radial directions of the axisymmetric orthotropic material were considered to be the principal directions. If this were not true, no component of the matrix of the material compliances **[S]** would be zero. In this case, normal strain components ε_r and ε_t and shear strain component γ_{rt} would be linked simultaneously to normal stresses σ_r and σ_t and to shear stress τ_{rt}, and a number of the simplifications that will be discussed below would not be possible. Note that in writing matrix **[S]** appearing in (1.19), the first term in the second line, i.e., $-\nu_{rt}/E_r$, is often substituted by $-\nu_{tr}/E_t$, as this matrix is symmetrical with respect to the principal diagonal.

Though we will be dealing chiefly with homogeneous isotropic materials in this text, generalizing for axisymmetric orthotropic materials presents no difficulties. The constitutive equations for homogeneous isotropic materials take the following form:

$$\begin{Bmatrix} \varepsilon_r \\ \varepsilon_t \\ \gamma_{rt} \end{Bmatrix} = \begin{bmatrix} \dfrac{1}{E} & \dfrac{-\nu}{E} & 0 \\ \dfrac{-\nu}{E} & \dfrac{1}{E} & 0 \\ 0 & 0 & \dfrac{1}{G} \end{bmatrix} \cdot \begin{Bmatrix} \sigma_r \\ \sigma_t \\ \tau_{rt} \end{Bmatrix}. \tag{1.20}$$

If the effect of a generic temperature variation deriving from a thermal load is to be taken into account, the strain components due to this thermal load must obviously be added to strains ε_r and ε_t caused by stresses $\{\sigma\}$. These additional strain components are given by $\alpha \cdot T$, where α and T are the material's coefficient of linear thermal expansion and the temperature respectively. If the material is homogeneous and isotropic, α does not depend on direction and, consequently, (1.20) becomes:

$$\begin{Bmatrix} \varepsilon_r \\ \varepsilon_t \\ \gamma_{rt} \end{Bmatrix} = \begin{bmatrix} \frac{1}{E} & \frac{-\nu}{E} & 0 \\ \frac{-\nu}{E} & \frac{1}{E} & 0 \\ 0 & 0 & \frac{1}{G} \end{bmatrix} \cdot \begin{Bmatrix} \sigma_r \\ \sigma_t \\ \tau_{rt} \end{Bmatrix} + \alpha \cdot T \cdot \begin{Bmatrix} 1 \\ 1 \\ 0 \end{Bmatrix}. \tag{1.21}$$

For an axisymmetric orthotropic material, (1.21) becomes:

$$\begin{Bmatrix} \varepsilon_r \\ \varepsilon_t \\ \gamma_{rt} \end{Bmatrix} = \begin{bmatrix} \frac{1}{E_r} & \frac{-\nu_{tr}}{E_t} & 0 \\ \frac{-\nu_{rt}}{E_r} & \frac{1}{E_t} & 0 \\ 0 & 0 & \frac{1}{G_{rt}} \end{bmatrix} \cdot \begin{Bmatrix} \sigma_r \\ \sigma_t \\ \tau_{rt} \end{Bmatrix} + T \cdot \begin{Bmatrix} \alpha_r \\ \alpha_t \\ 0 \end{Bmatrix}. \tag{1.22}$$

Taking the expressions of ε_r and ε_t from (1.21) and introducing them in the compatibility equation (1.17) gives the following compatibility equation in terms of stress:

$$(\sigma_t - \sigma_r) \cdot (1 + \nu) + r \cdot \frac{d\sigma_t}{dr} - \nu \cdot r \cdot \frac{d\sigma_r}{dr} + \alpha \cdot E \cdot r \frac{dT}{dr} = 0. \tag{1.23}$$

This equation applies assuming that quantities E, ν and α are constant throughout the disk. Were these quantities variable, (1.23) would become:

$$\begin{aligned} &(\sigma_t - \sigma_r) \cdot (1 + \nu) + r \cdot \frac{d\sigma_t}{dr} - \nu \cdot r \cdot \frac{d\sigma_r}{dr} + \cdot E \cdot r \cdot \sigma_t \frac{d}{dr}\left(\frac{1}{E}\right) \\ &\qquad - E \cdot r \cdot \sigma_r \cdot \frac{d}{dr}\left(\frac{\nu}{E}\right) + E \cdot r \cdot \frac{d}{dr}(\alpha \cdot T) = 0. \end{aligned} \tag{1.24}$$

1.3 General Differential Equation for Rotating Disk Subjected to Thermal Load

Here, analysis will be limited to the stress and strain state in an axisymmetric disk of arbitrary profile only rotating and subjected to a non-zero temperature gradient along the radius.

If the disk is only rotating ($\omega = const$), the only body force acting on it is that due to the centrifugal load. Accordingly, the tangential equilibrium equation given by the second relation (1.10) reduces to the identity $\tau_{rt} = 0$, which clearly satisfies the compatibility equation. Thus, if functions $h = h(r), \gamma = \gamma(r), E = E(r)$, $v = v(r), T = T(r)$ and $\alpha = \alpha(r)$ are known, the radial equilibrium equation given by the first relation (1.10) and compatibility equation (1.23) or (1.24) are sufficient to solve the problem completely, i.e., to determine how the two unknowns σ_r and σ_t vary with radius. From the two equilibrium and compatibility equations, which are first order differential equations in two unknowns σ_r and σ_t, we can obtain only one second order differential equation in one unknown.

Within certain arbitrary limits, two alternatives are usually considered in selecting this unknown: the first is to express stress components as a function of radial displacement u by means of the compatibility equation and then introduce these expressions in the equilibrium equation, while the second alternative is to first derive the expression of stress σ_t as a function of σ_r from the equilibrium equation and then introduce this expression in the compatibility equation. The two approaches are equivalent and lead to very similar solving equations. The first approach will be followed below, assuming that quantities γ, E, v and α are constant and thus independent of r. In dealing with the disk whose profile varies non-linearly according to a power of a linear function and which is loaded beyond yielding, the second approach will be used (see Chap. 12, Sect. 12.2).

Equation 1.21 give:

$$\begin{cases} \varepsilon_r = \dfrac{1}{E}(\sigma_r - v \cdot \sigma_t) + \alpha \cdot T \\ \varepsilon_t = \dfrac{1}{E}(\sigma_t - v \cdot \sigma_r) + \alpha \cdot T, \end{cases} \tag{1.25}$$

which, solved for σ_r and σ_t, yield the relationships:

$$\begin{cases} \sigma_r = \dfrac{E}{1 - v^2} \cdot [(\varepsilon_r - \alpha \cdot T) + v \cdot (\varepsilon_t - \alpha \cdot T)] \\ \sigma_t = \dfrac{E}{1 - v^2} \cdot [(\varepsilon_t - \alpha \cdot T) + v \cdot (\varepsilon_r - \alpha \cdot T)]. \end{cases} \tag{1.26}$$

Bearing the first two geometric relations (1.12) in mind, relations (1.26) can be written in the form:

$$\begin{cases} \sigma_r = \dfrac{E}{1 - v^2} \cdot \left[\left(\dfrac{du}{dr} - \alpha \cdot T\right) + v \cdot \left(\dfrac{u}{r} - \alpha \cdot T\right)\right] \\ \sigma_t = \dfrac{E}{1 - v^2} \cdot \left[\left(\dfrac{u}{r} - \alpha \cdot T\right) + v \cdot \left(\dfrac{du}{dr} - \alpha \cdot T\right)\right]. \end{cases} \tag{1.27}$$

Deriving the first of relations (1.27) by respect to r and introducing this derivative, together with relations (1.27) in the equilibrium equation, which in this case is the first (1.10), and carrying out a few other passages yields the following second order differential equation in u, which is the equation that solves the disk of arbitrary profile only rotating and subject to a non-zero temperature gradient along the radius:

$$\frac{d^2u}{dr^2}+\left(\frac{1}{h}\cdot\frac{dh}{dr}+\frac{1}{r}\right)\cdot\frac{du}{dr}+\left(\frac{\nu}{h\cdot r}\cdot\frac{dh}{dr}-\frac{1}{r^2}\right)\cdot u-(1+\nu)\cdot\alpha\cdot\left(\frac{dT}{dr}+\frac{T}{h}\cdot\frac{dh}{dr}\right)+\left(1-\nu^2\right)\cdot\frac{\gamma\cdot\omega^2\cdot r}{E}=0. \tag{1.28}$$

Whether this equation can be integrated analytically depends on the function $h = h(r)$. Closed form integration is possible, and relatively simple, in three special cases: constant thickness disk, uniform strength disk, and hyperbolic disk.[1]

In other cases, such as those of the linear tapered disk (the so-called conical profile disk) and the non-linearly variable thickness disk, a hypergeometric series solution is possible. Series solutions are also possible for the following other profiles (these will also be hypergeometric in all cases below except for the two-parameter exponential profile, whose solution involves the use of confluent hypergeometric series):

- Profile whose thickness varies according to an exponential function of the type characterizing the uniform strength disk, but with two parameters, and thus defined by relation $h = h_0 \cdot e^{-n\cdot\rho^k}$, where h_0 is the thickness at the axis, $\rho = r/r_e$ is the radius made dimensionless relative to the outside radius r_e, and n and k are the two geometric parameters controlling thickness at the outer edge relative to that at the axis and the profile shape respectively. This function makes it possible to describe solid and annular disks with concave, convex and inflection point profiles, but not conical disks;
- Profile whose thickness varies according to a generalization of Stodola's hyperbolic function, defined by the relation $h = h_0 \cdot (1+\rho)^a$, where a is a parameter controlling disk shape. This function does not give rise to singularity at the axis and thus, unlike Stodola's relationship, can be used to describe also the hyperbolic profile of a solid disk;
- Profile whose thickness varies according to an elliptical function defined by the relation $h = h_0 \cdot \left(1-n\cdot\rho^2\right)^{1/2}$, with one parameter n, whereby convex converging and concave diverging profiles can be described;
- Profile whose thickness varies according to the following two parabolic functions, both with two parameters, $h = h_0 \cdot \left(1-n\cdot\rho^k\right)$ and $h = h_0 \cdot \left[1-\left(\frac{r_e\cdot\rho}{r_e+n}\right)^k\right]$, whereby solid and annular convex and linear tapered disk profiles can be described.

---

[1] For a broad overview of these importacnt topics, see, in particular, Stodola [70], Love [44], Giovannozzi [29], Timoshenko and Goodier [74], Saada [62], Burr [7], Ugural and Fenster [76].

In all other cases, it is necessary to use numerical solutions which, however, pose no difficulties and yield results that, though approximate, are acceptable from the design standpoint.

In the following pages, the general (1.28) will be specialized and integrated for the three cases that can be solved in closed form, as well as for the two families consisting of the conical disk and the non-linearly variable thickness disk, with the disk subject to simple load conditions. In the linear elastic field (where stress is proportional to strain), the stress and strain states in a disk subjected to a complex load condition can obviously be determined by using the method of the superposition, that is as the superposition of the individual load conditions into which actual loading can be broken down, which are assumed to operate separately. The treatment used for the other disk profiles indicated above will not be discussed, as the analytical developments are formally similar to those for the two families we have just mentioned. These developments will thus be left to the reader.

Chapter 2
Constant Thickness Rotating Disk

2.1 General Equations

For the constant thickness disk ($h = const$; $dh/dr = 0$), the solving differential equation (1.28) is specialized in the following form:

$$\frac{d^2u}{dr^2} + \frac{1}{r} \cdot \frac{du}{dr} - \frac{u}{r^2} - (1 + \nu) \cdot \alpha \cdot \frac{dT}{dr} + \left(1 - \nu^2\right) \cdot \frac{\gamma \cdot \omega^2 \cdot r}{E} = 0. \tag{2.1}$$

This is a non-homogeneous second order Cauchy-Euler differential equation.

In the two particular cases of a disk only rotating and subjected to constant temperature and a non-rotating disk subject to a non-zero temperature gradient along the radius, (2.1) is further specialized in the two equations:

$$\frac{d^2u}{dr^2} + \frac{1}{r} \cdot \frac{du}{dr} - \frac{u}{r^2} = -\left(1 - \nu^2\right) \cdot \frac{\gamma \cdot \omega^2 \cdot r}{E}; \tag{2.2}$$

$$\frac{d^2u}{dr^2} + \frac{1}{r} \cdot \frac{du}{dr} - \frac{u}{r^2} = (1 + \nu) \cdot \alpha \cdot \frac{dT}{dr}. \tag{2.3}$$

General solutions of the three differential equation (2.1), (2.2) and (2.3) are obtained by adding the solution of the common associated homogeneous equation, given by

$$\frac{d^2u}{dr^2} + \frac{1}{r} \cdot \frac{du}{dr} - \frac{u}{r^2} = 0, \tag{2.4}$$

to particular solutions of the complete equations. Equation (2.4) does not contain the term for the body forces or the term for the temperature gradient. This is the equation for the load condition in which the disk is subjected only to surface forces

V. Vullo and F. Vivio, *Rotors: Stress Analysis and Design*,
Mechanical Engineering Series, DOI 10.1007/978-88-470-2562-2_2,
© Springer-Verlag Italia 2013

acting at the inner radius and/or the outer radius in the case of an annular disk, and only at the outer radius in the case of a solid disk.

Equation (2.5) can be written in the form:

$$\frac{d}{dr}\left[\frac{1}{r}\cdot\frac{d}{dr}(u\cdot r)\right]=0. \tag{2.5}$$

Integrating the latter equation twice and designating the two integration constants as C_1 and C_2 gives the following relation, which expresses radial displacement u as a function of radius r:

$$u=\frac{C_1}{2}\cdot r+\frac{C_2}{r}. \tag{2.6}$$

Deriving relation (2.6) by respect to r and introducing this derivative together with relation (2.6) in the first two geometric relations (1.14) yields:

$$\begin{cases}\varepsilon_r=\dfrac{du}{dr} & =\dfrac{C_1}{2}-\dfrac{C_2}{r^2}\\ \varepsilon_t=\dfrac{u}{r} & =\dfrac{C_1}{2}+\dfrac{C_2}{r^2}.\end{cases} \tag{2.7}$$

Substituting these relations in (1.27) from which temperature terms are omitted gives the following expressions of σ_r and σ_t as a function of radius r:

$$\begin{cases}\sigma_r=\dfrac{E\cdot C_1}{2\cdot(1-\nu)}-\dfrac{E\cdot C_2}{(1+\nu)}\cdot\dfrac{1}{r^2}\\ \sigma_t=\dfrac{E\cdot C_1}{2\cdot(1-\nu)}+\dfrac{E\cdot C_2}{(1+\nu)}\cdot\dfrac{1}{r^2}.\end{cases} \tag{2.8}$$

Finally, by multiplying and dividing the second term of the second members of expressions (2.8) by r_e^2, introducing the new dimensionless variable $\rho=r/r_e$, and using the notation

$$A=\frac{E\cdot C_1}{2\cdot(1-\nu)}\quad\text{and}\quad B=\frac{E\cdot C_2}{(1+\nu)\cdot r_e^2}, \tag{2.9}$$

we arrive at the following relations, which express radial and hoop stresses as a function of ρ:

$$\begin{cases}\sigma_r=A-B/\rho^2\\ \sigma_t=A+B/\rho^2.\end{cases} \tag{2.10}$$

These are Lamé's relations, from which it can be concluded that the sum $\sigma_r + \sigma_t = 2A$ is constant and invariant.

On the other hand, by introducing relations (2.10) in the second (1.25), from which the temperature term is omitted and bearing in mind that the second (1.14) gives $u = r \cdot \varepsilon_t$, we obtain the following relation expressing radial displacement $u = u(\rho)$:

$$u = \frac{r_e}{E} \cdot \rho \cdot \left[A \cdot (1 - \nu) + \frac{B}{\rho^2} \cdot (1 + \nu) \right] . \tag{2.11}$$

Relations (2.10) and (2.11) describe the distribution of radial and hoop stresses and radial displacement versus dimensionless variable ρ in a disk subjected only to surface forces acting at the inner and outer radii. In these relations, A and B are the integration constants, which can be determined by imposing boundary conditions.

For the annular disk, the boundary conditions to be imposed are:

$$\begin{cases} \sigma_r = \sigma_{r_i} & \text{for } \rho = \beta \text{, i.e., for } r = r_i \\ \sigma_r = \sigma_{r_e} & \text{for } \rho = 1 \text{, i.e., for } r = r_e; \end{cases} \tag{2.12}$$

$\beta = r_i / r_e$ indicates the minimum value assumed by ρ at the disk inner radius. For the solid disk, on the other hand, the boundary conditions to be imposed are:

$$\begin{cases} \sigma_r = \sigma_{r_e} & \text{for } \rho = 1 \\ u = 0 & \text{for } \rho = 0 \text{, i.e., for } r = 0. \end{cases} \tag{2.13}$$

The following general observations can be made on the basis of relations (2.10) and (2.11):

- Stresses σ_r and σ_t do not depend on the material's elastic properties, but only on the boundary conditions. This is in agreement with Michell's theorem, which states that in a multiply-connected body subjected to a plane stress state, stressed in the linear elastic field with neither body forces nor thermal loads and loaded by a system of external forces with a zero translation resultant on each boundary, separately considered, the stress state does not depend on the material's elastic constants. The assumptions on which Michell's theorem are based are all fulfilled here, including that of loads with zero resultant along all boundaries. The latter assumption, in fact, is subsumed in that of generalized axisymmetry, which requires that only uniformly distributed radial loads can act on each boundary. By contrast, strain components ε_r and ε_t, as well as radial displacement u are functions of the material's elastic properties;
- The sum $\sigma_r + \sigma_t$ is constant and invariant over the entire disk. This is true both for disks where a plane stress state is assumed, and for long cylindrical solids,

where a strain state characterized by $\varepsilon_z = const$ is assumed, provided that disks and long cylindrical solids are subjected only to surface force distributions on the inner and outer boundaries. In this case, we have the so-called generalized plane strain state. The sum $\varepsilon_r + \varepsilon_t$ is also an invariant;

- In the case of a solid disk subjected to surface force distribution at the outer radius, on the axis, i.e., for $r = 0$, we are dealing with a singularity. However, it can be said that, for reasons of symmetry, displacement u must become zero on the rotational axis. Near the axis, radial displacement u must thus be substantially equal to displacement at the axis: in other words, the limit of u for $r \rightarrow 0$ must be zero, i.e., $C_2 = B = 0$. In this case, as we will see below, the disk is subjected to radial and hoop stresses that are equal to each other at all points of the disk, as if the condition were one of uniform strength. This is no longer true if there is a hole at the axis, even if this hole is of infinitesimal diameter: in such cases, displacement u at the axis is not defined.

2.2 Non-Rotating Annular Disk, Loaded at the Outer Radius

We will now consider an annular disk loaded at the outer radius by uniformly distributed surface forces which thus give rise to a uniform distribution of stress σ_{re} at this radius. As regards design applications, both the case of positive stress σ_{re} (tensile stress) and of negative stress σ_{re} (compressive stress) are of interest. In both cases, the boundary conditions to be established in order to calculate integration constants A and B appearing in (2.10) and (2.11) given in the preceding section are:

$$\begin{cases} \sigma_r = 0 & \text{per} \quad \rho = \beta \\ \sigma_r = \sigma_{r_e} & \text{per} \quad \rho = 1 \end{cases}. \tag{2.14}$$

Accordingly, the system obtained from the first of relations (2.10) in the preceding paragraph gives:

$$\begin{cases} A = \sigma_{r_e} / \left(1 - \beta^2\right) \\ B = \sigma_{r_e} \cdot \beta^2 / \left(1 - \beta^2\right). \end{cases} \tag{2.15}$$

Substituting the values of the constants thus found in relations (2.10) and (2.11) in the preceding section gives the following expressions for σ_r, σ_t and u which provide a univocal solution of the problem:

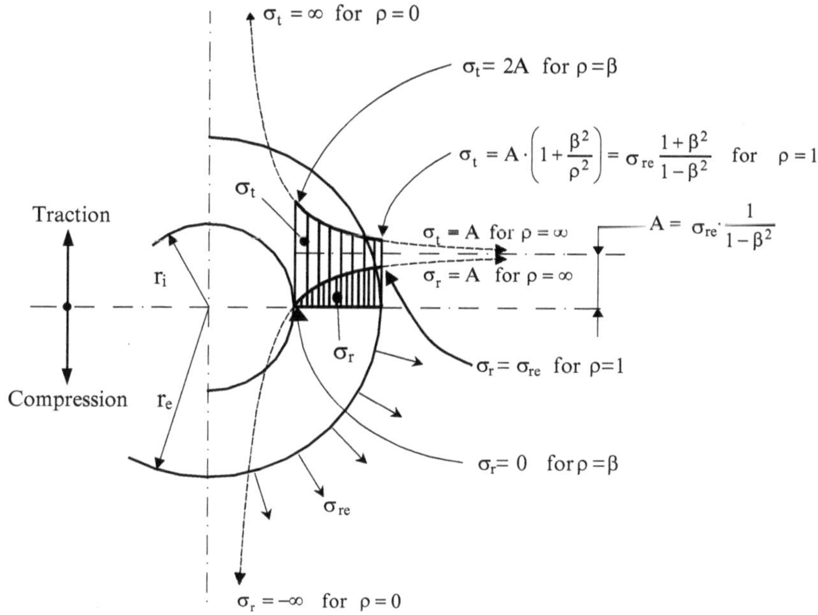

Fig. 2.1 Distribution curves of principal stresses σ_r and σ_t versus radius in an annular disk loaded at outer edge by a positive σ_{re}, and characteristic values of the quantities involved

$$\begin{cases} \sigma_r &= \sigma_{r_e} \cdot \dfrac{1}{1-\beta^2} \cdot \left(1 - \dfrac{\beta^2}{\rho^2}\right) \\ \sigma_t &= \sigma_{r_e} \cdot \dfrac{1}{1-\beta^2} \cdot \left(1 + \dfrac{\beta^2}{\rho^2}\right) \\ u &= \sigma_{r_e} \cdot \dfrac{r_e}{E} \cdot \dfrac{\rho}{1-\beta^2} \cdot \left[(1-\nu) + (1+\nu) \cdot \dfrac{\beta^2}{\rho^2}\right]. \end{cases} \tag{2.16}$$

As mentioned earlier, the first case applies to the loads from centrifugal forces caused by blades spaced at equal angles on the disk's outer surface. Figure 2.1, which refers to this case, shows radial and hoop stresses versus radius. As can be seen from the figure, the absolute maximum stress value is that reached by hoop stress at the inner edge, and constant A represents the asymptotic value that the two principal nonzero stresses (σ_r and σ_t) would assume for ρ tending to infinity.

The second case applies to the stresses deriving either from an external shrink-fit or from the direct application of an external pressure. In both circumstances, a stress $\sigma_{re} = -p_e$, where p_e denotes the shrink-fit pressure or direct pressure, will act on the outer radius. For this case, Fig. 2.2 shows radial and hoop stresses versus radius. Here again, it can be seen that the absolute maximum stress value is that reached by hoop stress at the inner edge, and constant A represents the asymptotic value that the two principal nonzero stresses (σ_r and σ_t) would assume for ρ tending to infinity.

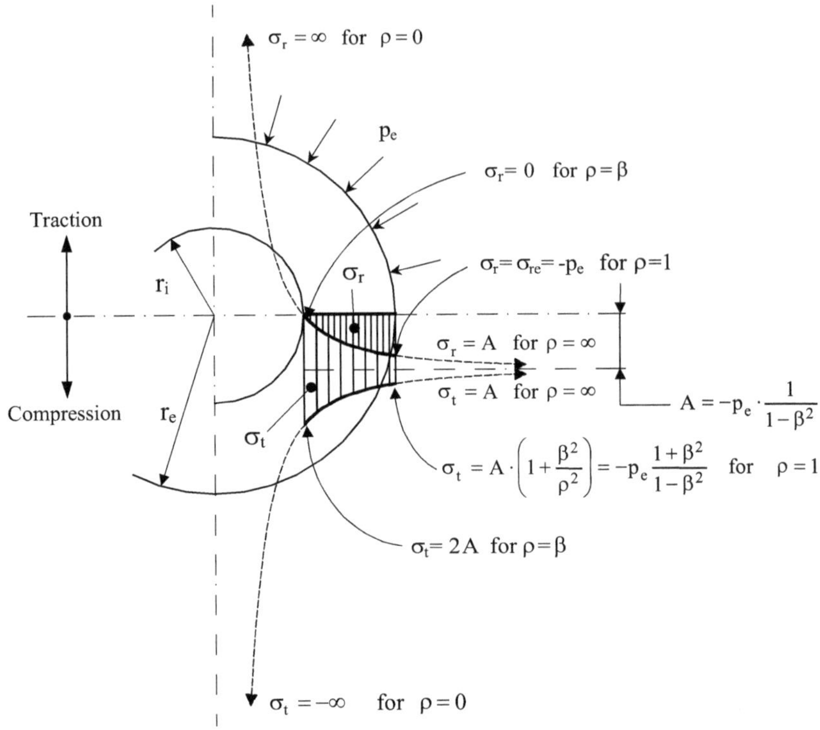

Fig. 2.2 Distribution curves of principal stresses σ_r and σ_t versus radius in an annular disk loaded at outer edge by a negative $\sigma_{re} = -p_e$, and characteristic values of the quantities involved

In both cases, there is a singularity for small finite values of β and for $\rho \to 0$. As the curves in Figs. 2.1 and 2.2 show, for $\rho \to 0$, radial stress and hoop stress would tend to ∞, whatever their sign (which is obviously correlated to that of σ_{re}). In reality, the radial stress at the inner radius, which is not loaded, will be zero, while the hoop stress at the inner radius will be $2A$; this, in fact, is the value obtained from the second relation (2.16), for $\beta^2/\rho^2 = 1$. Given the singularity, the ratio β^2/ρ^2 must be considered as unity for $\beta \to 0$.

2.3 Non-Rotating Annular Disk, Loaded at the Inner Radius

We will now consider an annular disk loaded at the inner radius by evenly distributed surface forces which thus give rise to a uniform distribution of stress σ_{ri} at this radius. Here again, both the case of positive stress σ_{ri} (tensile stress) and of negative stress σ_{ri} (compressive stress) are of interest from the design standpoint.

In both cases, the boundary conditions to be established in order to calculate integration constants A and B appearing in (2.10) and (2.11) are:

$$\begin{cases} \sigma_r = \sigma_{r_i} & \text{per} \quad \rho = \beta \\ \sigma_r = 0 & \text{per} \quad \rho = 1 \end{cases} \tag{2.17}$$

Accordingly, the system obtained from the first of relations (2.10) gives:

$$A = B = -\sigma_{r_i} \cdot \beta^2 / \left(1 - \beta^2\right) . \tag{2.18}$$

Substituting the values of the constants found in relations (2.10) and (2.11) gives the following expressions for σ_r, σ_t and u which provide a univocal solution of the problem:

$$\begin{cases} \sigma_r &= \sigma_{r_i} \cdot \dfrac{\beta^2}{1-\beta^2} \cdot \left(\dfrac{1}{\rho^2} - 1\right) \\ \sigma_t &= -\sigma_{r_i} \cdot \dfrac{\beta^2}{1-\beta^2} \cdot \left(\dfrac{1}{\rho^2} + 1\right) \\ u &= -\dfrac{\sigma_{r_i}}{E} \cdot r_e \cdot \rho \cdot \dfrac{\beta^2}{1-\beta^2} \cdot \left[(1-\nu) + (1+\nu) \cdot \dfrac{1}{\rho^2}\right] . \end{cases} \tag{2.19}$$

The first case applies to the stresses from centrifugal forces which arise at a given radius, which is assumed to be the inner radius, as a result of joining the disk by means of screw bolts to the flanged end of a shaft. Figure 2.3, which refers to this case, shows radial and hoop stresses versus radius. As can be seen from the figure, the absolute maximum stress value is that reached by hoop stress at the inner edge, and constant A represents the asymptotic value that the two principal nonzero stresses (σ_r and σ_t) would assume for ρ tending to infinity.

The second case applies to the stresses deriving either from internal shrink-fit or from the direct application of an internal pressure. In both circumstances, a stress $\sigma_{ri} = -p_i$, where p_i denotes the shrink-fit pressure or direct pressure, will act on the inner radius. For this case, Fig. 2.4 shows radial and hoop stresses versus radius. Here again, it can be seen that the absolute maximum stress value is that reached by hoop stress at the inner edge, and constant A represents the asymptotic value that the two principal nonzero stresses (σ_r and σ_t) would assume for ρ tending to infinity.

Here again, there is a singularity in both cases for small, finite values of β and for $\rho \to 0$. As the curves in Figs. 2.3 and 2.4 show, for $\rho \to 0$, radial stress and hoop stress would tend to ∞, whatever their sign (which is obviously correlated to that of σ_{ri}). In reality, the radial stress at the inner radius will assume the value deriving from the boundary condition ($\sigma_r = \sigma_{ri}$ for $\rho = \beta$), while the hoop stress at the inner radius will assume the value given by the second relation (2.19), for $\beta^2/\rho^2 = 1$; in

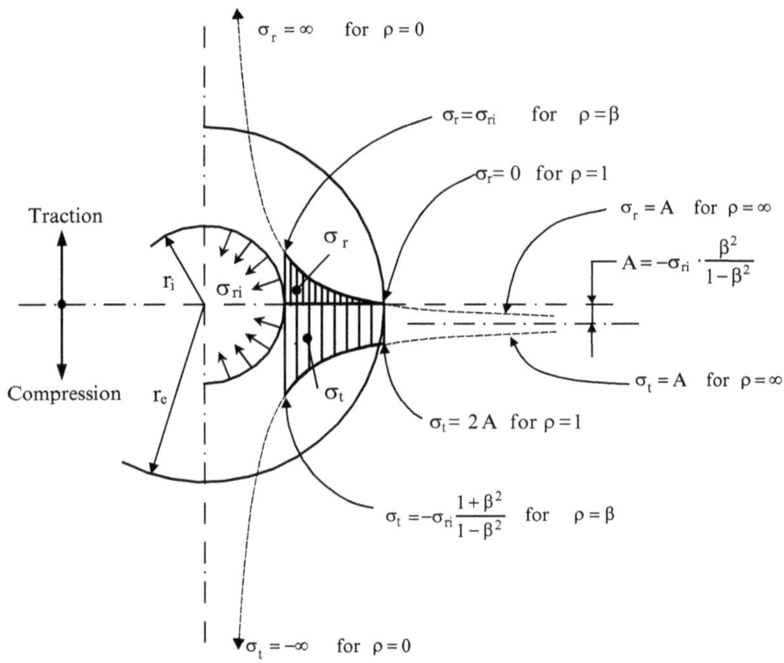

Fig. 2.3 Distribution curves of principal stresses σ_r and σ_t versus radius in an annular disk loaded at inner edge by a positive σ_{ri}, and characteristic values of the quantities involved

other words, here again the ratio β^2/ρ^2 must be considered as unity for $\beta \rightarrow 0$ given the singularity.

2.4 Non-Rotating Solid Disk, Loaded at the Outer Radius

Except that the disk is solid, the considerations given in Sect. 2.1 for the annular disk loaded at the outer radius also apply here. Naturally, σ_{re} will be positive in the case where the uniformly distributed surface forces on the outer radius simulate the stresses from centrifugal forces caused by blades spaced at equal angles on the disk's outer surface, and will be negative and equal to $-p_e$ in the case where the effect of an external shrink-fit is examined.

Whatever the sign of σ_{re}, for the relations (2.10) to continue to be valid even at the centre of the disk, where $\rho = 0$, we must necessarily have $B = 0$. The other condition to be imposed for calculating the remaining integration constant A is:

$$\sigma_r = \sigma_{r_e} \qquad \text{for} \qquad \rho = 1 \; , \tag{2.20}$$

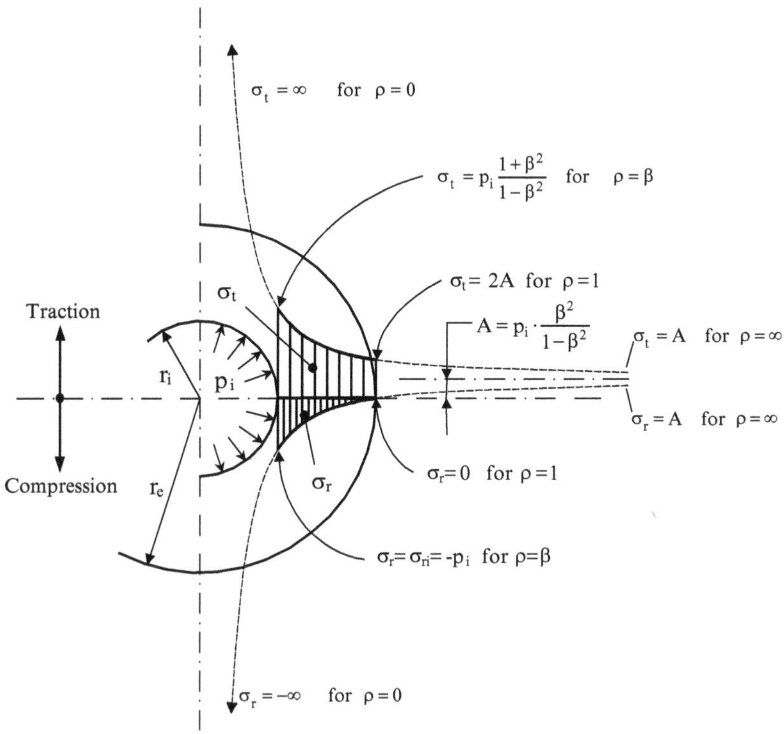

Fig. 2.4 Distribution curves of principal stresses σ_r and σ_t versus radius in an annular disk loaded at inner edge by a negative $\sigma_{ri} = -p_i$, and characteristic values of the quantities involved

From the first relation (2.10), we have

$$A = \sigma_{re}. \tag{2.21}$$

Expressions (2.10) and (2.11) then give:

$$\begin{cases} \sigma_r = \sigma_t = \sigma_{r_e} \\ u = \sigma_{r_e} \cdot \dfrac{r_e}{E} \cdot \rho \cdot (1 - \nu). \end{cases} \tag{2.22}$$

Consequently, stress is constant throughout the disk and positive or negative respectively in the cases of load from centrifugal forces caused by blades on the outer radius, or from an external shrink-fit. It is thus confirmed that the disk is subjected to radial and hoop stresses that are equal to each other at all points of the disk, as if the condition were one of uniform strength. Equation (2.22) can also be obtained from relations (2.16), with $\beta = 0$ in the latter.

2.5 Rotating Annular Disk

The general solution of (2.2) can be found by adding the solution of the associated homogeneous (2.5), which is already known, to a particular solution u_p of the complete equation. The latter can be readily expressed through the following relation:

$$u_p = -\left(1 - \nu^2\right) \cdot \frac{\gamma \cdot \omega^2 \cdot r^3}{8E}. \tag{2.23}$$

This is obtained by taking $u = C \cdot r^3$, substituting this relation together with its first and second derivatives in (2.2) and calculating constant C, once the coefficient of the powers of the variable r has been equalled to zero.

However, given the particularly simple structure of (2.2), which can be written in the form:

$$\frac{d}{dr}\left[\frac{1}{r} \cdot \frac{d}{dr}(u \cdot r)\right] = -\left(1 - \nu^2\right) \cdot \frac{\gamma \cdot \omega^2 \cdot r}{E}, \tag{2.24}$$

we have chosen for demonstration purposes to integrate it directly. With (2.24) in this form, performing two successive integrations followed by a derivation operation yields:

$$\begin{aligned} u &= -(1 - \nu^2) \cdot \frac{\gamma \cdot \omega^2}{E} \cdot \frac{r^3}{8} + C_1 \cdot \frac{r}{2} + \frac{C_2}{r} \\ \frac{du}{dr} &= -(1 - \nu^2) \cdot \frac{\gamma \cdot \omega^2}{E} \cdot \frac{3}{8} \cdot r^2 + \frac{C_1}{2} - \frac{C_2}{r^2}. \end{aligned} \tag{2.25}$$

Integrating equation (2.24) directly enables us to calculate radial displacement $\mathrm{u} = \mathrm{u}(r)$ as the sum of two contributions: the first, represented by the first term of the second member of the first relation (2.25), and thus coinciding with (2.23), regards centrifugal loading, while the second, given by the remaining two terms of the second member of the first relation (2.25), and thus coinciding with expression (2.6), is the contribution of the surface forces acting on the disk's inner and outer radii.

Substituting relations (2.25) in (1.27) from which temperature terms are omitted gives the following expressions of σ_r and σ_t as a function of radius r:

$$\begin{cases} \sigma_r = \dfrac{E \cdot C_1}{2 \cdot (1 - \nu)} - \dfrac{E \cdot C_2}{(1 + \nu)} \cdot \dfrac{1}{r^2} - \gamma \cdot \omega^2 \cdot \dfrac{3 + \nu}{8} \cdot r^2 \\ \sigma_t = \dfrac{E \cdot C_1}{2 \cdot (1 - \nu)} + \dfrac{E \cdot C_2}{(1 + \nu)} \cdot \dfrac{1}{r^2} - \gamma \cdot \omega^2 \cdot \dfrac{1 + 3\nu}{8} \cdot r^2. \end{cases} \tag{2.26}$$

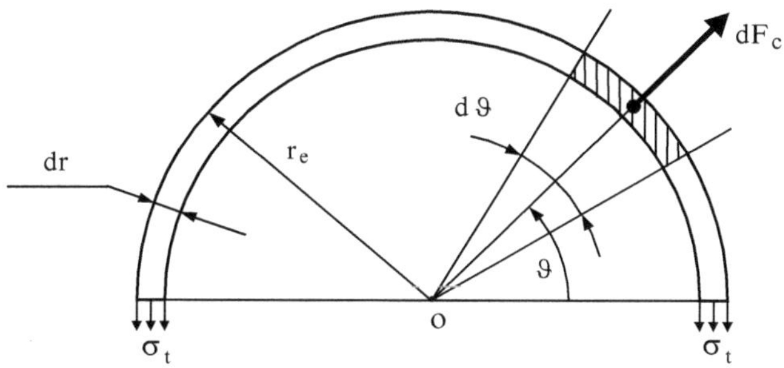

Fig. 2.5 Equilibrium of a peripheral semi-fiber, schematized as an elementary half-ring of radius r_e, radial thickness dr and axial dimension dz, subjected only to centrifugal loading

Finally, by multiplying and dividing the second and third term of the second members of expressions (2.26) by r_r^2, introducing the dimensionless variable ρ and taking

$$\sigma_0 = \gamma \cdot \omega^2 \cdot r_e^2; \quad C = \frac{3+\nu}{8} \cdot \sigma_0; \quad D = \frac{1+3\nu}{8} \cdot \sigma_0\,, \tag{2.27}$$

we arrive at the following relations, which express radial and hoop stresses as a function of ρ:

$$\begin{cases} \sigma_r = A - B/\rho^2 - C \cdot \rho^2 \\ \sigma_t = A + B/\rho^2 - D \cdot \rho^2. \end{cases} \tag{2.28}$$

The reference stress σ_0 given by the first (2.27) has a precise physical meaning: it is the value assumed by hoop stress σ_t in an infinitely thin ring of radius r_e (the outside, inside and mean radii in this case coincide) rotating around its axis at angular velocity ω, or in other words having peripheral velocity $\omega \cdot r_e$. For the half-ring shown in Fig. 2.5, obtained by sectioning a ring of radial thickness dr and axial length dz along a diametral plane, imposing equilibrium (d'Alembert's principle) in the direction along the ring's section plane gives:

$$2 \cdot \sigma_t \cdot dr \cdot dz = \int_0^{\pi} dF_c \cdot \sin\vartheta, \tag{2.29}$$

where dF_c is the elementary centrifugal force acting on the generic ring element, of angular width $d\vartheta$ and position ϑ relative to the section plane.

Bearing in mind that $dF_c = dm \cdot \omega^2 \cdot r_e = \gamma \cdot dV \cdot \omega^2 \cdot r_e = \gamma \cdot \omega^2 \cdot r_e^2 \cdot dr \cdot dz \cdot d\vartheta$ and then integrating the second member of expression (2.29), we arrive at relation:

$$\sigma_t = \gamma \cdot \omega^2 \cdot r_e^2 = \sigma_0\,. \tag{2.30}$$

The area $dA = dr \cdot dz$ of the elementary ring's section does not appear in this relation: as in the case of a catenary, each tangential fiber making up the ring bears the centrifugal load associated with it. The same relation shows that the hoop stress σ_t in an infinitely thin ring depends only on the materials density γ and on peripheral velocity $v = \omega \cdot r_e$. For any given material, there will always be a limit value of this velocity that cannot be exceeded without jeopardizing the ring's mechanical strength.

It should also be noted that for a disk rotating at angular velocity ω, stress σ_t, quantified with reference to its outermost peripheral fiber is the maximum value that σ_t can reach in the entire disk, assuming that the latter consists of fibers that are all independent. This justifies taking this stress as the reference stress σ_0 used to make the stresses due to centrifugal loading dimensionless. It also justifies the usual limitation on peripheral velocity in sizing rotating disks, given that it is equivalent to setting a limit on reference stress σ_0, leading to safe conditions in the entire disk.

On the basis of these considerations, it is clear that the rotor cannot be configured as a thin ring if high peripheral velocities are to be reached. The configuration with hub and crown ring connected by means of a predetermined number of spokes spaced at equal angles is not always suitable for ensuring sufficient structural strength for intense centrifugal loading. When peripheral velocity is very high (up to 400 to 500 m/s), the rotor should be configured as a disk with continuous walls of appropriate profile.

To complete our analysis of the rotating annular disk, the relation expressing radial displacement $u = u(\rho)$ must be determined. To this end, by introducing expressions (2.28) in the second (1.25), from which the temperature term is omitted, and bearing in mind that the second relation (1.14) gives $u = r \cdot \varepsilon_t$, we obtain:

$$u = \frac{r_e}{E} \cdot \rho \cdot \left[A \cdot (1 - v) + \frac{B}{\rho^2} \cdot (1 + v) - (D - v \cdot C) \cdot \rho^2 \right]. \qquad (2.31)$$

Relations (2.28) and (2.31) describe the distribution of radial and hoop stresses and radial displacement versus ρ in a disk subjected to centrifugal loading. In these relations, C and D are known terms, while A and B are the integration constants to be determined by imposing boundary conditions.

For the annular disk subjected only to centrifugal loading, the boundary conditions to be imposed are:

$$\begin{cases} \sigma_r = 0 & \text{per} \quad \rho = \beta \\ \sigma_r = 0 & \text{per} \quad \rho = 1 \end{cases}. \qquad (2.32)$$

Accordingly, the system obtained from the first of relations (2.28) gives:

$$\begin{aligned} A &= \frac{3 + v}{8} \cdot \sigma_0 \cdot \left(1 + \beta^2\right) \\ B &= \frac{3 + v}{8} \cdot \sigma_0 \cdot \beta^2 . \end{aligned} \qquad (2.33)$$

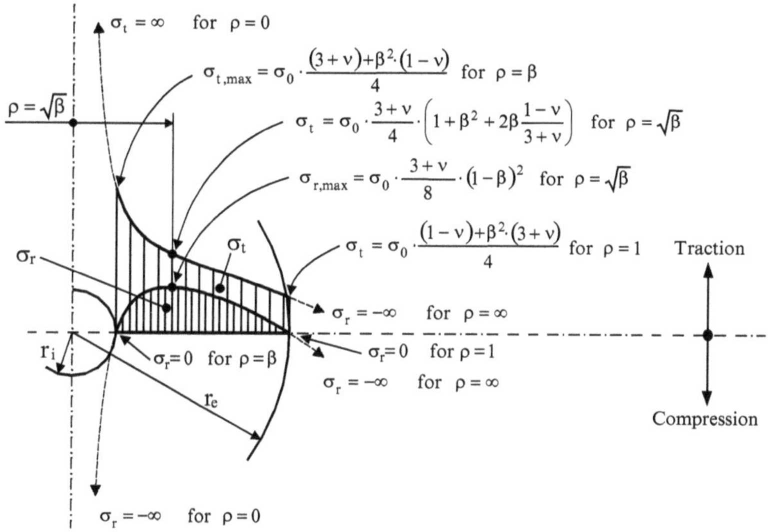

Fig. 2.6 Distribution curves of principal stresses σ_r and σ_t versus radius in a rotating annular disk and characteristic values of the quantities involved

Substituting the values of the constants thus found in relations (2.28) and (2.31) gives the following expressions for σ_r, σ_t and u which provide a univocal solution of the problem:

$$\begin{cases} \sigma_r = \dfrac{3+\nu}{8}\cdot\sigma_0\cdot\left(1+\beta^2-\dfrac{\beta^2}{\rho^2}-\rho^2\right) \\ \sigma_t = \dfrac{3+\nu}{8}\cdot\sigma_0\cdot\left(1+\beta^2+\dfrac{\beta^2}{\rho^2}-\dfrac{1+3\nu}{3+\nu}\cdot\rho^2\right) \\ u = \dfrac{r_e}{E}\cdot\rho\cdot\dfrac{3+\nu}{8}\cdot\sigma_0\cdot\left[\left(1+\beta^2\right)\cdot(1-\nu)+(1+\nu)\cdot\dfrac{\beta^2}{\rho^2}-\rho^2\cdot\left(\dfrac{1-\nu^2}{3+\nu}\right)\right]. \end{cases} \tag{2.34}$$

The radial and hoop stresses in the rotating disk are thus proportional to σ_0, so the square of peripheral velocity $v = \omega \cdot r_e$. Figure 2.6 shows the distribution of these radial and hoop stresses with radius in the annular disk subjected only to centrifugal loading. As can be seen from the figure, the absolute maximum value of the stresses characterizing the stress state is once again that reached by hoop stress at the inner edge. For finite β and for $\rho \to 0$, the hoop stress would tend to $+\infty$, while the radial stress would tend to $-\infty$. Here again, however, for $\beta \to 0$, we have a singularity whereby the ratio β^2/ρ^2 must be considered as unity in calculating hoop stress at the inner radius. Given the boundary condition, the radial stress at this radius will be zero, as is also confirmed by the first (2.34) for $\rho = \beta$ and β^2/ρ^2 considered as unity.

Radial stress, which is zero at the outer radius as well as the inner radius (see the boundary condition given by (2.32)), assumes its maximum value when $d\sigma_r/d\rho = 0$ or, equivalently, when the sum $[(\beta^2/\rho^2) + \rho^2]$ is minimum, i.e., when $\beta^2/\rho^2 = \rho^2$, given that the product of the two addends is constant. Maximum radial stress, which thus occurs for $\rho = \sqrt{\beta}$, or in other words for a radius $r = \sqrt{r_i \cdot r_e}$ equal to the geometric mean of the outer radius, is:

$$\sigma_{r,\max} = \frac{3+\nu}{8} \cdot (1-\beta)^2 \cdot \sigma_0. \tag{2.35}$$

By contrast, hoop stress σ_t increases continuously from the outer edge to the inner edge, where it respectively assumes the minimum value and the maximum value given by the relation:

$$\sigma_{t,\,\min,\max} = \left[\frac{1+\beta^2}{2} \mp \frac{1+\nu}{4} \cdot \left(1-\beta^2\right)\right] \cdot \sigma_0, \tag{2.36}$$

where the minus sign applies for the outer edge and the plus sign for the inner edge. The absolute maximum value of the stress field for an annular disk is thus that of the hoop stress at the inner boundary, where it is:

$$\sigma_{t,\,\max} = \sigma_0 \cdot \frac{(3+\nu) + \beta^2 \cdot (1-\nu)}{4}; \tag{2.37}$$

In fact, $\sigma_{t,max}$ is always greater than $\sigma_{r,max}$ and the ratio σ_t/σ_r is always greater than 1.

Figure 2.7a shows the curves for ratios $\sigma_{t,max}/\sigma_0$ and $\sigma_{t,min}/\sigma_0$ given by (2.36) as a function of β^2 for $\nu = 0.3$. Point S, for the case examined above of an elementary ring of radial thickness dr and axial length dz (see Fig. 2.5), is characterized by the equality $\sigma_{t,max} = \sigma_{t,min} = \sigma_0$. For any given outer radius r_e, the hoop stresses at the inner edge and at the outer edge decrease (the latter more rapidly than the former) as the inner radius r_i decreases, dropping to 0.825 and 0.175 respectively of the value for the elementary ring.

Figure 2.7b shows curves for the expressions appearing in parentheses in the first two relations (2.34) as a function of ρ for $\nu = 0.3$; if we do not consider the constant factor $\sigma_0 \cdot (3+\nu)/8$, these relations represent for stresses σ_r and σ_t. The solid-line curves apply for $\beta = 0.25$, i.e., the case where $r_i = r_e/4$. The dashed-line σ_t curves apply for other values of β (0.50; 0.75). Relation (2.37) indicates that stress $\sigma_{t,max}$ at the inner boundary is a parabolic function of β; curve *m-n* highlights this variation.

For a very small inner radius, i.e., for $\beta \to 0$, stress σ_t near the hole varies significantly by comparison with the solid disk. This circumstance is clearly shown by curve *m–p–q* (Fig. 2.7b), where the maximum value corresponds to the relation

$$\sigma_{t,\,\max} \cong \sigma_0 \cdot \frac{(3+\nu)}{4}, \tag{2.38}$$

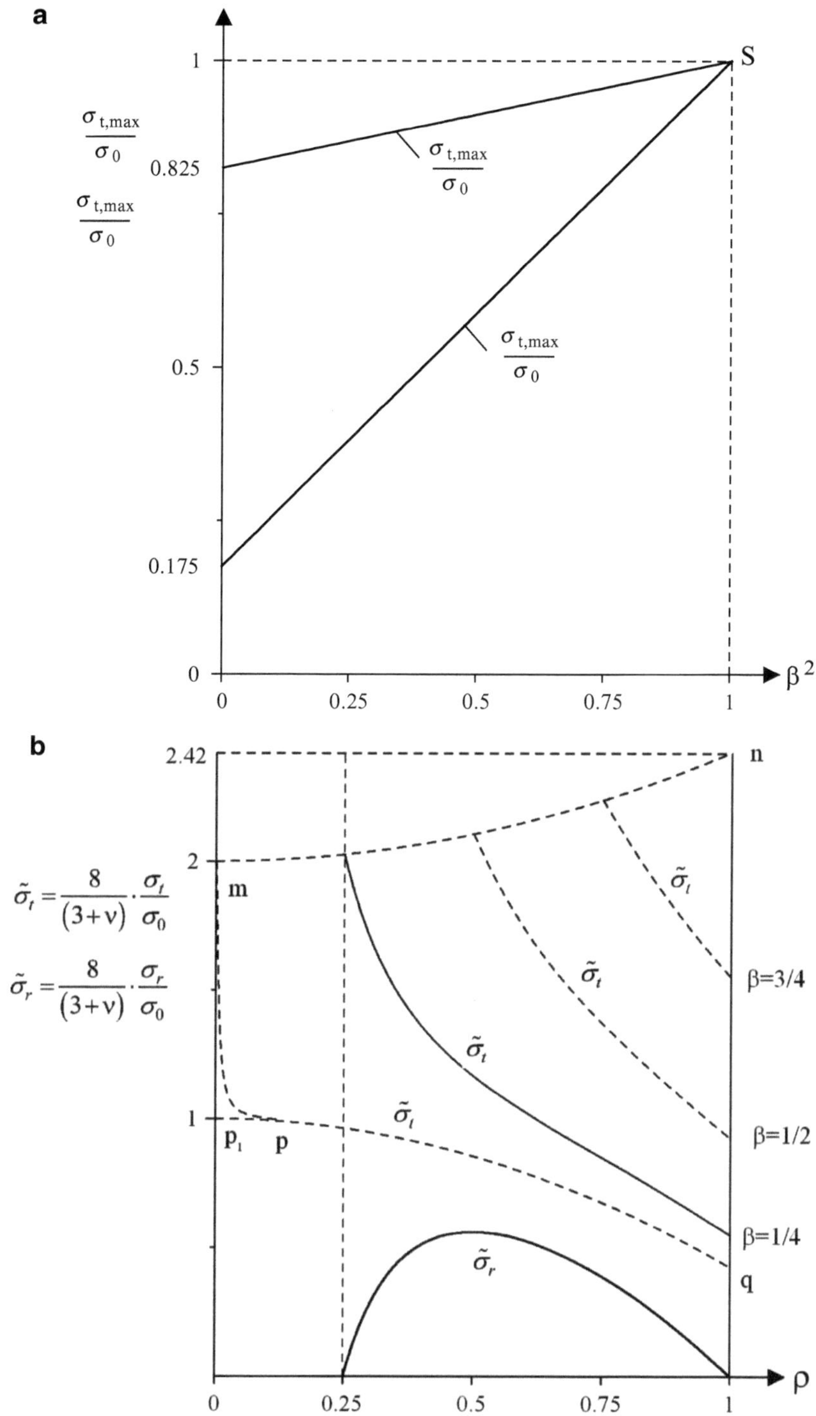

Fig. 2.7 (**a**) Distribution curves of dimensionless stresses $\sigma_{t,max}/\sigma_0$ and $\sigma_{t,min}/\sigma_0$ as a function of β^2, for $\nu = 0.3$; (**b**) distribution curves of dimensionless stresses $\tilde{\sigma}_t$ and $\tilde{\sigma}_r$ versus dimensionless radius, for different values of β

which is obtained from (2.37) for $\beta \rightarrow 0$. However, the presence of a central hole of this kind ($\beta \rightarrow 0$) has no effect on the displacement field.

For $r_i \rightarrow r_e$, i.e., for $\beta \rightarrow 1$, as indicated earlier, relation (2.37) yields $\sigma_{t,max} = \sigma_0$, which coincides with the stress obtained for an infinitely thin disk of radius r_e rotating around its axis at angular velocity ω.

It can be concluded from Fig. 2.7b that stress $\sigma_{t,max}$ for an annular disk shows little variation with hole radius, as its value for an infinitely thin disk is only 20 % higher than that obtained for with an axial hole of very small diameter.

In mechanical disk design, it is obvious that the most dangerous stress state must be considered, i.e., that arising at the inner radius. For $\rho = \beta$, the Tresca's criterion, i.e. the maximum-shear-stress criterion (whose results, in this case, coincide with those obtained by using the maximum-normal-stress criterion and the maximum-distortion-energy criterion) leads to the expression (see Sect. 9.1):

$$\tau_{\max} = (\sigma_t/2)_{\rho=\beta} = \sigma_0 \cdot \frac{(3+\nu) + \beta^2 \cdot (1-\nu)}{8}. \tag{2.39}$$

By contrast, using the maximum octahedral-shear-stress criterion, which is one of the most reliable criteria for predicting yield, and hence the onset of the plastic flow, and is directly correlated with the maximum-distortion-energy criterion, leads to the following relation:

$$\tau_{oct} = \frac{\sqrt{2}}{3} \cdot \sigma_t = \frac{\sqrt{2}}{3} \cdot \sigma_0 \cdot \frac{(3+\nu) + \beta^2 \cdot (1-\nu)}{4}. \tag{2.40}$$

At the inner edge of the disk where the material is at greatest risk, it is thus clear that both criteria indicate that the danger of yielding depends on the higher principal stress, which is the hoop stress.

With either the maximum-shear-stress criterion or the maximum octahedral-shear-stress criterion, the equivalent or ideal stress σ_e depends on σ_0, β and ν. For steels ($\nu = 0.3$), we have: $\sigma_e = \tau_{max} = 0.413{\cdot}\sigma_0$, for $\beta = 0.1$ ($r_e = 10{\cdot}r_i$), and $\sigma_e = \tau_{oct} = 0.389{\cdot}\sigma_0$, again for $\beta = 0.1$. For $\beta \rightarrow 1$, or in other words for disks of infinitely small radial thickness ($r_e \approx r_i$), we have: $\sigma_e = \tau_{max} = 0.500{\cdot}\sigma_0$ and $\sigma_e = \tau_{oct} = 0.471{\cdot}\sigma_0$.

To conclude this section it should be emphasized that, as Figs 2.6 and 2.7 clearly show, the stress state along the radius is not such that optimal use can be made of the material's strength properties. From the design standpoint, the use of constant thickness disks should be limited to low operating speeds, which lead to equally low maximum stresses. As will be demonstrated below, other types of disk profile offer undeniable advantages over constant thickness disks: examples of such profiles include those used in turbine disks, which feature larger thicknesses at the inner radius and smaller thicknesses at the outer radius, so that the profile tapers from a thick centre to a thinner edge.

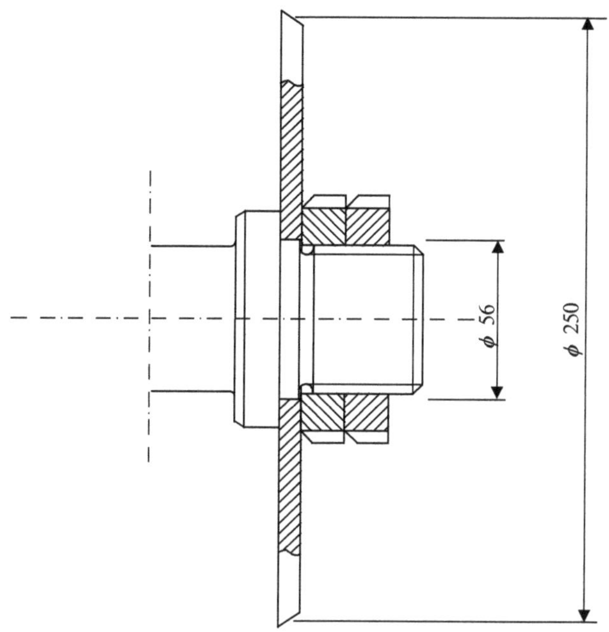

Fig. 2.8 Geometry of a rotating cutter for reams of paper

2.5.1 *Example 1*

A thin steel disk (Fig. 2.8) is used at a rotation velocity of $n = 14{\cdot}10^3$ rpm as a rotary cutter for reams of paper. Except for its cutting edge, the disk can be considered as being of constant thickness. It features a 250 mm outside diameter and a 56 mm inside diameter and is made of unquenched AISI 1060 steel, with yield strength (or yield stress) $\sigma_y = 480$ MPa. The disk is installed on the end of a shaft, where it is retained axially against the shaft shoulder by means of a nut and locknut. Disk rotation is thus driven by friction, though the effects of friction are neglected in calculating stresses and displacements. We will calculate the stress state and radial displacement at the inner and outer radii.

From the problem data, we have: $\nu = 0.30$; $\beta = r_i/r_e = 28/125 = 0.224$; $\sigma_0 = \gamma \cdot \omega^2 \cdot r_e{}^2 = 7{,}800{\cdot}(2{\cdot}\pi{\cdot}14{,}000/60)^2{\cdot}\ 0.125^2 = 262$ MPa; $E = 210$ GPa.

Equation (2.34) yield:

$$\sigma_r = 108 \cdot \left(1 + \beta^2 - \frac{\beta^2}{\rho^2} - \rho^2\right) = 108 \cdot \left(1.05 - \frac{0.05}{\rho^2} - \rho^2\right)$$

$$\sigma_t = 108 \cdot \left(1 + \beta^2 + \frac{\beta^2}{\rho^2} - \frac{1 + 3\nu}{3 + \nu} \cdot \rho^2\right) = 108 \cdot \left(1.05 + \frac{0.05}{\rho^2} - 0.58 \cdot \rho^2\right)$$

$$u = 108 \cdot \frac{r_e}{E} \cdot \rho \cdot \left[(1 + \beta^2) \cdot (1 - \nu) + (1 + \nu) \cdot \frac{\beta^2}{\rho^2} - \rho^2 \cdot \left(\frac{1 - \nu^2}{3 + \nu}\right)\right]$$

$$= 6.43 \cdot 10^{-5} \cdot \rho \cdot \left[0.735 + \frac{0.065}{\rho^2} - 0.275 \cdot \rho^2\right].$$

Radial stress is zero at the outer radius ($\rho = 1$) and at the inner radius ($\rho = \beta$); it is maximum for $\rho = \sqrt{\beta}$, or in other words for $\rho = 0.47$ ($r = 59.2$ mm), and, being a tensile stress, is $\sigma_{r,max} = 65$ MPa.

The values assumed by hoop stress at the inner and outer radii respectively are:

$$(\sigma_t)_{\rho=\beta} = 218 \text{ MPa}; \ (\sigma_t)_{\rho=1} = 56 \text{ MPa}.$$

Using Tresca's criterion, the maximum-shear-stress occurs at the inner radius and is:

$$\tau_{max} = \sigma_t/2 = 109\text{MPa}.$$

If the shear forces are relatively small and velocity is stationary, the load can be regarded as static. In this case, the safety factor s for the material according to the maximum-shear-stress criterion (which here coincides with the maximum-normal-stress criterion) is:

$$s = (\sigma_y/2)/\tau_{max} = 2.20.$$

Were the disk to start and stop frequently, the service conditions would be those of pulsating stress. The disk would thus be operating under fatigue and the safety factor would be lower.

The values for radial displacement at the inner and outer radii are as follows:

$$(u)_{\rho=\beta} = 2.91 \cdot 10^{-5} \text{ m}; \ (u)_{\rho=1} = 3.38 \cdot 10^{-5} \text{ m}.$$

2.5.2 *Example 2*

The mechanical component shown in Fig. 2.9, made of unquenched AISI 1060 steel with $\sigma_y = 480$ MPa, consists of a constant thickness disk, a hub and a crown ring, also of constant thickness, and featuring known geometry ($r_m = 50$ mm; $r_i = 90$ mm; $r_e = 260$ mm; $r_c = 300$ mm; $a = 100$ mm; $b = 130$ mm; $h = 20$ mm).

Assuming that the component operates at a steady-state velocity of $6 \cdot 10^3$ rpm and that the surface forces at radii r_m and r_c can be considered zero, we will determine the stress state in the disk connecting the hub to the crown ring (disk of thickness h), as well as the radial displacements at radii r_i and r_e.

To solve the problem, it is first necessary to determine the radial stresses in section A ($r = r_i$) and in section B ($r = r_e$), designated respectively as $\sigma_{r,A}$ and $\sigma_{r,B}$, which are two hyperstatic unknowns. To do so, it is sufficient to write two compatibility equations establishing on the one hand that the radial displacement of the outer edge of the disk is equal to the displacement of the inner edge of the ring, and, on the other hand, that the radial displacement of the inner edge of the disk is equal to the radial displacement of the outer edge of the hub.

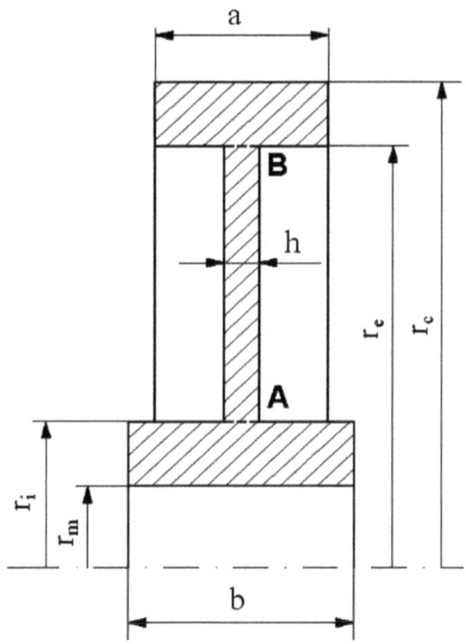

Fig. 2.9 Geometry of a disk with hub and crown ring, not subject to surface forces at radii r_m and r_c

Because of the principle of superposition, however, the radial displacement $u_{B,d}$ of interface B, regarded as part of the central disk, will be the sum of three contributions due to the centrifugal load, the stress $\sigma_{r,B}$ applied to radius $r = r_e$ and the stress $\sigma_{r,A}$ applied to radius $r = r_i$. Consequently, bearing the third relations of (2.34), (2.16) and (2.19) in mind, we can write:

$$u_{B,d}=\frac{r_e}{E}\cdot\rho\cdot\frac{3+v}{8}\cdot\sigma_0\cdot\left[(1+\beta^2)\cdot(1-v)+(1+v)\cdot\frac{\beta^2}{\rho^2}-\rho^2\cdot\left(\frac{1-v^2}{3+v}\right)\right]+$$
$$+\sigma_{r,B}\cdot\frac{r_e}{E}\cdot\frac{\rho}{1-\beta^2}\cdot\left[(1-v)+(1+v)\cdot\frac{\beta^2}{\rho^2}\right]-\sigma_{r,A}\cdot\frac{r_e}{E}\cdot\rho\cdot\frac{\beta^2}{1-\beta^2}\cdot\left[(1-v)+(1+v)\cdot\frac{1}{\rho^2}\right]; \quad (2.41)$$

this is to be calculated for $\rho = 1$, $\beta = r_i/r_e = 90/260$, $\sigma_0 = \gamma \cdot \omega^2 \cdot r_e^2 = 7{,}800\cdot(2\cdot\pi\cdot6{,}000/60)^2\cdot 0.125^2 = 208$ MPa, $v = 0.3$ and $E = 210$ GPa.

Again because of the principle of superposition, when interface B is regarded as part of the crown ring, its radial displacement $u_{B,c}$ will be the sum of three contributions due to the centrifugal load, the stress $\sigma_{r,B}\cdot(h/a)$ applied to radius r_e and the stress at radius r_c due to the blades (if present). In this example, the latter will be zero, because the surface force at radius r_e is supposed equal to zero. The stress $\sigma_{r,B}\cdot(h/a)$ acting at crown ring radius r_e is justified by the fact that, for there to be equilibrium, the product of radial stress by thickness at the disk-ring interface must remain constant. It is obvious that Saint Venant's principle does not hold at this interface. Nevertheless, this approximation, though rough, is indispensable if we wish to use the relations of the mono-dimensional thin disk theory (also see Chap. 5: Disk of Uniform Strength).

On the basis of the foregoing considerations, and taking into account the third relations of (2.34) and (2.19), we can write:

$$u_{B,c} = \frac{r_c}{E} \cdot \rho \cdot \frac{3+\nu}{8} \cdot \sigma_{0,c} \cdot \left[(1+\beta_c^2) \cdot (1-\nu) + (1+\nu) \cdot \frac{\beta_c^2}{\rho^2} - \rho^2 \cdot \left(\frac{1-\nu^2}{3+\nu} \right) \right] + \\ -\sigma_{r,B} \cdot \frac{h}{a} \cdot \frac{r_c}{E} \cdot \rho \cdot \frac{\beta_c^2}{1-\beta_c^2} \cdot \left[(1-\nu) + (1+\nu) \cdot \frac{1}{\rho^2} \right]. \tag{2.42}$$

This relation is to be calculated for $\rho = \beta_c = r_e/r_c = 260/300$ and $\sigma_{0,c} = \gamma \cdot \omega^2 \cdot r_c^2 = 277$ MPa, as well as for $\nu = 0.3$ and $E = 210$ GPa.

If a surface force distribution due to the presence of blades acts on the outer radius of the crown ring, the second member of (2.42) would include a third term deriving from specializing the third relation (2.16).

Likewise, the radial displacement $u_{A,d}$ of interface A, regarded as part of the central disk, will be the sum of three contributions due to the centrifugal load, the stress $\sigma_{r,B}$ applied to radius $r = r_e$ and the stress $\sigma_{r,A}$ applied to radius $r = r_i$. Here again, bearing the third relations of (2.34), (2.16) and (2.19) in mind, we thus obtain:

$$u_{A,d} = \frac{r_e}{E} \cdot \rho \cdot \frac{3+\nu}{8} \cdot \sigma_0 \cdot \left[(1+\beta^2) \cdot (1-\nu) + (1+\nu) \cdot \frac{\beta^2}{\rho^2} - \rho^2 \cdot \left(\frac{1-\nu^2}{3+\nu} \right) \right] + \\ + \sigma_{r,B} \cdot \frac{r_e}{E} \cdot \frac{\rho}{1-\beta^2} \cdot \left[(1-\nu) + (1+\nu) \cdot \frac{\beta^2}{\rho^2} \right] - \sigma_{r,A} \cdot \frac{r_e}{E} \cdot \rho \cdot \frac{\beta^2}{1-\beta^2} \cdot \left[(1-\nu) + (1+\nu) \cdot \frac{1}{\rho^2} \right]; \tag{2.43}$$

this is the same as relation (2.41), with the difference that it is to be calculated for $\rho = \beta = r_i/r_e = 90/260$. The values for σ_0, ν and E are unchanged with respect to (2.41).

When interface A is regarded as part of the hub, its radial displacement $u_{A,m}$ will be the sum of three contributions due to the centrifugal load, the stress $\sigma_{r,A} \cdot (h/b)$ applied to radius r_i and the stress at radius r_m due to the fact that the hub is shrink-fit on the shaft. In this example, the latter stress, which is negative because it is a compression stress, is again assumed to be zero. Stress $\sigma_{r,A} \cdot (h/b)$ is justified by considerations similar to those which made it possible to calculate stress $\sigma_{r,B} \cdot (h/a)$ at the disk-ring interface.

Here again, bearing in mind the third relations (2.34) and (2.16), we can write:

$$u_{A,m} = \frac{r_i}{E} \cdot \rho \cdot \frac{3+\nu}{8} \cdot \sigma_{0,m} \cdot \left[(1+\beta_m^2) \cdot (1-\nu) + (1+\nu) \cdot \frac{\beta_m^2}{\rho^2} - \rho^2 \cdot \left(\frac{1-\nu^2}{3+\nu} \right) \right] + \\ + \sigma_{r,A} \cdot \frac{h}{b} \cdot \frac{r_i}{E} \cdot \frac{\rho}{1-\beta_m^2} \cdot \left[(1-\nu) + (1+\nu) \cdot \frac{\beta_m^2}{\rho^2} \right]. \tag{2.44}$$

In addition to the usual values of v and E, this relation is to be calculated for $\rho = 1$, $\beta_m = r_m/r_i = 50/90$ and $\sigma_{0,m} = \gamma \cdot \omega^2 \cdot r_i^2 = 25$ MPa.

If a negative surface force distribution resulting from the shrink-fit acts on the inner radius of the hub, the second member of (2.44) would include a third term which can be readily expressed by specializing the third relation (2.19).

Equating (2.41) with (2.42) and (2.43) with (2.44), substituting the values of the known quantities in these equalities and adding the common terms, we obtain the following linear system with two hyperstatic unknowns $\sigma_{r,A}$ and $\sigma_{r,B}$:

$$\begin{cases} 0.636 \cdot \sigma_{r,B} - 0.071 \cdot \sigma_{r,A} & = 54.08 \\ 0.204 \cdot \sigma_{r,B} - 0.163 \cdot \sigma_{r,A} & = -14.87. \end{cases} \tag{2.45}$$

Solving this system makes it possible to calculate the following values of the unknowns:

$$\begin{cases} \sigma_{r,A} & = 229.7 \text{ MPa} \\ \sigma_{r,B} & = 110.9 \text{ MPa} . \end{cases} \tag{2.46}$$

Substituting these values of $\sigma_{r,A}$ and $\sigma_{r,B}$ in either of relations (2.43) or (2.44) and (2.41) or (2.42) gives the following values for radial displacement at interfaces A ($r = r_i$) and B ($r = r_e$):

$$\begin{cases} u_A & = 0.029 \cdot 10^{-3} \text{ m} \\ u_B & = 0.127 \cdot 10^{-3} \text{ m} . \end{cases} \tag{2.47}$$

Next to be determined is the stress state in the disk of thickness h, for which boundary conditions have been found. We are thus dealing with a rotating disk of given geometry, loaded at the inner and outer edges by known surface forces $\sigma_{r,A}$ and $\sigma_{r,B}$.

Thus, using the first relation (2.45), imposing the two boundary conditions:

$$\begin{cases} \sigma_r = \sigma_{r,A} & = 229.7 \text{ MPa, for } \rho = \beta = r_i/r_e \\ \sigma_r = \sigma_{r,B} & = 110.9 \text{ MPa, for } \rho = 1 \end{cases} \tag{2.48}$$

and remembering that $C = [(3+v)/8] \cdot \sigma_0$, with $\sigma_0 = \gamma \cdot \omega^2 \cdot r_e^2 = 208$ MPa, yields the following system of two equations:

$$\begin{cases} A - 8.35 \cdot B - 10.29 & = 229.7 \\ A - B - 85.87 & = 110.9. \end{cases} \tag{2.49}$$

Solving this system makes it possible to determine the following values of the integration constants:

$$\begin{cases} A & = 190.92 \text{ MPa} \\ B & = -5.88 \text{ MPa}. \end{cases} \tag{2.50}$$

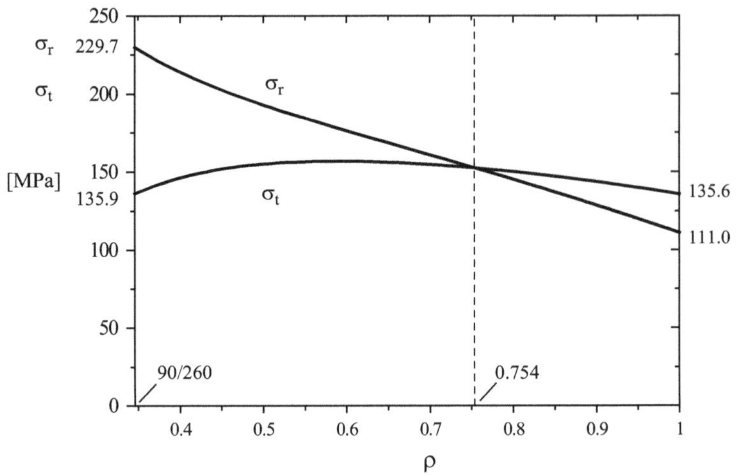

Fig. 2.10 Distribution curves of stresses σ_r and σ_t versus dimensionless radius in the disk shown in Fig. 2.9

Given relations (2.28), and noting that $D = [(1+3\cdot\nu)/8]\cdot\sigma_0$, the stress state in the disk of thickness h is given by:

$$\begin{cases} \sigma_r &= 190.92 + 5.88/\rho^2 - 85.87 \cdot \rho^2 \\ \sigma_t &= 190.92 - 5.88/\rho^2 - 49.44 \cdot \rho^2. \end{cases} \tag{2.51}$$

Figure 2.10 shows distribution curves of stresses σ_r and σ_t in the disk versus dimensionless radius ρ. Radial stress decreases nonlinearly from the inner radius to the outer radius, while hoop stress, though also varying nonlinearly, increases from the inner radius until it reaches a maximum value ($\sigma_{t,max} = 156.8$ MPa) for $\rho = 0.587$, subsequently dropping until it assumes a value at the outer radius which is almost equal to that at the inner radius.

The three failure criteria based on τ_{max}, σ_{max} and maximum-distortion-energy all indicate that the critical section is interface A, i.e., that at the disk's inner radius ($r = r_i$), where $\sigma_{r,A} = 229.7$ MPa and $\sigma_{t,A} = 135.9$ MPa.

If Tresca's criterion is used, the maximum-shear-stress at radius r_i is:

$$\tau_{max} = \sigma_r/2 = 114.9 \text{ MPa}.$$

Here again, in stationary conditions, the load can be regarded as static. In this case, the safety factor s for the material according to the maximum-shear-stress criterion (which here coincides with the maximum-normal-stress criterion) is:

$$s = (\sigma_y/2)/\tau_{max} = 2.09.$$

Were the disk to start and stop frequently, the service conditions would be those of pulsating stress. The disk would thus be operating under fatigue and the safety factor would be lower.

If the maximum-distortion-energy criterion is used, the equivalent stress in interface A would be $\sigma_e = 200{,}04$ MPa. In stationary conditions, the resulting safety factor would be:

$$s = \sigma_y/\sigma_e = 2.40.$$

2.5.3 Example 3

The mechanical component in the previous example (Fig. 2.9) is also subjected to surface forces $\sigma_{rm} = -\ 10$ MPa and $\sigma_{rc} = 40$ MPa applied to hub radius r_m and crown ring radius r_c respectively. These surface forces simulate respectively the stresses due to the fact that the hub is shrink-fit on the shaft, and the presence of blades. We will determine the stress state in the disk connecting the hub to the crown ring (disk of thickness h) and evaluate the influence that disk thickness h has on this stress state.

As in the example illustrated above, in solving the problem it is first necessary to determine the radial stresses in section A ($r = r_i$) and in section B ($r = r_e$), designated respectively as $\sigma_{r,A}$ and $\sigma_{r,B}$, which are two hyperstatic unknowns. To do so, it is necessary to write two compatibility equations, establishing on the one hand that the radial displacement of the outer edge of the disk is equal to the displacement of the inner edge of the ring, and, on the other hand, that the radial displacement of the inner edge of the disk is equal to the radial displacement of the outer edge of the hub.

Because of the principle of superposition, the radial displacement $u_{B,d}$ of interface B, regarded as part of the central disk, will be the sum of three contributions due to the centrifugal load, the stress $\sigma_{r,B}$ applied to radius $r = r_e$ and the stress $\sigma_{r,A}$ applied to radius $r = r_i$. Consequently, bearing the third relations of (2.34), (2.16) and (2.19) in mind, we can write:

$$u_{B,d} = \frac{r_e}{E} \cdot \rho \cdot \frac{3+\nu}{8} \cdot \sigma_0 \cdot \left[(1+\beta^2) \cdot (1-\nu) + (1+\nu) \cdot \frac{\beta^2}{\rho^2} - \rho^2 \cdot \left(\frac{1-\nu^2}{3+\nu} \right) \right] +$$
$$+ \sigma_{r,B} \cdot \frac{r_e}{E} \cdot \frac{\rho}{1-\beta^2} \cdot \left[(1-\nu) + (1+\nu) \cdot \frac{\beta^2}{\rho^2} \right] - \sigma_{r,A} \cdot \frac{r_e}{E} \cdot \rho \cdot \frac{\beta^2}{1-\beta^2} \cdot \left[(1-\nu) + (1+\nu) \cdot \frac{1}{\rho^2} \right]. \quad (2.52)$$

This is to be calculated for $\rho = 1$, $\beta = r_i/r_e = 90/260$, $\sigma_0 = \gamma \cdot \omega^2 \cdot r_e^2 = 7{,}800 \cdot (2 \cdot \pi \cdot 6{,}000/60)^2 \cdot 0.125^2 = 208$ MPa, $\nu = 0.3$ and $E = 210$ GPa.

Again because of the principle of superposition, when interface B is regarded as part of the crown ring, its radial displacement $u_{B,c}$ will be the sum of three contributions due to the centrifugal load, the stress $\sigma_{r,B}\cdot(h/a)$ applied to radius r_e and the stress at radius r_c due to the blades.

On the basis of the foregoing considerations, and taking into account the third relations of (2.34), (2.16) and (2.19), we can write:

$$u_{B,c}=\frac{r_c}{E}\cdot\rho\cdot\frac{3+v}{8}\cdot\sigma_{0,c}\cdot\left[\left(1+\beta_c^2\right)\cdot(1-v)+(1+v)\cdot\frac{\beta_c^2}{\rho^2}-\rho^2\cdot\left(\frac{1-v^2}{3+v}\right)\right]+$$
$$+\sigma_{rc}\cdot\frac{r_c}{E}\frac{\rho}{1-\beta_c^2}\cdot\left[(1-v)+(1+v)\cdot\frac{\beta_c^2}{\rho^2}\right]-\sigma_{r,B}\cdot\frac{h}{a}\frac{r_c}{E}\cdot\rho\cdot\frac{\beta_c^2}{1-\beta_c^2}\cdot\left[(1-v)+(1+v)\cdot\frac{1}{\rho^2}\right].\quad(2.53)$$

This is to be calculated for $\rho=\beta_c=r_e/r_c=260/300$, $\sigma_{rc}=40$ MPa and $\sigma_{0,c}=\gamma\cdot\omega^2\cdot r_c^2=277$ MPa, as well as for $v=0.3$ and $E=210$ GPa.

Likewise, the radial displacement $u_{A,d}$ of interface A, regarded as part of the central disk, will be the sum of three contributions due to the centrifugal load, the stress $\sigma_{r,B}$ applied to radius $r=r_e$ and the stress $\sigma_{r,A}$ applied to radius $r=r_i$. Here again, bearing the third relations of (2.34), (2.16) and (2.19) in mind, we thus obtain:

$$u_{A,d}=\frac{r_e}{E}\cdot\rho\cdot\frac{3+v}{8}\cdot\sigma_0\cdot\left[\left(1+\beta^2\right)\cdot(1-v)+(1+v)\cdot\frac{\beta^2}{\rho^2}-\rho^2\cdot\left(\frac{1-v^2}{3+v}\right)\right]+$$
$$+\sigma_{r,B}\cdot\frac{r_e}{E}\frac{\rho}{1-\beta^2}\cdot\left[(1-v)+(1+v)\cdot\frac{\beta^2}{\rho^2}\right]-\sigma_{r,A}\cdot\frac{r_e}{E}\cdot\rho\cdot\frac{\beta^2}{1-\beta^2}\cdot\left[(1-v)+(1+v)\cdot\frac{1}{\rho^2}\right].\quad(2.54)$$

This is the same as relation (2.52), with the difference that it is to be calculated for $\rho=\beta=r_i/r_e=90/260$. The values for σ_0, v and E are unchanged with respect to (2.52).

When interface A is regarded as part of the hub, its radial displacement $u_{A,m}$ will be the sum of three contributions due to the centrifugal load, the stress $\sigma_{r,A}\cdot(h/b)$ applied to radius r_i and the stress σ_{rm} at radius r_m due to the fact that the hub is shrink-fit on the shaft, which is clearly negative because it is a compression stress. Here again, bearing in mind the third relations (2.34), (2.16) and (2.19), we can write:

$$u_{A,m}=\frac{r_i}{E}\cdot\rho\cdot\frac{3+v}{8}\cdot\sigma_{0,m}\cdot\left[\left(1+\beta_m^2\right)\cdot(1-v)+(1+v)\cdot\frac{\beta_m^2}{\rho^2}-\rho^2\cdot\left(\frac{1-v^2}{3+v}\right)\right]+$$
$$+\sigma_{r,A}\cdot\frac{h}{b}\frac{r_i}{E}\frac{\rho}{1-\beta_m^2}\cdot\left[(1-v)+(1+v)\cdot\frac{\beta_m^2}{\rho^2}\right]-\sigma_{r,m}\cdot\frac{r_i}{E}\cdot\rho\cdot\frac{\beta_m^2}{1-\beta_m^2}\cdot\left[(1-v)+(1+v)\cdot\frac{1}{\rho^2}\right].\quad(2.55)$$

In addition to the usual values of v and E, this relation is to be calculated for $\rho=1$, $\beta_m=r_m/r_i=50/90$, $\sigma_{rm}=-10$ MPa and $\sigma_{0,m}=\gamma\cdot\omega^2\cdot r_i^2=25$ MPa.

Equating (2.52) with (2.53) and (2.54) with (2.55), substituting the values of the known quantities in these equalities and adding the common terms, we obtain a linear system with two unknowns $\sigma_{r,A}$ and $\sigma_{r,B}$. Solving this system makes it possible to calculate the following values of the unknowns:

$$\begin{cases} \sigma_{r,A} & = 235.9 \text{ MPa} \\ \sigma_{r,B} & = 119.8 \text{ MPa} . \end{cases} \tag{2.56}$$

Substituting these values of $\sigma_{r,A}$ and $\sigma_{r,B}$ in either of relations (2.54) or (2.55) and (2.52) or (2.53) gives the following values for radial displacement at interfaces A ($r = r_i$) and B ($r = r_e$):

$$\begin{cases} u_A & = 0.033 \cdot 10^{-3} \text{ m} \\ u_B & = 0.135 \cdot 10^{-3} \text{ m} . \end{cases} \tag{2.57}$$

Next to be determined is the stress state in the disk of thickness h, for which boundary conditions have been found. We are thus dealing with a rotating disk of given geometry, loaded at the inner and outer edges by known surface forces $\sigma_{r,A}$ and $\sigma_{r,B}$.

Thus, using the first relation (2.28), establishing the two boundary conditions:

$$\begin{cases} \sigma_r = \sigma_{r,A} & = 235.9 \text{ MPa, per } \rho = \beta = r_i/r_e \\ \sigma_r = \sigma_{r,B} & = 119.8 \text{ MPa, per } \rho = 1 \end{cases} \tag{2.58}$$

and remembering that $C = [(3+v)/8]\cdot\sigma_0$, with $\sigma_0 = \gamma \cdot \omega^2 \cdot r_e^{\,2} = 208$ MPa, yields the following system of two equations:

$$\begin{cases} A - 8.35 \cdot B - 10.29 & = 235.9 \\ A - B - 85.87 & = 119.8. \end{cases} \tag{2.59}$$

Solving this system makes it possible to determine the following values of the integration constants:

$$\begin{cases} A & = 200.13 \text{ MPa} \\ B & = -5.51 \text{ MPa}. \end{cases} \tag{2.60}$$

Given relations (2.28), and noting that $D = [(1 + 3\cdot v)/8]\cdot\sigma_0$, the stress state in the disk of thickness h is given by:

$$\begin{cases} \sigma_r = 200.13 + 5.51/\rho^2 - 85.87 \cdot \rho^2 \\ \sigma_t = 200.13 - 5.51/\rho^2 - 49.44 \cdot \rho^2. \end{cases} \tag{2.61}$$

Figure 2.11 shows distribution curves of stresses σ_r and σ_t in the disk versus dimensionless radius ρ. Here again, radial stress decreases nonlinearly from the

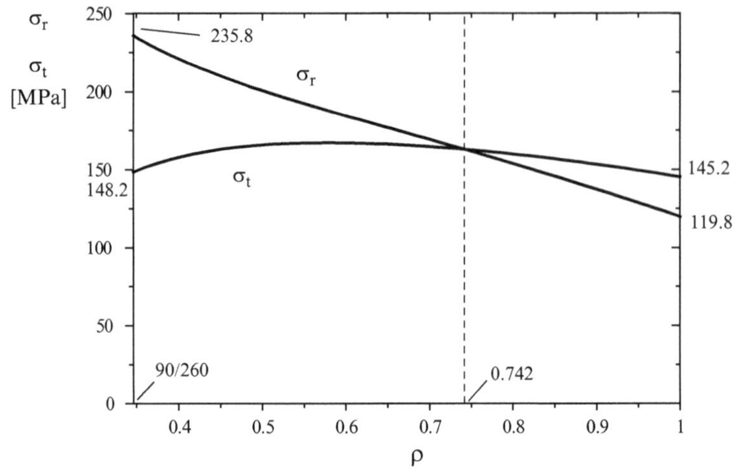

Fig. 2.11 Distribution curves of stresses σ_r and σ_t versus dimensionless radius in the disk shown in Fig. 2.9, subjected to surface forces at radii r_m and r_c

inner radius to the outer radius, while hoop stress, though also varying nonlinearly, increases from the inner radius until it reaches a maximum value ($\sigma_{t,max} = 167.1$ MPa) for $\rho = 0.578$, subsequently dropping until it assumes a value at the outer radius which is almost equal to that at the inner radius.

The three failure criteria based on τ_{max}, σ_{max} and maximum-distortion-energy all indicate that the critical section is interface A, i.e., that at the disk's inner radius ($r = r_i$), where $\sigma_{r,A} = 235.9$ MPa and $\sigma_{t,A} = 148.2$ MPa.

Here again, in stationary conditions, the load can be regarded as static. In this case, the safety factor s for the material according to the maximum-shear-stress criterion (which here coincides with the maximum-normal-stress criterion) is:

$$s = (\sigma_y/2)/\tau_{max} = 240/118.0 = 2.03.$$

If the maximum-distortion-energy criterion is used, the equivalent stress in interface A would be $\sigma_e = 206.4$ MPa. In stationary conditions, the resulting safety factor would be:

$$s = \sigma_y/\sigma_e = 2.33.$$

Finally, to evaluate the influence of the central disk's thickness h on its stress state, Fig. 2.12 shows radial stress σ_r and hoop stress σ_t versus radius ρ for thicknesses h varying from $h = 20$ mm (that used in this exercise) to $h = 80$ mm, i.e., a value near that of the outer ring. The curves not only show that h has a major impact on the stress state, but also that the trend is reversed as h increases. As thickness goes from $h = 20$ mm to $h = 80$ mm, the concavity of the curves

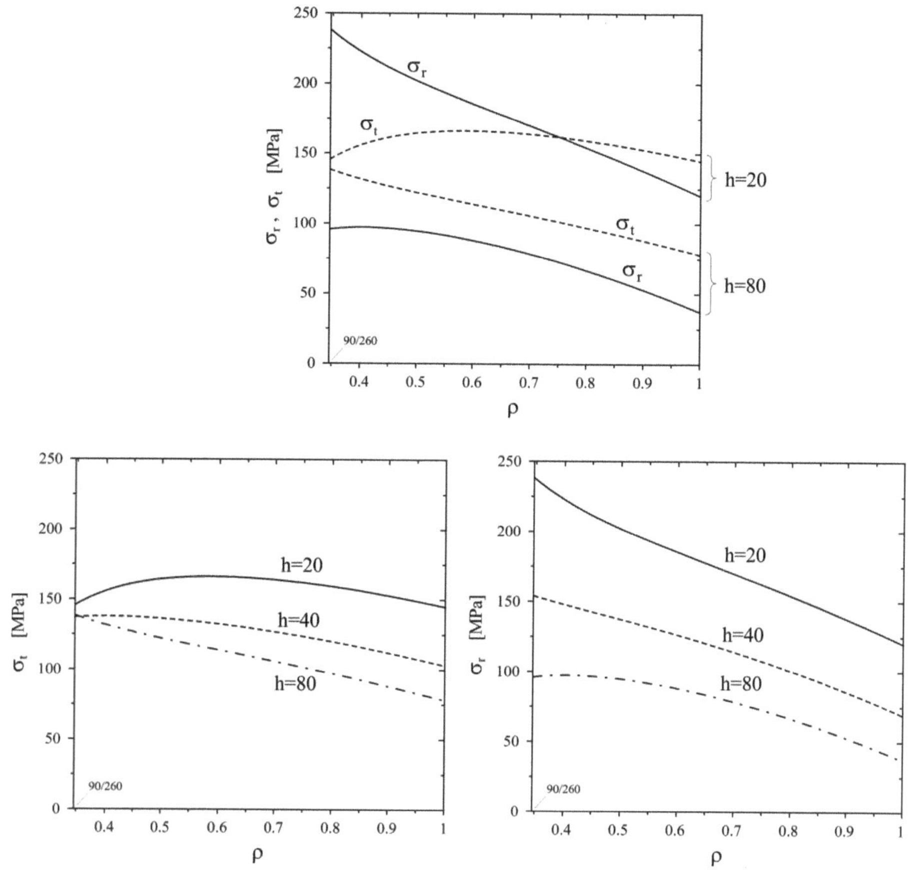

Fig. 2.12 Distribution curves of stresses σ_r and σ_t versus dimensionless radius in the disk shown in Fig. 2.9, for different thicknesses h

representing radial and hoop stresses is inverted, with the radial stress for the 80 mm thick disk assuming values that are constantly below those for hoop stress.

2.5.4 *Example 4*

We will now consider the rotor shown in Fig. 2.13, which is a constant thickness disk with radial slots on the peripheral surface that serve as seats for blades. The disk is made of unquenched AISI 1060 steel, with $\sigma_y = 480$ MPa, has a steady-state rotating velocity of $1.8{\cdot}10^3$ rpm, and features known geometry ($r_i = 100$ mm; $r_e = 400$ mm; $R = 660$ mm). Assuming that the surface force at radius $r = r_i$ is zero and that the mass of the blades is that of the material removed in

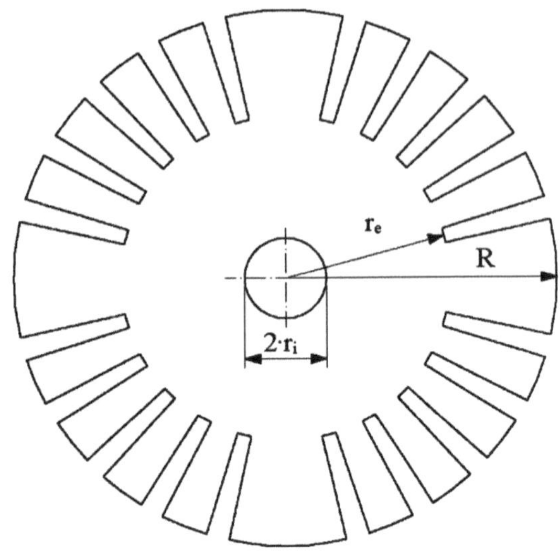

Fig. 2.13 Geometry of a rotor with peripheral radial slots

machining the slots, we will determine the total hoop stress acting on the disk's inner radius.

To solve the problem, it is first necessary to determine the radial stress acting at radius $r = r_e$. Because of the peripheral radial slots, the part of the rotor between radius r_e and radius R is not capable of withstanding hoop stresses.

As the radial stress at radius r_e is due to the centrifugal force to which the mass between radii r_e and R is subjected, we will have:

$$\sigma_{re} = \frac{1}{2\pi \cdot h \cdot r_e} \cdot \int_{r_e}^{R} dF_c, \tag{2.62}$$

where h is disk thickness and $dF_c = r \cdot \omega^2 \cdot dm = r \cdot \omega^2 \cdot \gamma \cdot dV$ is the elementary centrifugal force on mass dm of the annular section between radii r and $r + dr$, whose volume dV is equal to $2\pi \cdot r \cdot h \cdot dr$. We will thus have:

$$\sigma_{re} = \frac{\gamma \cdot \omega^2}{r_e} \cdot \int_{r_e}^{R} r^2 \cdot dr = \frac{\gamma \cdot \omega^2}{3 \cdot r_e} \cdot \left(R^3 - r_e^3\right) = 51.6 \text{ MPa}. \tag{2.63}$$

As the principle of superposition applies in the linear elastic field, the total hoop stress at radius r_i can be expressed by the following sum:

$$\sigma_{t,tot} = (\sigma_t)_{\sigma_{re}} + (\sigma_t)_{\omega}, \tag{2.64}$$

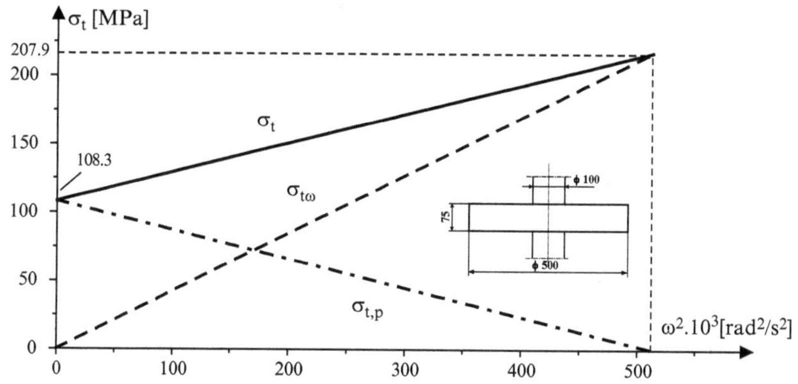

Fig. 2.14 Geometry of the shrink-fit shaft/disk assembly and stresses $\sigma_{t,p}$, $\sigma_{t,\omega}$ and σ_t versus ω^2

where the two terms in the second member represent the contribution of the stationary disk loaded at radius r_e by surface force σ_{re} and that of the disk only rotating and thus subjected only to the associated centrifugal load.

As the first contribution is given by the second relation (2.16) for $\rho = \beta = 100/400$, we thus obtain:

$$(\sigma_t)_{\sigma_{re}} = 110.1 \text{ MPa}. \tag{2.65}$$

The second contribution is obtained from the second relation (2.34) for $\rho = 1$ and for $\sigma_0 = \gamma \cdot \omega^2 \cdot r_e^2 = 44.34$ MPa. Accordingly, we have:

$$(\sigma_t)_\omega = 27.2 \text{ MPa}. \tag{2.66}$$

The total hoop stress acting at disk radius r_i will thus be:

$$\sigma_{t,tot} = 137.3 \text{ MPa}. \tag{2.67}$$

2.5.5 Example 5

An annular steel disk ($\gamma = 7.8 \cdot 10^3$ kg/m^3, $E = 200$ GPa and $\nu = 0.3$) of constant thickness $h = 75$ mm, inside diameter $2r_i = 0.1$ m and outside diameter $2r_e = 0.5$ m is assembled on a solid shaft made of the same material (Fig. 2.14) with a shrink-fit capable of generating a contact pressure $p_c = 100$ MPa at the disk/shaft interface, when the connection is stationary and at an assembly temperature coinciding with ambient temperature. We will determine the interference i needed to generate the assigned contact pressure p_c at ambient temperature; the maximum hoop stress in the disk caused by contact pressure when the connection is stationary at ambient temperature, as well as its variation with the square of angular velocity ω; the maximum hoop stress in the disk arising from rotation alone which is capable of bringing contact pressure p_c to zero at ambient temperature, and its variation with the square of angular velocity ω; and the maximum total hoop stress at the disk's inner radius when the connection is rotating at ambient temperature and a velocity of $n = 3{,}000$ rpm.

With the connection stationary and at ambient temperature, radial displacement u_d of the interface, regarded as part of the disk and subjected to a surface force distribution $\sigma_{ri} = -p_c$ at radius r_i is given by the third relation (2.19) for $\rho = \beta = r_i/r_e = 0.2$. We thus have:

$$u_d = \frac{p_c}{E} \cdot r_e \cdot \frac{\beta^3}{1-\beta^2} \cdot \left[(1-\nu) + \frac{(1+\nu)}{\beta^2}\right] = 3.46 \cdot 10^{-5}\text{m}.$$

Again with the connection stationary and at ambient temperature, when the interface is regarded as part of the shaft and subjected to a surface force distribution $\sigma_{re} = -p_c$ at the outside radius of the shaft, which is equal to r_i, its radial displacement u_a is given by the second relation (2.22) for $\rho = 1$. We thus have:

$$u_a = -p_c \cdot \frac{r_i}{E} \cdot (1-\nu) = -1.75 \cdot 10^{-5}\text{m}.$$

Consequently, the radial interference i needed at ambient temperature to generate the assigned contact pressure p_c at the disk/shaft interface will be:

$$i = u_d - u_a = (3.46 + 1.75) \cdot 10^{-5}\text{m} = 5.21 \cdot 10^{-5}\text{m}.$$

With the connection stationary and at ambient temperature, the maximum hoop stress in the disk resulting from contact pressure p_c is calculated using the second (2.19) for $\sigma_{ri} = -p_c$ and $\rho = \beta = 0.2$. This stress occurs at the inner radius and is:

$$\sigma_{t,p} = p_c \cdot \frac{1+\beta^2}{1-\beta^2} = 108.33 \text{ MPa}.$$

With the connection rotating at ambient temperature, the radial displacement u'_d of the interface regarded as part of the disk and due to rotation alone can be calculated using the third (2.34) for $\rho = \beta = 0.2$, noting that $\sigma_0 = \gamma \cdot \omega^2 \cdot r_e^2$. We thus have:

$$u'_d = \frac{\gamma \cdot \omega^2 \cdot r_e^3}{E} \cdot \beta \cdot \frac{3+\nu}{8} \cdot \left[(1+\beta^2) \cdot (1-\nu) + (1+\nu) - \beta^2 \cdot \left(\frac{1-\nu^2}{3+\nu}\right)\right]$$
$$= 1.025 \cdot 10^{-10} \cdot \omega^2.$$

Again with the connection rotating at ambient temperature, when the interface is regarded as part of the shaft, its radial displacement u'_a due to rotation alone is calculated using the third (2.34) for $\rho = 1$ and $\sigma_0 = \gamma \cdot \omega^2 \cdot r_i^2$, where, as indicated earlier, r_i is the shaft outer radius. We will thus have:

$$u'_a = \frac{\gamma \cdot \omega^2 \cdot r_i^3}{E} \cdot \frac{3+v}{8}\left[(1-v) - \frac{1-v^2}{3+v}\right] = 8.53 \cdot 10^{-13} \cdot \omega^2.$$

As u'_d and u'_a are both in the same direction, the contact pressure p_c will drop to zero when ω reaches a value such that assembly interference is zero, i.e., when $u'_d - u'_a = i$; this condition, which expressed in explicit form is:

$$(1.025 \cdot 10^{-10} - 8.53 \cdot 10^{-10}) \cdot \omega^2 = 5.21 \cdot 10^{-5},$$

yields $\omega = 716$ rad/s. At this rotation velocity, there is no interference at the shaft/disk connection, which is thus loose. In this case, it should be noted that the contribution of displacement u'_a relative to displacement u'_d could be neglected, as it is less than 1 %.

The maximum hoop stress in the disk resulting only from angular velocity $\omega = 716$ rad/s is calculated using the second (2.34) for $\sigma_0 = \gamma \cdot \omega^2 \cdot r_e^2 = 7.8 \cdot 10^3 \cdot 716^2 \cdot 0.25^2 = 249.92$ MPa and for $\rho = \beta = 0.2$. This stress occurs at the inner radius and is:

$$\sigma_{t,\omega} = \sigma_0 \cdot \frac{(3+v) + \beta^2 \cdot (1-v)}{4} = 207.93 \text{ MPa}.$$

As the maximum hoop stress resulting from simple rotation given by the second (2.34) is an increasing linear function of the square of angular velocity ω and thus drops to zero for $\omega = 0$, its variation with ω^2 in the interval $0 \leq \omega \leq 716$ rad/s of design interest will be that represented by the dashed line in Fig. 2.14.

On the other hand, assembly interference, which is proportional to contact pressure, decreases linearly with the square of ω. Consequently, the maximum hoop stress resulting from contact pressure will also be a decreasing linear function of ω^2; in the same interval $0 \leq \omega \leq 716$ rad/s, it will be represented by the dashed and dotted line in Fig. 2.14.

The maximum total hoop stress σ_t at the disk's inner radius with the connection at operating angular velocity ω is obtained as the superposition of the effects of the two previous contributions, or in other words as the sum of the stress $\sigma_{t,p}$ due to contact pressure and the stress $\sigma_{t,\omega}$ due to rotation only, both calculated at velocity ω. The curve thus obtained is represented with a solid line in Fig. 2.14.

From this distribution curve of σ_t, it can be concluded (though it is obviously also possible to proceed analytically) that the maximum total hoop stress at the disk's inner radius at a rotating velocity of 3,000 rpm ($\omega = 314$ rad/s) is 127.48 MPa.

2.6 Rotating Solid Disk

As indicated earlier, the condition to be imposed for the solid disk is that radial displacement u be zero at the centre ($r = \rho = 0$). From the first relation (2.25), it follows that $C_2 = 0$ and thus, given relation (2.9), $B = 0$. As a result, expressions (2.28) become:

$$\begin{cases} \sigma_r = A - C \cdot \rho^2 \\ \sigma_t = A - D \cdot \rho^2. \end{cases} \tag{2.68}$$

At the centre of the disk, we will thus have $\sigma_r = \sigma_t = A$. But as stress σ_r is zero at the outer radius ($\rho = 1$) we can conclude from the first relation (2.68) that $A = C$. Accordingly, at the centre of the disk we have:

$$\sigma_r = \sigma_t = A = C = \frac{3+\nu}{8} \cdot \sigma_0. \tag{2.69}$$

For the solid disk, (2.34) become:

$$\begin{cases} \sigma_r & = & \frac{3+\nu}{8} \sigma_0 . (1 - \rho^2). \\ \sigma_t & = & \frac{3+\nu}{8} \sigma_0 . \left(1 - \frac{1+3\nu}{3+\nu} \cdot \rho^2\right). \\ u & = & \frac{r_e}{E} \cdot \rho \cdot \frac{3+\nu}{8} \sigma_0 . \left[(1-\nu) - \rho^2 \cdot \frac{1-\nu^2}{3+\nu}\right]. \end{cases} \tag{2.70}$$

These equations express the radial and hoop stresses and radial displacement as a function of ρ. Equation (2.70) can also be derived from (2.34) by setting $\beta = 0$ in the latter.

Figure 2.15a shows these stresses' variation with the radius in a solid disk subjected to centrifugal loading alone. It can be concluded from this figure that the maximum stress value is that reached simultaneously by radial stress and hoop stress at the centre of the disk. Here as for the constant thickness annular disk, the stress state along the radius is not such that optimal use can be made of the material's strength properties. However, comparison with the equivalent annular disk shows that, all other conditions remaining equal, the solid disk is much stronger. In fact, the first two relations (2.70) indicate that at the axis, stresses σ_r and σ_t are maximum and equal to each other, and are:

$$\sigma_{t,\max} = \sigma_{r,\max} = \frac{3+\nu}{8} \cdot \sigma_0 . \tag{2.71}$$

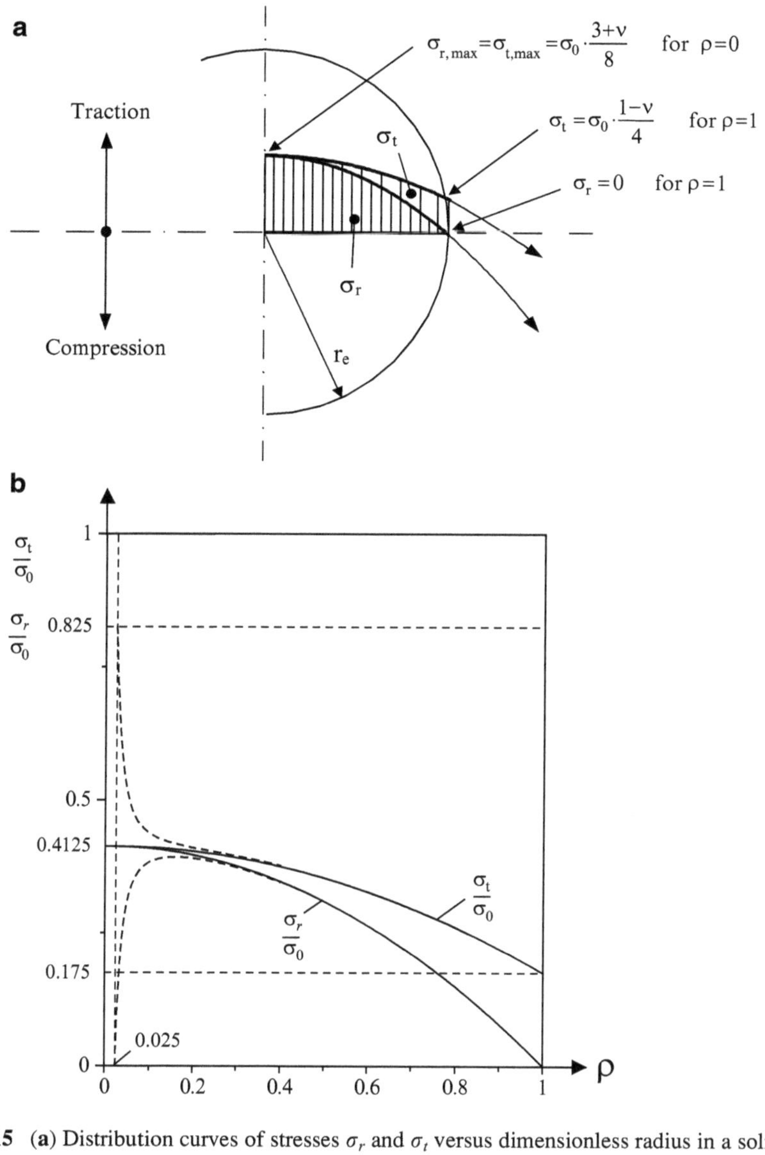

Fig. 2.15 (**a**) Distribution curves of stresses σ_r and σ_t versus dimensionless radius in a solid disk only rotating and characteristic values of the quantities involved; (**b**) variation of ratios σ_t/σ_0 and σ_r/σ_0 versus ρ for $\nu = 0.3$

Comparing this relation with (2.37), we can conclude that, because of the stress concentration, the stress at the boundary of a central hole of very small radius is twice that occurring on the axis of a solid disk. Again not considering the constant factor $\sigma_0 \cdot (3+\nu)/8$, the dashed–line p_1-p-q curve in Fig. 2.7b shows variation of stress σ_t along the radius of a solid disk.

Table 2.1 Equivalent stresses calculated with the τ_{max} and τ_{oct} failure criteria for the annular disk with two different values of β and for the corresponding solid disk, for $\nu = 0.3$

	Annular disk		
Failure criterion	$\beta = 0.1$	$\beta \approx 1$	Solid disk
τ_{max}	$\sigma_e = 0.413 \cdot \sigma_0$	$\sigma_e = 0.500 \cdot \sigma_0$	$\sigma_e = 0.206 \cdot \sigma_0$
τ_{oct}	$\sigma_e = 0.389 \cdot \sigma_0$	$\sigma_e = 0.471 \cdot \sigma_0$	$\sigma_e = 0.194 \cdot \sigma_0$

Figure 2.15b shows the numerical values of ratios σ_t/σ_0 e σ_r/σ_0 given by the first two relations (2.70) as a function of ρ for $\nu = 0.3$ (solid-line curves). The dashed-line curves apply to the numerical values assumed by the same ratios for an annular disk with very small β ($\beta = 0.025$). These curves indicate that for $\beta \to 0$, the ratio σ_r/σ_0 also tends to zero, whereas the ratio σ_t/σ_0 tends to a value twice that of the corresponding value for the solid disk.

Here again, in mechanical disk design, it is necessary to consider the most dangerous stress state, which is that occurring at the rotational axis. For $\rho = 0$, the maximum-shear-stress criterion (whose results again coincide with those obtained by using the maximum-normal-stress criterion and the maximum-distortion-energy criterion) leads to the expression:

$$\tau_{\max} = (\sigma_t/2)_{\rho=0} = \sigma_0 \cdot \frac{3+\nu}{16}. \tag{2.72}$$

By contrast, using the maximum octahedral-shear-stress criterion, which was introduced earlier for the annular disk, leads to the following relation:

$$\tau_{oct} = \frac{\sqrt{2}}{3} \cdot \sigma_t = \frac{\sqrt{2}}{3} \cdot \sigma_0 \cdot \frac{3+\nu}{8}. \tag{2.73}$$

At the rotational axis where the material is at greatest risk, it is clear that both criteria indicate that the hazard of structural failure depends on the higher principal stress, i.e., $(\sigma_t)_{\rho=0} = (\sigma_r)_{\rho=0}$. With both failure criteria, the equivalent stress σ_e thus depends on σ_0 and ν. For steels ($\nu = 0.3$), we have: $\sigma_e = \tau_{max} = 0.206 \cdot \sigma_0$, for $\rho = 0$, and $\sigma_e = \tau_{oct} = 0.194 \cdot \sigma_0$, again for $\rho = 0$.

For a more immediate comparison of the strength of rotating solid and annular disks having the same outer radius r_e and made of the same material, Table 2.1 summarizes the results obtained using the two failure criteria based on τ_{max} and τ_{oct}.

The table, which applies to $\nu = 0.3$, indicates that the equivalent stress in the solid disk is around half that which occurs in the annular disk with a small value of β ($\beta = 0.1$) and that the difference becomes more pronounced as β increases, becoming greatest for a rotor having the form of a thin disk ($\beta \approx 1$). In this latter case, the equivalent stress calculated with both failure criteria is over 240 % higher in the rotor configured as a thin ring than in the corresponding solid rotor.

Though the table also shows that the τ_{max} criterion is somewhat more prudential than the τ_{oct} criterion, experimental findings confirm that the latter is a better interpretation of the behaviour of ductile materials.

2.7 Constant Thickness Disk Subjected to Angular Acceleration

When axisymmetry is assumed, as was mentioned earlier, the stresses appearing in the second (1.10), which expresses dynamic equilibrium in the direction perpendicular to the radius, will be only the shear stresses, and the normal stresses do not appear. The second (1.10) is written in terms of dimensionless variable ρ as follows:

$$\frac{d}{d\rho}(\tau_{rt} \cdot \rho \cdot h) + \tau_{rt} \cdot h + \gamma \cdot \dot{\omega} \cdot r_e^2 \cdot \rho^2 \cdot h = 0 \,. \tag{2.74}$$

As shear stress is not linked to normal strain components ε_t and ε_r, but only to shear strain component γ_{rt}, it can be concluded from compatibility (1.17) that any stress distribution τ_{rt} is compatible. Consequently, the problem can be solved though simple integration of the equilibrium equation.

Equation (2.74) can be rewritten as follows:

$$\frac{d(\tau_{rt} \cdot \rho \cdot h)}{d\rho} + \frac{1}{\rho} \cdot (\tau_{rt} \cdot \rho \cdot h) + \gamma \cdot \dot{\omega} \cdot r_e^2 \cdot \rho^2 \cdot h = 0 \,, \tag{2.75}$$

which is a linear first order differential equation. Setting $\tau_{rt} \cdot \rho \cdot h = x \cdot y$ in this equation, with $x = x(\rho)$ and $y = y\,(\rho)$, and integrating, gives:

$$\tau_{rt} = \frac{C}{\rho^2 \cdot h} - \frac{\dot{\omega} \cdot r_e^2 \cdot \int_\rho^1 \gamma \cdot h \cdot \rho^3 \cdot d\rho}{\rho^2 \cdot h} \,, \tag{2.76}$$

where C is an integration constant that can be determined by establishing that the shear stress at the outer radius assumes the value $(\tau_{rt})_e$, deriving from the application of a driving or braking torque at the rim of the disk. By setting this boundary condition, we obtain the relation:

$$\tau_{rt} = \frac{(\tau_{rt})_e \cdot h_e}{\rho^2 \cdot h} - \dot{\omega} \cdot r_e^2 \cdot \frac{\int_\rho^1 \gamma \cdot h \cdot \rho^3 \cdot d\rho}{\rho^2 \cdot h} \,, \tag{2.77}$$

which is valid regardless of whether or not the material is isotropic. In general, integrating (2.77) poses no difficulties.

For a disk with constant thickness and density, (2.77) can be reduced to the simple relation:

$$\tau_{rt} = \frac{(\tau_{rt})_e}{\rho^2} - \dot{\omega} \cdot r_e^2 \cdot \gamma \cdot \frac{(1 - \rho^4)}{4\rho^2} \,. \tag{2.78}$$

If no driving or braking torque is applied at the outer radius of a disk with constant thickness and density, relation (2.76) gives $C = 0$. Consequently, as $(\tau_{rt})_e = 0$, we will have:

$$\tau_{rt} = -\dot{\omega} \cdot r_e^2 \cdot \gamma \cdot \frac{(1 - \rho^4)}{4 \cdot \rho^2} \ . \tag{2.79}$$

The maximum-shear-stress occurs at the inner radius, while the distribution of this shear stress is independent of the boundary conditions at the inner radius.

2.8 Summary of Results for Constant Thickness Disk Loaded at the Inner and Outer Radius and Subjected to Centrifugal Load

In summary form, Table 2.2 shows all results obtained for annular and solid disks of constant thickness subjected to different types of load. For each geometry and each load type, the table indicates boundary conditions and the general relations for σ_r, σ_t and u as a function of dimensionless variable ρ. The maximum values of stresses σ_r and σ_t and where they occur are also shown.

Table 2.2 Constant thickness disks; geometry (annular disk or solid disk), type of loading, boundary conditions, general relations for calculating stresses σ_r and σ_t and radial displacement u, maximum values of σ_r and σ_t and their location; reference stress $\sigma_0 = \gamma \cdot \omega^2 \cdot r_e^2$; γ = density; ω = angular velocity; ν = Poisson's ratio; E = Young's modulus

Loading	Boundary conditions	Stresses σ_r and σ_t and radial displacement u
σ_{ri} at inner radius 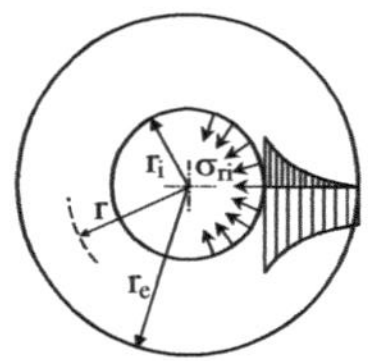 	$\sigma_r = \sigma_{ri}$ for $\rho = \beta$ $\sigma_r = 0$ for $\rho = 1$	$\sigma_r = \sigma_{r_i} \cdot \frac{\beta^2}{1-\beta^2} \cdot \left(\frac{1}{\rho^2} - 1\right); \quad (\sigma_r)_{\max} = \sigma_{r_i}$ for $\rho = \beta$ $\sigma_t = -\sigma_{r_i} \cdot \frac{\beta^2}{1-\beta^2} \cdot \left(\frac{1}{\rho^2} + 1\right); \quad (\sigma_t)_{\max} = -\sigma_{r_i} \cdot \frac{1+\beta^2}{1-\beta^2}$ for $\rho = \beta$ $u = -\sigma_{r_i} \cdot \frac{r_e}{E} \cdot \rho \cdot \frac{\beta^2}{1-\beta^2} \cdot \left[(1-\nu) + (1+\nu) \cdot \frac{1}{\rho^2}\right].$
σ_{re} at outer radius 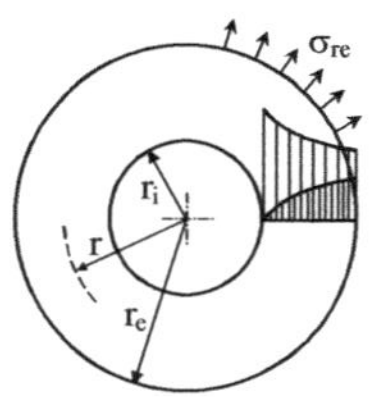 	$\sigma_r = 0$ for $\rho = \beta$ $\sigma_r = \sigma_{re}$ for $\rho = 1$	$\sigma_r = \sigma_{r_e} \cdot \frac{1}{1-\beta^2} \cdot \left(1 - \frac{\beta^2}{\rho^2}\right); \quad (\sigma_r)_{\max} = \sigma_{r_e}$ for $\rho = 1$ $\sigma_t = \sigma_{r_e} \cdot \frac{1}{1-\beta^2} \cdot \left(1 + \frac{\beta^2}{\rho^2}\right); \quad (\sigma_t)_{\max} = \frac{2 \cdot \sigma_{r_e}}{1-\beta^2}$ for $\rho = \beta$ $u = \sigma_{r_e} \cdot \frac{r_e}{E} \cdot \frac{\rho}{1-\beta^2} \cdot \left[(1-\nu) + (1+\nu) \cdot \frac{\beta^2}{\rho^2}\right].$
Centrifugal loading only 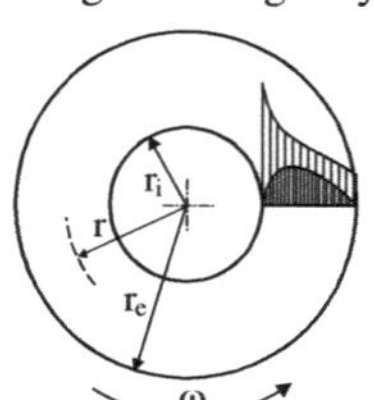 	$\sigma_r = 0$ for $\rho = \beta$ $\sigma_r = 0$ for $\rho = 1$	$\sigma_r = \frac{3+\nu}{8} \cdot \sigma_0 \cdot \left(1 + \beta^2 - \frac{\beta^2}{\rho^2} - \rho^2\right); (\sigma_r)_{\max} = \frac{3+\nu}{8} \cdot (1-\beta)^2 \cdot \sigma_0$ for $\rho = \sqrt{\beta}$ $\sigma_t = \frac{3+\nu}{8} \cdot \sigma_0 \cdot \left(1 + \beta^2 + \frac{\beta^2}{\rho^2} - \frac{1+3\nu}{3+\nu} \cdot \rho^2\right); \quad (\sigma_t)_{\max} = \frac{(3+\nu) + \beta^2(1-\nu)}{4} \cdot \sigma_0$ for $\rho = \beta$ $u = \frac{r_e}{E} \cdot \rho \cdot \frac{3+\nu}{8} \cdot \sigma_0 \cdot \left[(1+\beta^2) \cdot (1-\nu) + (1+\nu) \cdot \frac{\beta^2}{\rho^2} - \left(\frac{1-\nu^2}{3+\nu}\right) \cdot \rho^2\right].$

(continued)

Table 2.2 (continued)

Loading	Boundary conditions	Stresses σ_r and σ_t and radial displacement u
σ_{re} at outer radius 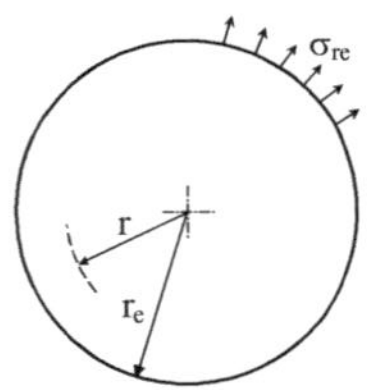	$\sigma_r = \sigma_{re}$ for $\rho = 1$ $u = 0$ for $\rho = 0$	$\sigma_r = \sigma_t = \sigma_{r_e}$ $u = \sigma_{r_e} \cdot \frac{r_e}{E} \cdot \rho \cdot (1 - \nu)$
Centrifugal loading only 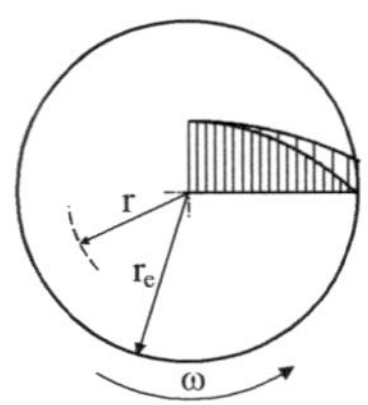	$\sigma_r = 0$ for $\rho = 1$ $u = 0$ for $\rho = 0$	$\sigma_r = \frac{3+\nu}{8} \cdot \sigma_0 \cdot (1 - \rho^2); (\sigma_r)_{\max} = \frac{3+\nu}{8} \cdot \sigma_0 \quad \text{for} \quad \rho = 0$ $\sigma_t = \frac{3+\nu}{8} \cdot \sigma_0 \cdot \left(1 - \frac{1+3\nu}{3+\nu} \cdot \rho^2\right); \quad (\sigma_t)_{\max} = (\sigma_r)_{\max} = \frac{3+\nu}{8} \cdot \sigma_0 \quad \text{for} \quad \rho = 0$ $u = \frac{r_e}{E} \cdot \rho \cdot \frac{3+\nu}{8} \cdot \sigma_0 \cdot \left[(1 - \nu) - \frac{1-\nu^2}{3+\nu} \cdot \rho^2\right].$

Chapter 3
Thermal Loads and Fictitious Density Variation Along the Radius

3.1 Annular Disk, Subjected to Thermal Load

Here again, the general solution of (2.3) can be found by adding the solution of the associated homogeneous equation (2.4) to a particular solution of the complete equation. As the general solution of the associated homogeneous equation (2.4) is already known, the problem of the disk subject to a non-zero temperature gradient along the radius is reduced to determining a particular integral of (2.3). Note that the derivative dT/dr appears in this latter equation. It follows that a constant temperature resulting from slow, uniform heating or cooling of the disk does not cause stresses, but only expansion. This observation is valid in general, regardless of the disk's shape, provided the material is isotropic and its elastic and thermophysical properties are independent of the radius.

To calculate this particular integral, it is necessary to know the function $T = T(r)$ of temperature distribution along the radius. Frequently, the temperature does not depend only on radius r, but varies according to a more complex function. In many current design applications (disks for gas and steam turbines, for example), however, the assumption that temperature varies only as a function of the radius is a sufficiently close approximation from the engineering standpoint, and can thus be profitably used by the structural designer because of the simplifications that it permits.

In this context, we will consider three functions of temperature variation with the radius, viz.:

$$T = T_0 + k \cdot r^n \; ; \tag{3.1}$$

$$T = \sum_{i=0}^{n} k_i \cdot r^i \; ; \tag{3.2}$$

$$T = T(r) \; . \tag{3.3}$$

V. Vullo and F. Vivio, *Rotors: Stress Analysis and Design*,
Mechanical Engineering Series, DOI 10.1007/978-88-470-2562-2_3,
© Springer-Verlag Italia 2013

The first two, which are taken from the literature, express $T = T(r)$ through an *n-th* degree function and a polynomial of degree n respectively, while with the third, any function $T = T(r)$ whatsoever is assumed. In the latter case, it should be noted that the function $T = T(r)$ need not necessarily be restricted to functions that can be integrated analytically; for those that cannot, it is possible to proceed with numerical integration, as the results thus obtained, though approximate, are acceptable for design purposes.

In function (3.1), k and n are constant, with n being any real exponent, whether positive or negative, an integer or a fraction, while T_0, which is also constant, is the reference temperature, and in general coincides with the ambient temperature or the assembly temperature. In function (3.2), k_i are constant and $i = 0, 1, 2, \ldots, n$ (for this function, $k_0 = T_0$ is the reference temperature, again coinciding with the ambient temperature or the assembly temperature). In even more general terms, the function $T = T(r)$ given by (3.2) can be expressed in the form $T = T_0 + \sum_{i=1}^{n} k_i \cdot r^{m_i}$, where k_i are constants and m_1, m_2, ..., m_n are exponents that are not necessarily integers.

3.1.1 *Function* T *Given by an* n-th *Degree Function*

Where function $T = T(r)$ is expressed by (3.1), relation (3.4) written in the following form can be used:

$$\frac{d}{dr}\left[\frac{1}{r}\cdot\frac{d}{dr}(u\cdot r)\right] = (1+\nu)\cdot\alpha\cdot\frac{dT}{dr}\,. \tag{3.4}$$

Given that $dT/dr = k \cdot n \cdot r^{n-1}$, performing two successive integrations (here again, we have chosen for demonstration purposes to proceed with direct integration) followed by a derivation operation yields:

$$\begin{cases} u = (1+\nu)\cdot\alpha\cdot k\cdot\dfrac{r^{n+1}}{n+2} + C_1\cdot\dfrac{r}{2} + \dfrac{C_2}{r} \\ \dfrac{du}{dr} = (1+\nu)\cdot\alpha\cdot k\cdot\dfrac{n+1}{n+2}\cdot r^n + \dfrac{C_1}{2} - \dfrac{C_2}{r^2}\,. \end{cases} \tag{3.5}$$

It should be noted that the first term in the second member of the first of these relations is the particular integral of the non-homogeneous second order differential equation (3.4); also it can be deduced with the usual procedure that applies to this type of differential equation, i.e., by setting $u = C\cdot r^{n+1}$, substituting this relation together with its first and second derivatives in (3.4), where $dT/dr = k \cdot n \cdot r^{n-1}$ and then calculating constant C, once the coefficient of the power function r^{n-1} has been equated to zero.

Substituting expressions (3.5) in relations (1.27), where it is established that $\alpha \cdot T = \alpha \cdot k \cdot r^n$, given that, as indicated above, $\alpha \cdot T_0$ makes no contribution to stresses, yields the following expressions of σ_r and σ_t as a function of radius r:

$$\begin{cases} \sigma_r = \dfrac{E \cdot C_1}{2 \cdot (1-\nu)} - \dfrac{E \cdot C_2}{(1+\nu)} \cdot \dfrac{1}{r^2} - \dfrac{E \cdot \alpha \cdot k}{n+2} \cdot r^n \\ \sigma_t = \dfrac{E \cdot C_1}{2 \cdot (1-\nu)} + \dfrac{E \cdot C_2}{(1+\nu)} \cdot \dfrac{1}{r^2} - E \cdot \alpha \cdot k \cdot \dfrac{n+1}{n+2} \cdot r^n. \end{cases} \tag{3.6}$$

Subsequently, by multiplying and dividing the second and third term of the second member of expressions (3.6) by r_e^2 and by r_e^n respectively, introducing the dimensionless variable ρ, using relations (2.9) and taking

$$C' = \frac{E \cdot \alpha \cdot k}{n+2} \cdot r_e^n \text{ e } D' = E \cdot \alpha \cdot k \cdot \frac{n+1}{n+2} \cdot r_e^n, \tag{3.7}$$

we arrive at the following relations, which express radial and hoop stresses as a function of ρ:

$$\begin{cases} \sigma_r = A - B/\rho^2 - C' \cdot \rho^n \\ \sigma_t = A + B/\rho^2 - D' \cdot \rho^n. \end{cases} \tag{3.8}$$

Introducing relations (3.8) in the second (1.25) and bearing in mind that, given the second relation (1.14), $u = r \cdot \varepsilon_t$, yields the following expression of radial displacement $u = u(\rho)$ for the configuration at reference temperature $T = T_0$, which in general is assumed to be the assembly temperature T_a:

$$\begin{aligned} u &= \frac{r_e}{E} \cdot \rho \cdot \left[A \cdot (1-\nu) + \frac{B}{\rho^2} \cdot (1+\nu) - \left(D' - \nu \cdot C' - E \cdot \alpha \cdot k \cdot r_e^n \right) \cdot \rho^n \right] = \\ &= \frac{r_e}{E} \cdot \rho \cdot \left[A \cdot (1-\nu) + \frac{B}{\rho^2} \cdot (1+\nu) + \frac{E \cdot \alpha \cdot k \cdot (1+\nu)}{n+2} \cdot r_e^n \cdot \rho^n \right]. \end{aligned} \tag{3.9}$$

Obviously, where the assembly temperature to differ from the reference temperature, the further term $\alpha \cdot E \cdot (T_a - T_0)$ would appear within the square brackets in expression (3.9). Relations (3.8) and (3.9) describe the distribution of radial and hoop stresses and radial displacement versus ρ in a disk subjected to a non-zero temperature gradient along the radius expressed by (3.1). In these relations, as the temperature variation along the radius is known, C' and D' represent known terms for a given material and for an assigned outer radius r_e, while A and B are the integration constants to be determined by imposing boundary conditions.

It should be noted that if the function of temperature variation with the radius expressed by (3.1) were to be characterized by exponent $n = -1$, we would have

$dT/dr = -k{\cdot}r^{-2}$ at the second member of relation (3.4). As a result, the solution of this equation would no longer be independent of that of the associated non-homogeneous equation, as two terms in r^{-2} would appear in relation (3.4). Nevertheless, a solution is still possible, but it will not be given here as it is of little design interest. The difficulty is purely mathematical and can be readily circumvented, with results that provide an excellent approximation, by using an exponent n which is close to but not equal to -1 (for example, $n = -0.999$ or $n = -1.001$).

In the case considered here of an annular disk subjected only to thermal load, the boundary conditions to be imposed are:

$$\begin{cases} \sigma_r = 0 & \text{for} \quad \rho = 1 \\ \sigma_r = 0 & \text{for} \quad \rho = \beta. \end{cases} \tag{3.10}$$

Accordingly, the system obtained from the first of relations (3.8) gives:

$$\begin{aligned} A &= \frac{E \cdot k \cdot \alpha}{n+2} \cdot r_e^n \cdot \frac{1-\beta^{n+2}}{1-\beta^2} \\ B &= \frac{E \cdot k \cdot \alpha}{n+2} \cdot r_e^n \cdot \frac{1-\beta^n}{1-\beta^2} \cdot \beta^2. \end{aligned} \tag{3.11}$$

Substituting the values of the constants thus found in relations (3.8) and (3.9) gives the following expressions for σ_r, σ_t and u which provide a univocal solution of the problem:

$$\begin{cases} \sigma_r = \dfrac{E \cdot \alpha \cdot k}{n+2} \cdot r_e^n \cdot \left[\dfrac{1-\beta^{n+2}}{1-\beta^2} - \dfrac{1-\beta^n}{1-\beta^2} \cdot \dfrac{\beta^2}{\rho^2} - \rho^n\right] \\ \sigma_t = \dfrac{E \cdot \alpha \cdot k}{n+2} \cdot r_e^n \cdot \left[\dfrac{1-\beta^{n+2}}{1-\beta^2} + \dfrac{1-\beta^n}{1-\beta^2} \cdot \dfrac{\beta^2}{\rho^2} - (n+1) \cdot \rho^n\right] \\ u = \rho \cdot \dfrac{\alpha \cdot k}{n+2} \cdot r_e^{n+1} \cdot \left[\dfrac{1-\beta^{n+2}}{1-\beta^2} \cdot (1-\nu) + \dfrac{1-\beta^n}{1-\beta^2} \cdot (1+\nu) \cdot \dfrac{\beta^2}{\rho^2} + (1+\nu) \cdot \rho^n\right]. \end{cases} \tag{3.12}$$

3.1.2 *Function* T *Given by an* n *Degree Polynomial*

By contrast, where function $T = T(r)$ is expressed by (3.2), relation (3.4) can again be used, given that $dT/dr = \sum_{i=1}^{n} i \cdot k_i \cdot r^{i-1}$, thus, performing two successive integrations followed by a derivation operation yields:

$$
\begin{aligned}
u &= \alpha \cdot (1+\nu) \cdot \sum_{i=1}^{n} k_i \cdot \frac{r^{i+1}}{i+2} + C_1 \cdot \frac{r}{2} + \frac{C_2}{r} \\
\frac{du}{dr} &= \alpha \cdot (1+\nu) \cdot \sum_{i=1}^{n} \frac{i+1}{i+2} \cdot k_i \cdot r^i + \frac{C_1}{2} - \frac{C_2}{r^2}.
\end{aligned} \tag{3.13}
$$

It should also be noted that integrating relation (3.4) with $dT/dr = \sum_{i=1}^{n} i \cdot k_i \cdot r^{i-1}$ is also possible using the principle of superposition: in this case, the total stress state in the disk will be given by the sum of the stress states for the n terms of the summation appearing in function (3.2), each calculated by imposing the same boundary conditions. It should also be observed that if function (3.2) were characterized by a term with exponent $i = -1$, the latter would give rise to the same integration problem indicated for function (3.1), to which the reader is referred.

Substituting expressions (3.13) in relations (1.27), where it is established that $\alpha \cdot T = \alpha \cdot \sum_{i=0}^{n} k_i \cdot r^i = \alpha \cdot \sum_{i=1}^{n} k_i \cdot r^i$, given that $\alpha{\cdot}k_0$ is a constant term and thus makes no contribution to stresses, yields the following expressions of σ_r and σ_t as a function of radius r:

$$
\begin{cases}
\sigma_r = \dfrac{E \cdot C_1}{2 \cdot (1-\nu)} - \dfrac{E \cdot C_2}{(1+\nu)} \cdot \dfrac{1}{r^2} - \alpha \cdot E \cdot \displaystyle\sum_{i=1}^{n} k_i \cdot \frac{r^i}{i+2} \\
\sigma_t = \dfrac{E \cdot C_1}{2 \cdot (1-\nu)} + \dfrac{E \cdot C_2}{(1+\nu)} \cdot \dfrac{1}{r^2} - \alpha \cdot E \cdot \displaystyle\sum_{i=1}^{n} \frac{i+1}{i+2} \cdot k_i \cdot r^i.
\end{cases} \tag{3.14}
$$

Subsequently, by multiplying and dividing the second and third term of the second member of expressions (3.14) by $r_e^{\,2}$ and by $r_e^{\,i}$ respectively, introducing the dimensionless variable ρ as well as the constants A and B given by relations (2.9), we arrive at the following relations, which express radial and hoop stresses as a function of ρ:

$$
\begin{cases}
\sigma_r = A - \dfrac{B}{\rho^2} - \alpha \cdot E \cdot \displaystyle\sum_{i=1}^{n} k_i \cdot r_e^i \cdot \frac{\rho^i}{i+2} \\
\sigma_t = A + \dfrac{B}{\rho^2} - \alpha \cdot E \cdot \displaystyle\sum_{i=1}^{n} \frac{i+1}{i+2} \cdot k_i \cdot r_e^i \cdot \rho^i.
\end{cases} \tag{3.15}
$$

Relations (3.15) are more general than relations (3.8), which are a particular case thereof. It is sufficient to consider the *n-th* term of polynomial (3.2) alone, or in other words, it is sufficient to set $i = n$, $k_1 = k_2 = \ldots\ k_{n-1} = 0$ in the latter polynomial to reduce relations (3.15) to relations (3.8).

Introducing relations (3.15) in the second (1.25) and noting that, given the second relation (1.14), $u = r{\cdot}\varepsilon_t$, we obtain the following expression for radial displacement $u = u(\rho)$ for the configuration at reference temperature $T_0 = k_0$ (this coincides with the first term of the series expansion of relation (3.2)), which here again is in general assumed to be the assembly temperature T_a:

$$u = \frac{r_e}{E} \cdot \rho \cdot \left[A \cdot (1 - \nu) + \frac{B}{\rho^2} \cdot (1 + \nu) + \alpha \cdot E \cdot (1 + \nu) \cdot \sum_{i=1}^{n} \frac{k_i \cdot r_e^i \cdot \rho^i}{i + 2} \right]. \quad (3.16)$$

Once again, it is obvious that if the assembly temperature were to differ from the reference temperature $T_0 = k_0$, the further term $\alpha \cdot E \cdot (T_a - T_0)$ would appear within the square brackets in expression (3.16). Relation (3.16) is more general than relation (3.9), which is a particular case thereof, and it is again sufficient to consider the *n-th* term of polynomial (3.2) alone, for (3.16) to be reduced to (3.9). Relations (3.15) and (3.16) describe the distribution of radial and hoop stresses and radial displacement versus ρ in a disk subjected to a non-zero temperature gradient along the radius expressed by (3.2), while A and B are the integration constants to be determined by imposing boundary conditions.

With the boundary conditions for the annular disk subjected only to thermal load ($\sigma_r = 0$ for both $\rho = 1$ and for $\rho = \beta$), the resulting system obtained from the first (3.15) yields:

$$\begin{aligned} A &= \frac{\alpha \cdot E}{1 - \beta^2} \cdot \sum_{i=1}^{n} \frac{k_i \cdot r_e^i}{i + 2} \cdot \left(1 - \beta^{i+2}\right) \\ B &= \frac{\alpha \cdot E}{1 - \beta^2} \cdot \beta^2 \cdot \sum_{i=1}^{n} \frac{k_i \cdot r_e^i}{i + 2} \cdot \left(1 - \beta^i\right). \end{aligned} \quad (3.17)$$

Substituting the values of the constants thus found in relations (3.15) and (3.16) gives the following expressions for σ_r, σ_t and u:

$$\begin{cases} \sigma_r = \alpha \cdot E \cdot \sum_{i=1}^{n} \frac{k_i \cdot r_e^i}{i + 2} \cdot \left[\frac{1 - \beta^{i+2}}{1 - \beta^2} - \frac{1 - \beta^i}{1 - \beta^2} \cdot \frac{\beta^2}{\rho^2} - \rho^i \right] \\ \sigma_t = \alpha \cdot E \cdot \sum_{i=1}^{n} \frac{k_i \cdot r_e^i}{i + 2} \cdot \left[\frac{1 - \beta^{i+2}}{1 - \beta^2} + \frac{1 - \beta^i}{1 - \beta^2} \cdot \frac{\beta^2}{\rho^2} - (i + 1) \cdot \rho^i \right] \\ u = \rho \cdot \alpha \cdot \sum_{i=1}^{n} \frac{k_i \cdot r_e^{i+1}}{i + 2} \cdot \left[\frac{1 - \beta^{i+2}}{1 - \beta^2} \cdot (1 - \nu) + \frac{1 - \beta^i}{1 - \beta^2} \cdot (1 + \nu) \cdot \frac{\beta^2}{\rho^2} + (1 + \nu) \cdot \rho^i \right]. \end{cases} \quad (3.18)$$

3.1.3 General Function T

Finally, in the most general possible case in which the function of temperature variation with the radius is expressed by (3.3), relation (3.4) can again be used, and performing two successive integrations followed by a derivation operation yields:

$$
\begin{aligned}
u &= \alpha \cdot (1+\nu) \cdot \frac{1}{r} \cdot \int_{r_i}^{r} T \cdot r \cdot dr + C_1 \cdot \frac{r}{2} + \frac{C_2}{r} \\
\frac{du}{dr} &= -\alpha \cdot (1+\nu) \cdot \frac{1}{r^2} \cdot \int_{r_i}^{r} T \cdot r \cdot dr + \alpha \cdot (1+\nu) \cdot T + \frac{C_1}{2} - \frac{C_2}{r^2}.
\end{aligned}
\tag{3.19}
$$

Substituting expression (3.19) in relations (1.27) yields the following expressions of σ_r and σ_t as a function of radius r:

$$
\begin{cases}
\sigma_r = \dfrac{E \cdot C_1}{2 \cdot (1-\nu)} - \dfrac{E \cdot C_2}{(1+\nu)} \cdot \dfrac{1}{r^2} - \alpha \cdot E \cdot \dfrac{1}{r^2} \cdot \displaystyle\int_{r_i}^{r} T \cdot r \cdot dr \\
\sigma_t = \dfrac{E \cdot C_1}{2 \cdot (1-\nu)} + \dfrac{E \cdot C_2}{(1+\nu)} \cdot \dfrac{1}{r^2} + \alpha \cdot E \cdot \left(\dfrac{1}{r^2} \cdot \displaystyle\int_{r_i}^{r} T \cdot r \cdot dr - T \right).
\end{cases}
\tag{3.20}
$$

Subsequently, proceeding as for the passage from relations (3.14) to (3.15), yields the following relations which express radial and hoop stresses as a function of ρ:

$$
\begin{cases}
\sigma_r = A - \dfrac{B}{\rho^2} - \alpha \cdot E \cdot \dfrac{1}{\rho^2} \cdot \displaystyle\int_{\beta}^{\rho} T \cdot \rho \cdot d\rho \\
\sigma_t = A + \dfrac{B}{\rho^2} + \alpha \cdot E \cdot \left(\dfrac{1}{\rho^2} \cdot \displaystyle\int_{\beta}^{\rho} T \cdot \rho \cdot d\rho - T \right).
\end{cases}
\tag{3.21}
$$

Introducing relations (3.21) in the second (1.25) and noting that, given the second relation (1.14), $u = r \cdot \varepsilon_t$, we obtain the following expression for radial displacement $u = u(\rho)$:

$$
u = \frac{r_e}{E} \cdot \rho \cdot \left[A \cdot (1-\nu) + \frac{B}{\rho^2} \cdot (1+\nu) + \frac{\alpha \cdot E \cdot (1+\nu)}{\rho^2} \cdot \int_{\beta}^{\rho} T \cdot \rho \cdot d\rho \right]. \tag{3.22}
$$

Relations (3.21) and (3.22) describe the distribution of radial and hoop stresses and radial displacement versus ρ in a disk subjected to a non-zero temperature gradient along the radius expressed by (3.3), while A and B are the integration constants to be determined by imposing boundary conditions.

With the usual boundary conditions for the annular disk subjected only to thermal load ($\sigma_r = 0$ for both $\rho = 1$ and for $\rho = \beta$), the resulting system obtained from the first relation (3.21) yields:

$$
\begin{aligned}
A &= \frac{\alpha \cdot E}{1-\beta^2} \cdot \int_{\beta}^{1} T \cdot \rho \cdot d\rho \\
B &= \frac{\alpha \cdot E \cdot \beta^2}{1-\beta^2} \cdot \int_{\beta}^{1} T \cdot \rho \cdot d\rho.
\end{aligned}
\tag{3.23}
$$

Substituting the values of the constants thus found in relations (3.21) and (3.22) gives the following expressions for σ_r, σ_t and u which provide a univocal solution of the problem:

$$
\begin{cases}
\sigma_r = \alpha \cdot E \cdot \left[\dfrac{1}{1-\beta^2} \cdot \left(1 - \dfrac{\beta^2}{\rho^2}\right) \cdot \displaystyle\int_{\beta}^{1} T \cdot \rho \cdot d\rho - \dfrac{1}{\rho^2} \cdot \displaystyle\int_{\beta}^{\rho} T \cdot \rho \cdot d\rho \right] \\
\sigma_t = \alpha \cdot E \cdot \left[\dfrac{1}{1-\beta^2} \cdot \left(1 + \dfrac{\beta^2}{\rho^2}\right) \cdot \displaystyle\int_{\beta}^{1} T \cdot \rho \cdot d\rho + \dfrac{1}{\rho^2} \cdot \displaystyle\int_{\beta}^{\rho} T \cdot \rho \cdot d\rho - T \right] \\
u = \dfrac{\alpha \cdot r_e}{\rho} \cdot \left[\dfrac{\rho^2 \cdot (1-\nu) + \beta^2 \cdot (1+\nu)}{(1-\beta^2)} \cdot \displaystyle\int_{\beta}^{1} T \cdot \rho \cdot d\rho + (1+\nu) \cdot \displaystyle\int_{\beta}^{\rho} T \cdot \rho \cdot d\rho \right].
\end{cases}
\tag{3.24}
$$

In the literature, (3.24) are also found written in terms of variable r, i.e., in the form:

$$
\begin{cases}
\sigma_r = \dfrac{\alpha \cdot E}{r_e^2 - r_i^2} \cdot \left(1 - \dfrac{r_i^2}{r^2}\right) \cdot \displaystyle\int_{r_i}^{r_e} T \cdot r \cdot dr - \dfrac{\alpha \cdot E}{r^2} \cdot \displaystyle\int_{r_i}^{r} T \cdot r \cdot dr \\
\sigma_t = \dfrac{\alpha \cdot E}{r_e^2 - r_i^2} \cdot \left(1 + \dfrac{r_i^2}{r^2}\right) \cdot \displaystyle\int_{r_i}^{r_e} T \cdot r \cdot dr + \alpha \cdot E \cdot \left(\dfrac{1}{r^2} \cdot \displaystyle\int_{r_i}^{r} T \cdot r \cdot dr - T \right) \\
u = \dfrac{\alpha}{r \cdot \left(r_e^2 - r_i^2\right)} \left[r^2 \cdot (1-\nu) + r_i^2 \cdot (1+\nu) \right] \cdot \displaystyle\int_{r_i}^{r_e} T \cdot r \cdot dr + \dfrac{\alpha \cdot (1+\nu)}{r} \cdot \displaystyle\int_{r_i}^{r} T \cdot r \cdot dr.
\end{cases}
\tag{3.25}
$$

3.1.4 Example

We will now consider an annular disk of constant thickness made of unquenched AISI 1060 steel, with $\sigma_y = 480$ MPa and having outside and inside radius $r_e = 1.0$ m and $r_i = 0.5$ m. Let the disk be subjected to centripetal heat flow characterized by a linear temperature distribution along the radius described by function $T = T_0 + kr$, with T_0 (reference temperature) and k constant, and let $T_e = 120°$C and $T_i = T_0 = 20°$C be the temperatures at the outer and inner radii.

We will also consider another disk, again annular and of constant thickness, having the same outer radius as the previous disk, but with $r_i = 0.1$ m and subjected to the same temperature differential $\Delta T = T_e - T_i$ across the outer and inner radii, and a linear temperature distribution along the radius.

Assuming that the material's thermophysical and mechanical properties remain unchanged up to temperature T_e, we will calculate the distribution of stresses due to thermal loading in the two disks when stationary, and we will determine the maximum values and their locations, discussing how variation in β influences the stress field.

From the problem data, we thus have: $\alpha = 12 \cdot 10^{-6}\,°\text{C}^{-1}$; $\nu = 0.3$; $E = 210$ GPa; $\beta = 0.5$ and $k = (T_e - T_i)/(r_e - r_i) = 200°$C/m for the first disk; $\beta = 0.1$ and $k = (T_e - T_i)/(r_e - r_i) = 111.11°$C/m for the second disk. Data for constant k appearing in linear function $T = T(r)$ are deduced from relation $T = T_i + [(T_e - T_i)/(r_e - r_i)] \cdot (r - r_i)$ which expresses the temperature function in explicit form.

Using the first two relations (3.12) which express the stress field due to heat flow in the annular disk, in which $n = 1$, we obtain the two pairs of dashed-line and solid-line curves shown in Fig. 3.1, which apply respectively to the disk with the larger diameter hole ($\beta = 0.5$) and the disk with the smaller diameter hole ($\beta = 0.1$). As can be seen from these curves, the radial stress for any value of β remains positive at all times, with a maximum value that, if the temperature variation function is (3.1), occurs at $\rho = \left[2\beta^2(1-\beta^n)/n(1-\beta^2)\right]^{1/(n+2)}$, and, in the case of interest to us here with $n = 1$, occurs at $\rho = \left[2\beta^2/(1+\beta)\right]^{1/3}$. Hoop stress assumes a positive absolute maximum value at the inner radius, decreases from the inner radius to the outer radius as ρ increases, becoming null for the value of ρ at which the term in square brackets appearing in the second (3.12) is zero. It then changes sign, becoming a compression stress, and again increases in absolute value until reaching its negative absolute maximum value at the outer radius.

It can thus be concluded that the material is at greatest risk at the inner radius, and that the absolute maximum value of the stress field at the inner radius is heavily influenced by ratio β and, for any given temperature difference $\Delta T = T_e - T_i$ and with all other conditions remaining equal, increases as β decreases. Conversely, the absolute value of compressive hoop stress at the outer radius drops as β decreases.

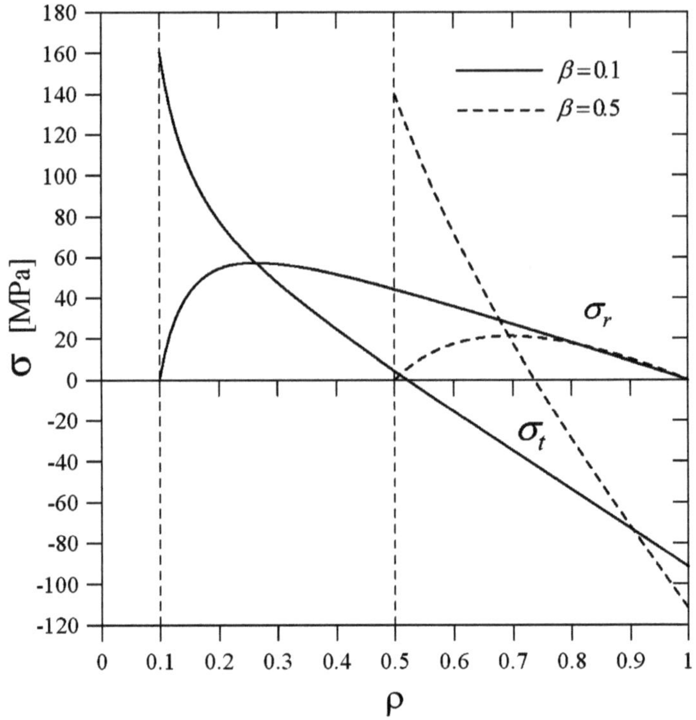

Fig. 3.1 Distribution curves of principal stresses σ_r and σ_t in two annular disks with the same outer radius and different inner radii, subjected to the same temperature differential across inner radius ($T_i = 20°C$) and outer radius ($T_e = 120°C$). Reference temperature $T_0 = 20°C$

3.2 Solid Disk, Subjected to Thermal Load

For the solid disk subjected to a temperature gradient along the radius, we will again consider the three functions $T = T(r)$ given by (3.1), (3.2) and (3.3) respectively.

3.2.1 *Function* T *Given by an* n-th *Degree Function*

If the function of temperature variation with the radius is expressed by (3.1), we must impose the condition that radial displacement at the centre is zero, as was done for the rotating solid disk. From the first relation (3.5), for $r = 0$, we obtain $C_2 = 0$ and thus, given the second relation (2.9), $B = 0$. As a result, expressions (3.8) become:

$$\begin{aligned} \sigma_r &= A - C' \cdot \rho^n \\ \sigma_t &= A - D' \cdot \rho^n. \end{aligned} \tag{3.26}$$

Here again, we will have $\sigma_r = \sigma_t = A$ at the disk axis, where $\rho = 0$. But as stress σ_r is zero at the outer radius, where $\rho = 1$, we can conclude from the first relation (3.26) that $A = C'$. Accordingly, at the centre of the disk we have:

$$\sigma_r = \sigma_t = A = C' = \frac{E \cdot \alpha \cdot k}{n+2} \cdot r_e^n. \tag{3.27}$$

For the solid disk, (3.12) become:

$$\begin{cases} \sigma_r = \dfrac{E \cdot \alpha \cdot k}{n+2} \cdot r_e^n \cdot (1 - \rho^n) \\ \sigma_t = \dfrac{E \cdot \alpha \cdot k}{n+2} \cdot r_e^n \cdot [1 - (n+1) \cdot \rho^n] \\ u = \dfrac{\alpha \cdot k \cdot r_e^{n+1}}{n+2} \cdot \rho \cdot [(1-\nu) + (1+\nu) \cdot \rho^n]. \end{cases} \tag{3.28}$$

These equations express the radial and hoop stresses and radial displacement as a function of ρ. Equation (3.28) can also be derived from (3.12) by setting $\beta = 0$ in the latter.

3.2.2 *Function* T *Given by an* n *Degree Polynomial*

If function $T = T(r)$ is expressed by (3.2), and again imposing the condition that radial displacement at the centre is zero, the first relation (3.13) yields $C_2 = 0$ for $r = 0$, and thus, given the second relation (2.9), $B = 0$. Accordingly, relations (3.15) become:

$$\begin{cases} \sigma_r = A - \alpha \cdot E \cdot \displaystyle\sum_{i=1}^{n} k_i \cdot r_e^i \cdot \frac{\rho^i}{i+2} \\ \sigma_t = A - \alpha \cdot E \cdot \displaystyle\sum_{i=1}^{n} \frac{i+1}{i+2} \cdot k_i \cdot r_e^i \cdot \rho^i. \end{cases} \tag{3.29}$$

Obviously, at the disk axis, where $\rho = 0$, we will again have $\sigma_r = \sigma_t = A$. But as stress σ_r is zero at the outer radius, where $\rho = 1$, we can conclude from the first relation (3.29) that:

$$A = \alpha \cdot E \cdot \sum_{i=1}^{n} \frac{k_i \cdot r_e^i}{i+2}. \tag{3.30}$$

Accordingly, at the centre of the disk we have:

$$\sigma_r = \sigma_t = A = \alpha \cdot E \cdot \sum_{i=1}^{n} \frac{k_i \cdot r_e^i}{i+2}. \tag{3.31}$$

For the solid disk, relations (3.18) become:

$$\begin{cases} \sigma_r = \alpha \cdot E \cdot \sum_{i=1}^{n} \frac{k_i \cdot r_e^i}{i+2} \cdot \left(1 - \rho^i\right) \\ \sigma_t = \alpha \cdot E \cdot \sum_{i=1}^{n} \frac{k_i \cdot r_e^i}{i+2} \cdot \left[1 - (i+1) \cdot \rho^i\right] \\ u = \rho \cdot \alpha \cdot \sum_{i=1}^{n} \frac{k_i \cdot r_e^{i+1}}{i+2} \cdot \left[(1-\nu) + (1+\nu) \cdot \rho^i\right]. \end{cases} \tag{3.32}$$

These equations can be derived from (3.18) by setting $\beta = 0$ in the latter. Here again, relations (3.32) are more general than relations (3.28), which are a particular case thereof. It is sufficient to consider the *n-th* term of polynomial (3.2) alone, or in other words, it is sufficient to set $i = n$, $k_1 = k_2 = \ldots k_{n-1} = 0$ in the latter polynomial to reduce relations (3.32) to relations (3.28).

3.2.3 General Function T

Finally, in the most general possible case in which the function of temperature variation with the radius is expressed by (3.3), we again impose the condition that radial displacement at the centre is zero; thus, for $r = r_i = 0$, we obtain $C_2 = 0$ from the first relation (3.19), and consequently, given the second relation (2.9), $B = 0$.

However, rather than calculating stresses σ_r and σ_t and radial displacement u by specializing expressions (3.21) to the case considered here ($B = 0$), calculating the integration constant A from the first of these expressions (3.21) thus specialized, with the condition that $\sigma_r = 0$ at the outer radius, i.e., for $\rho = 1$ and then substituting the constant found in this way in expressions (3.21) and in relation (3.22), we prefer here to start directly from (3.24) and (3.25), establishing that ratio β tends to zero in (3.24) and inner radius r_i tends to zero in (3.25). In this way, the following relations are obtained from (3.24) and (3.25) respectively:

$$\begin{cases} \sigma_r = \alpha \cdot E \cdot \left[\int_0^1 T \cdot \rho \cdot d\rho - \frac{1}{\rho^2} \cdot \int_0^\rho T \cdot \rho \cdot d\rho \right] \\ \sigma_t = \alpha \cdot E \cdot \left[\int_0^1 T \cdot \rho \cdot d\rho + \frac{1}{\rho^2} \cdot \int_0^\rho T \cdot \rho \cdot d\rho - T \right] \\ u = \frac{\alpha \cdot r_e}{\rho} \cdot \left[\rho^2 \cdot (1 - \nu) \cdot \int_0^1 T \cdot \rho \cdot d\rho + (1 + \nu) \cdot \int_0^\rho T \cdot \rho \cdot d\rho \right]. \end{cases} \tag{3.33}$$

$$\begin{cases} \sigma_r = \alpha \cdot E \cdot \left[\frac{1}{r_e^2} \cdot \int_0^{r_e} T \cdot r \cdot dr - \frac{1}{r^2} \cdot \int_0^r T \cdot r \cdot dr \right] \\ \sigma_t = \alpha \cdot E \cdot \left[\frac{1}{r_e^2} \cdot \int_0^{r_e} T \cdot r \cdot dr + \frac{1}{r^2} \cdot \int_0^r T \cdot r \cdot dr - T \right] \\ u = \frac{\alpha \cdot (1 - \nu) \cdot r}{r_e^2} \cdot \int_0^{r_e} T \cdot r \cdot dr + \frac{\alpha \cdot (1 + \nu)}{r} \cdot \int_0^r T \cdot r \cdot dr. \end{cases} \tag{3.34}$$

Relations (3.34) can be used to calculate stresses and displacement at the centre of the disk. In fact, with a Taylor series expansion of $T = T(r)$, such that

$$T(r) = T(0) + T'(0) \cdot r + \frac{1}{2} T''(0) \cdot r^2 + \ldots \tag{3.35}$$

and noting that

$$\begin{aligned} \lim_{r \to 0} \frac{1}{r^2} \int_0^r T \cdot r \cdot dr &= \frac{1}{2} \cdot T(0) \\ \lim_{r \to 0} \frac{1}{r} \int_0^r T \cdot r \cdot dr &= 0, \end{aligned} \tag{3.36}$$

we arrive at the conclusion, which is in any case fairly intuitive and has already been demonstrated using other methods, that at the centre of the disk we have:

$$(\sigma_r)_{r=0} = (\sigma_t)_{r=0}; (u)_{r=0} = 0. \tag{3.37}$$

3.3 Summary of Results for Constant Thickness Disk Subjected to Thermal Load

In summary form, Table 3.1 shows all results obtained for annular and solid disks of constant thickness subjected to a temperature gradient along the radius described by three different functions $T = T(r)$. For each geometry and each of the three functions describing thermal load, the table indicates boundary conditions and the general relations for σ_r, σ_t and u as a function of dimensionless variable ρ.

3.4 Constant Thickness Disk Subjected to Centrifugal and Thermal Loads

The stress state in a constant thickness annular disk subjected simultaneously to centrifugal and thermal loads is given by the following three pairs of general relations:

$$\begin{cases} \sigma_r = A - B/\rho^2 - C \cdot \rho^2 - C' \cdot \rho^n \\ \sigma_t = A + B/\rho^2 - D \cdot \rho^2 - D' \cdot \rho^n. \end{cases} \tag{3.38}$$

$$\begin{cases} \sigma_r = A - \dfrac{B}{\rho^2} - C \cdot \rho^2 - \alpha \cdot E \cdot \displaystyle\sum_{i=1}^{n} k_i \cdot r_e^i \cdot \frac{\rho^i}{i+2} \\ \sigma_t = A + \dfrac{B}{\rho^2} - D \cdot \rho^2 - \alpha \cdot E \cdot \displaystyle\sum_{i=1}^{n} \frac{i+1}{i+2} \cdot k_i \cdot r_e^i \cdot \rho^i. \end{cases} \tag{3.39}$$

$$\begin{cases} \sigma_r = A - \dfrac{B}{\rho^2} - C \cdot \rho^2 - \alpha \cdot E \cdot \dfrac{1}{\rho^2} \cdot \displaystyle\int_\beta^\rho T \cdot \rho \cdot d\rho \\ \sigma_t = A + \dfrac{B}{\rho^2} - D \cdot \rho^2 + \alpha \cdot E \cdot \left(\dfrac{1}{\rho^2} \cdot \displaystyle\int_\beta^\rho T \cdot \rho \cdot d\rho - T \right). \end{cases} \tag{3.40}$$

These relations apply in cases where the function $T = T(r)$ is expressed by relations (3.1), (3.2) and (3.3).

In these relations, the coefficients C, D, C' and D', as well as the terms in the summations and integrals, are zero in load conditions resulting from surface forces acting on the inner edge and/or the outer edge. If the disk is only rotating, the coefficients C' and D' and the terms in the summations and integrals are zero, while if the disk is stationary but subject to a non-zero temperature gradient along the radius, the coefficients C and D are zero.

Table 3.1 Constant thickness disks; geometry (annular disk or solid disk), thermal load represented by three different functions $T = T(r)$, boundary conditions and general relations for calculating stresses σ_r and σ_t and radial displacement u

Thermal load	Boundary conditions	Stresses σ_r and σ_t and radial displacement u
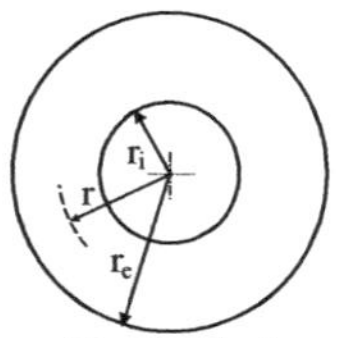 $T = T_0 + k \cdot r^n$	$\sigma_r = 0$ for $\rho = \beta$ $\sigma_r = 0$ for $\rho = 1$	$\sigma_r = \frac{E \cdot \alpha \cdot k}{n+2} \cdot r_e^n \cdot \left[\frac{1-\beta^{n+2}}{1-\beta^2} - \frac{1-\beta^n}{1-\beta^2} \cdot \frac{\beta^2}{\rho^2} - \rho^n\right]$ $\sigma_t = \frac{E \cdot \alpha \cdot k}{n+2} \cdot r_e^n \cdot \left[\frac{1-\beta^{n+2}}{1-\beta^2} + \frac{1-\beta^n}{1-\beta^2} \cdot \frac{\beta^2}{\rho^2} - (n+1) \cdot \rho^n\right]$ $u = \rho \cdot \frac{\alpha \cdot k}{n+2} \cdot r_e^{n+1} \cdot \left[\frac{1-\beta^{n+2}}{1-\beta^2} \cdot (1-\nu) + \frac{1-\beta^n}{1-\beta^2} \cdot (1+\nu) \cdot \frac{\beta^2}{\rho^2} + (1+\nu) \cdot \rho^n\right]$
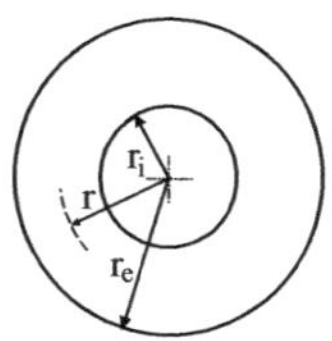 $T = \sum_{i=0}^{n} k_i \cdot r^i$	$\sigma_r = 0$ for $\rho = \beta$ $\sigma_r = 0$ for $\rho = 1$	$\sigma_r = \alpha \cdot E \cdot \sum_{i=1}^{n} \frac{k_i \cdot r_e^i}{i+2} \cdot \left[\frac{1-\beta^{i+2}}{1-\beta^2} - \frac{1-\beta^i}{1-\beta^2} \cdot \frac{\beta^2}{\rho^2} - \rho^i\right]$ $\sigma_t = \alpha \cdot E \cdot \sum_{i=1}^{n} \frac{k_i \cdot r_e^i}{i+2} \cdot \left[\frac{1-\beta^{i+2}}{1-\beta^2} + \frac{1-\beta^i}{1-\beta^2} \cdot \frac{\beta^2}{\rho^2} - (i+1) \cdot \rho^i\right]$ $u = \rho \cdot \alpha \cdot \sum_{i=1}^{n} \frac{k_i \cdot r_e^{i+1}}{i+2} \cdot \left[\frac{1-\beta^{i+2}}{1-\beta^2} \cdot (1-\nu) + \frac{1-\beta^i}{1-\beta^2} \cdot (1+\nu) \cdot \frac{\beta^2}{\rho^2} + (1+\nu) \cdot \rho^i\right]$
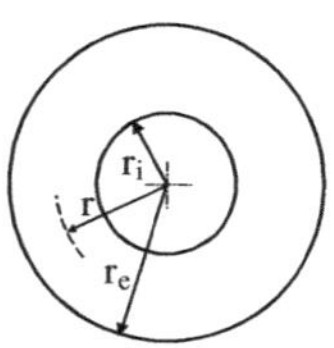 $T = T(r)$	$\sigma_r = 0$ for $\rho = \beta$ $\sigma_r = 0$ for $\rho = 1$	$\sigma_r = \alpha \cdot E \cdot \left[\frac{1}{1-\beta^2} \cdot \left(1 - \frac{\beta^2}{\rho^2}\right) \cdot \int_\beta^1 T \cdot \rho \cdot d\rho - \frac{1}{\rho^2} \cdot \int_\beta^\rho T \cdot \rho \cdot d\rho\right]$ $\sigma_t = \alpha \cdot E \cdot \left[\frac{1}{1-\beta^2} \cdot \left(1 + \frac{\beta^2}{\rho^2}\right) \cdot \int_\beta^1 T \cdot \rho \cdot d\rho + \frac{1}{\rho^2} \cdot \int_\beta^\rho T \cdot \rho \cdot d\rho - T\right]$ $u = \frac{\alpha \cdot r_e}{\rho} \cdot \left[\frac{\rho^2 \cdot (1-\nu) + \beta^2 \cdot (1+\nu)}{(1-\beta^2)} \cdot \int_\beta^1 T \cdot \rho \cdot d\rho + (1+\nu) \cdot \int_\beta^\rho T \cdot \rho \cdot d\rho\right]$.

(continued)

Table 3.1 (continued)

Thermal load	Boundary conditions	Stresses σ_r and σ_t and radial displacement u
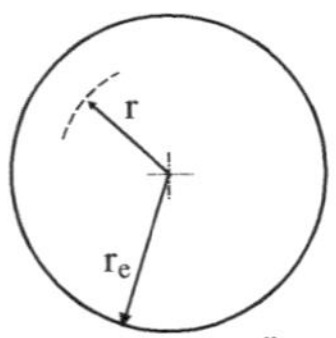 $T = T_0 + k \cdot r^n$	$\sigma_r = 0$ for $\rho = 1$ $u = 0$ for $\rho = 0$	$\left\{\begin{array}{lll} \sigma_r & = & \frac{E \cdot \alpha \cdot k}{n+2} \cdot r_e^n \cdot (1-\rho^n) \\ \sigma_t & = & \frac{E \cdot \alpha \cdot k}{n+2} \cdot r_e^n \cdot [1-(n+1)\cdot \rho^n] \\ u & = & \frac{\alpha \cdot k \cdot r_e^{n+1}}{n+2} \cdot \rho \cdot [(1-v)+(1+v)\cdot \rho^n] \end{array}\right.$
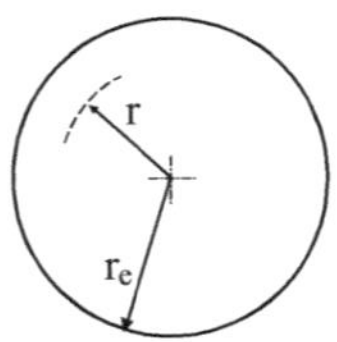 $T = \sum_{i=0}^{n} k_i \cdot r^i$	$\sigma_r = 0$ for $\rho = 1$ $u = 0$ for $\rho = 0$	$\left\{\begin{array}{lll} \sigma_r & = & \alpha \cdot E \cdot \sum\limits_{i=1}^{n} \frac{k_i \cdot r_e^i}{i+2} \cdot (1-\rho^i) \\ \sigma_t & = & \alpha \cdot E \cdot \sum\limits_{i=1}^{n} \frac{k_i \cdot r_e^i}{i+2} \cdot [1-(i+1)\cdot \rho^i] \\ u & = & \rho \cdot \alpha \cdot \sum\limits_{i=1}^{n} \frac{k_i \cdot r_e^{i+1}}{i+2} \cdot [(1-v)+(1+v)\cdot \rho^i] \end{array}\right.$
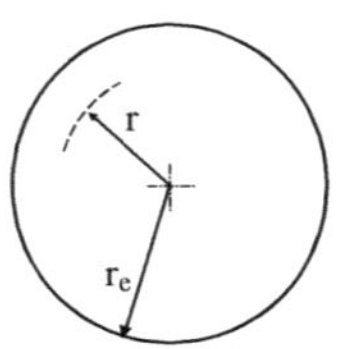 $T = T(r)$	$\sigma_r = 0$ for $\rho = 1$ $u = 0$ for $\rho = 0$	$\left\{\begin{array}{lll} \sigma_r & = & \alpha \cdot E \cdot \left[\int\limits_0^1 T \cdot \rho \cdot d\rho - \frac{1}{\rho^2} \cdot \int\limits_0^\rho T \cdot \rho \cdot d\rho\right] \\ \sigma_t & = & \alpha \cdot E \cdot \left[\int\limits_0^1 T \cdot \rho \cdot d\rho + \frac{1}{\rho^2} \cdot \int\limits_0^\rho T \cdot \rho \cdot d\rho - T\right] \\ u & = & \frac{\alpha \cdot r_e}{\rho} \cdot \left[\rho^2 \cdot (1-v) \cdot \int\limits_0^1 T \cdot \rho \cdot d\rho + (1+v) \cdot \int\limits_0^\rho T \cdot \rho \cdot d\rho\right] \end{array}\right.$

If C, D, C', D' and the terms making up the summations and the integrals are all zero, strain ε_z in the axial direction will in accordance with the generalized form of Hooke's law be:

$$\varepsilon_z = -\frac{\nu}{E} \cdot (\sigma_r + \sigma_t) = -2 \cdot \frac{\nu}{E} \cdot A = const. \tag{3.41}$$

For the disk subjected to surface forces acting on the inner and/or outer edges, axial expansion will be constant, and the strain state can be considered as a generalized plane strain state. In this case, the assumption of a plane stress state, and that of an uniform axial translation of one generic cross section, will all lead to the same result. It was mentioned earlier that the stress state due to surface forces in annular disks coincides with that in thick-walled tubes; however, this is no longer true when C, D, C', D' and the terms making up the summations and the integrals are not zero (for example, in disks subjected to thermal load or to centrifugal load).

The stress state in a constant thickness solid disk subjected simultaneously to centrifugal and thermal loads is also given by the three pairs of relations (3.38), (3.39) and (3.40), where we set $B = 0$.

3.5 Stresses in Rotating Disks Having a Fictitious Density Variation Along the Radius

As was indicated in the introduction, the peripheral surface of a disk featuring blades spaced at equal angles, the slots serving as seats for the blades and the material between each slot and the next can be simulated by means of a fictitious, discrete increase in the density of the material variously distributed therein. In other design applications (impellers for centrifugal compressors, impellers for centrifugal pumps and the like), the problem is that of evaluating the stress and strain states arising in the disk as a result of blades evenly distributed on the two side faces (5 and 6 in Fig. I.1). These blades produce a significant increase in centrifugal load stresses, without contributing appreciably to the disk's strength.

The body forces due to these blades are simulated by considering the disk without blades and introducing a fictitious variation in the disk's mass per unit volume (or density) along the radius. This fictitious variation will be discrete for peripheral blades and continuous for lateral blades. A continuous density variation function expressed in the following form is normally used:

$$\gamma(r) = \gamma_0 \cdot (1 + dm'/dm), \tag{3.42}$$

where γ_0 is the density of the basic disk material and dm' and dm are respectively the elementary masses of the blades and of the portion of the disk between two coaxial cylinders of radius r and $r + dr$.

Among the relations expressing density variation with radius that have been introduced in the literature, mention should be made of the exponential function, often uses in literature (see, as an example Güven 1992):

$$\gamma(r) = \gamma_e \cdot r^m, \tag{3.43}$$

where γ_e is the disk's density at its outer radius and m is any exponent, and of the most general relation, used by Giovannozzi [28], which expresses function $\gamma = \gamma(r)$ as follows:

$$\gamma(r) = \sum_{i=0}^{n} \gamma_i \cdot r^i, \tag{3.44}$$

or in other words through an *n-th* degree polynomial in *r*; as indicated earlier, in this latter function γ_0 represents the density of the rotor material, while $\gamma_1, \gamma_2, \ldots$ are constants and $i = 0, 1, 2, \ldots, n$. Here again, in even more general terms, the function $\gamma = \gamma(r)$ can be expressed in the form $\gamma = \gamma_0 + \sum_{i=1}^{n} \gamma_i \cdot r^{m_i}$, where γ_i are constants and $m_1, m_2, \ldots, m_n$ are exponents that are not necessarily integers. This function was introduced by Botto [6].

Using the exponential function (3.43) and considering a constant thickness disk, (2.2) becomes:

$$\frac{d}{dr}\left[\frac{1}{r} \cdot \frac{d}{dr}(u \cdot r)\right] = -\frac{(1 - \nu^2)}{E} \cdot \gamma_e \cdot \omega^2 \cdot r^{m+1}. \tag{3.45}$$

Integrating the latter equation directly in successive passages yields:

$$\begin{aligned} u &= -\frac{(1 - \nu^2) \cdot \gamma_e \cdot \omega^2 \cdot r^{m+3}}{E \cdot (m+2) \cdot (m+4)} + C_1 \cdot \frac{r}{2} + \frac{C_2}{r} \\ \frac{du}{dr} &= -\frac{(1 - \nu^2) \cdot \gamma_e \cdot \omega^2 \cdot (m+3) \cdot r^{m+2}}{E \cdot (m+2) \cdot (m+4)} + \frac{C_1}{2} - \frac{C_2}{r^2}. \end{aligned} \tag{3.46}$$

Substituting relations (3.46) in (1.27) from which temperature terms are omitted gives the following expressions of σ_r and σ_t as a function of radius *r*:

$$\begin{cases} \sigma_r = \dfrac{E \cdot C_1}{2 \cdot (1 - \nu)} - \dfrac{E \cdot C_2}{(1 + \nu)} \cdot \dfrac{1}{r^2} - \gamma_e \cdot \omega^2 \cdot \dfrac{m + 3 + \nu}{(m+2) \cdot (m+4)} \cdot r^{m+2} \\ \sigma_t = \dfrac{E \cdot C_1}{2 \cdot (1 - \nu)} + \dfrac{E \cdot C_2}{(1 + \nu)} \cdot \dfrac{1}{r^2} - \gamma_e \cdot \omega^2 \cdot \dfrac{1 + \nu \cdot (m+3)}{(m+2) \cdot (m+4)} \cdot r^{m+2}. \end{cases} \tag{3.47}$$

Continuing to consider radius r as an independent variable, setting $B = E \cdot C_2/(1 + \nu)$ and bearing the first relation (2.9) in mind, we arrive at the following relations expressing radial and hoop stresses as a function of *r*:

$$\begin{cases} \sigma_r = A - \dfrac{B}{r^2} - \gamma_e \cdot \omega^2 \cdot \dfrac{m+3+\nu}{(m+2)\cdot(m+4)} \cdot r^{m+2} \\ \sigma_t = A + \dfrac{B}{r^2} - \gamma_e \cdot \omega^2 \cdot \dfrac{1+\nu\cdot(m+3)}{(m+2)\cdot(m+4)} \cdot r^{m+2}. \end{cases} \tag{3.48}$$

Finally, introducing relations (3.48) in the expression for ε_t from which the temperature term has been omitted and noting that $u = r \cdot \varepsilon_t$ yields the following relation for displacement $u(r)$ at the generic radius r:

$$u = \frac{r}{E} \cdot \left[A \cdot (1-\nu) + \frac{B}{r^2} \cdot (1+\nu) - \frac{(1-\nu^2)\cdot\gamma_e\cdot\omega^2}{(m+2)\cdot(m+4)} \cdot r^{m+2} \right]. \tag{3.49}$$

Introducing the dimensionless variable ρ and noting that in this case the reference stress is $\sigma_0 = \gamma_e \cdot \omega^2 \cdot r_e^2$, relations (3.48) and (3.49) become:

$$\begin{cases} \sigma_r = A - \dfrac{B}{\rho^2} - \dfrac{m+3+\nu}{(m+2)\cdot(m+4)} \cdot \sigma_0 \cdot r_e^m \cdot \rho^{m+2} \\ \sigma_t = A + \dfrac{B}{\rho^2} - \dfrac{1+\nu\cdot(m+3)}{(m+2)\cdot(m+4)} \cdot \sigma_0 \cdot r_e^m \cdot \rho^{m+2} \\ u = \dfrac{r_e\cdot\rho}{E} \cdot \left[A\cdot(1-\nu) + \dfrac{B}{\rho^2}\cdot(1+\nu) - \dfrac{(1-\nu^2)}{(m+2)\cdot(m+4)} \cdot \sigma_0 \cdot r_e^m \cdot \rho^{m+2} \right], \end{cases} \tag{3.50}$$

where, however, B is given by the second relation (2.9).

From this point onwards, the procedure involves steps that are entirely similar to those described above, in the first of which integration constants A and B are calculated by imposing boundary conditions, which will obviously differ according to whether the disk is annular or solid. These steps will not be further illustrated here.

If the most general polynomial function (3.44) is used and a constant thickness disk is again considered, (2.2) becomes

$$\frac{d}{dr}\left[\frac{1}{r} \cdot \frac{d}{dr}(u \cdot r)\right] = -(1-\nu^2) \cdot \frac{\omega^2 \cdot r}{E} \cdot \sum_{i=0}^{n} \gamma_i \cdot r^i; \tag{3.51}$$

Integrating this equation directly in successive passages yields:

$$\begin{cases} u = -(1-\nu^2) \cdot \dfrac{\omega^2}{E} \cdot \displaystyle\sum_{i=0}^{n} \gamma_i \cdot \dfrac{r^{i+3}}{(i+2)\cdot(i+4)} + C_1 \cdot \dfrac{r}{2} + \dfrac{C_2}{r} \\ \dfrac{du}{dr} = -(1-\nu^2) \cdot \dfrac{\omega^2}{E} \cdot \displaystyle\sum_{i=0}^{n} \gamma_i \cdot \dfrac{(i+3)}{(i+2)\cdot(i+4)} \cdot r^{i+2} + \dfrac{C_1}{2} - \dfrac{C_2}{r^2}. \end{cases} \tag{3.52}$$

Substituting relations (3.52) in (1.27) from which temperature terms are omitted gives the following expressions of σ_r and σ_t as a function of radius r:

$$\begin{cases} \sigma_r = \dfrac{E \cdot C_1}{2 \cdot (1-v)} - \dfrac{E \cdot C_2}{(1+v)} \cdot \dfrac{1}{r^2} - \omega^2 \cdot \sum_{i=0}^{n} \gamma_i \cdot \dfrac{(i+3+v)}{(i+2) \cdot (i+4)} \cdot r^{i+2} \\ \sigma_t = \dfrac{E \cdot C_1}{2 \cdot (1-v)} + \dfrac{E \cdot C_2}{(1+v)} \cdot \dfrac{1}{r^2} - \omega^2 \cdot \sum_{i=0}^{n} \gamma_i \cdot \dfrac{1+(i+3) \cdot v}{(i+2) \cdot (i+4)} \cdot r^{i+2}. \end{cases} \tag{3.53}$$

Then, continuing to consider radius r as an independent variable, setting $B = E{\cdot}C_2/(1 + v)$, bearing in mind the first relation (2.9) and (2.27), which express the constant A and known terms σ_0, C and D respectively, and isolating the γ_0 term from the summations ($\sigma_0 = \gamma_0{\cdot}\omega^2{\cdot}r_e^2$), we arrive at the following relations expressing radial and hoop stresses as a function of r:

$$\begin{cases} \sigma_r = A - \dfrac{B}{r^2} - C \cdot r^2 + \omega^2 \cdot \sum_{i=1}^{n} \gamma_i \cdot \dfrac{(i+3+v)}{1-(i+3)^2} \cdot r^{i+2} \\ \sigma_t = A + \dfrac{B}{r^2} - D \cdot r^2 + \omega^2 \cdot \sum_{i=1}^{n} \gamma_i \cdot \dfrac{1+(i+3) \cdot v}{1-(i+3)^2} \cdot r^{i+2}. \end{cases} \tag{3.54}$$

Finally, introducing relations (3.54) in the expression for ε_t from which the temperature term has been omitted and noting that $u = r{\cdot}\varepsilon_t$ yields the following relation for displacement $u(r)$:

$$u = \frac{r}{E} \cdot \left[A \cdot (1-v) + \frac{B}{r^2} \cdot (1+v) - (D - v \cdot C) \cdot r^2 + \omega^2 \cdot (1-v^2) \cdot \sum_{i=1}^{n} \gamma_i \cdot \frac{r^{i+2}}{1-(i+3)^2} \right]. \tag{3.55}$$

Here again, from this point onwards, the procedure involves steps that are entirely similar to those described above, in the first of which integration constants A and B are calculated by imposing boundary conditions, which will obviously differ according to whether the disk is annular or solid. These steps will also not be further illustrated here.

Chapter 4
Hyperbolic Disks

As was indicated above, the general equation (1.28) can be integrated in closed form even if the disk features a hyperbolic profile, i.e., one defined by an equation of the following type (Stodola[1] [70]):

$$h = C \cdot r^a \tag{4.1}$$

where C and a are constants. In current design applications (turbine disks), disk thickness decreases from the centre towards the periphery, and we thus have $a < 0$. For any given thickness at the inner radius, a rotating disk with a profile of this kind, and thus tapering from the inner to the outer radius, will provide the undeniable advantage of lower weight and rotational inertia than a constant thickness disk, and, as we will see in a moment, a distribution of stresses along the radius that is more conducive to optimal use of the material's strength characteristics. In reality, an actual turbine disk also features a crown ring housing the blade attachments and an inner hub for connection to the shaft. Obviously, relation (4.1) is capable of describing only the portion of the disk with a hyperbolic profile, while crown ring and hub are usually considered as two constant thickness disks.

With function (4.1), for $a = 0$, we again have the case of the constant thickness disk. The same function can also be expressed in the following form:

$$h = h_e \cdot \rho^a \tag{4.2}$$

where the meaning of the symbols remains the same, and h_e is disk thickness at the outer radius r_e. Passing from magnitudes to logarithms, it can be concluded from

[1] Historically, Stodola's hyperbolic profile (1924) analyzed here was the second variable thickness profile introduced in rotating disk theory.

V. Vullo and F. Vivio, *Rotors: Stress Analysis and Design*, Mechanical Engineering Series, DOI 10.1007/978-88-470-2562-2_4,

© Springer-Verlag Italia 2013

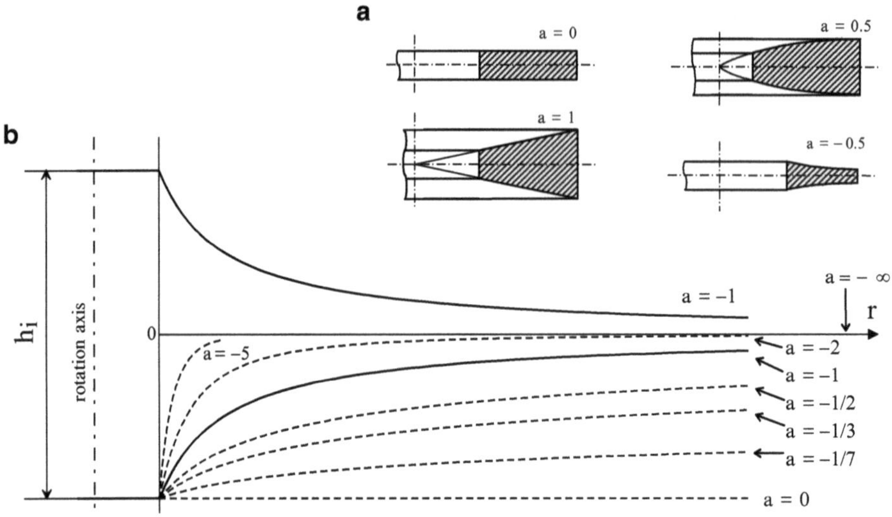

Fig. 4.1 Hyperbolic disks: (**a**) profiles obtained by varying exponent a; (**b**) profiles tapering towards the outer radius ($a < 0$) for a disk with $\beta = 0.1$

both of these relations that any value of $a \neq 0$ gives rise to a singularity[2] at the rotational axis, i.e., to an infinite thickness for $r = \rho = 0$. Consequently, the theory presented below can be used only for annular disks.

In the literature, the thickness at inner radius h_i is occasionally assumed as the reference thickness. In this case, the function of thickness variation with dimensionless radius ρ takes the following form:

$$h = h_i \cdot \left(\frac{\rho}{\beta}\right)^a. \tag{4.3}$$

It follows from relations (4.2) and (4.3) that $h_i/h_e = (r_e/r_i)^{-a}$; the latter makes it possible to determine exponent a once r_e, r_i, h_e and h_i have been established.

Figure 4.1a shows the different disk profiles that can be obtained, using the above relations, with different values of exponent a. As can be seen, relation (4.1) can not only describe constant thickness disks ($a = 0$) and disks whose profile tapers nonlinearly from the inner to the outer radius ($a < 0$), but also disks whose profile diverges from the inner to the outer radius, both nonlinearly ($a > 0$) and

[2] A variant of Stodola's hyperbolic function was also introduced, in which thickness is defined by the relation $h = h_0 \cdot (1 + \rho)^a$, where h_0 is thickness on the axis and a is a negative exponent. This does not give rise to a singularity for $\rho = 0$, and can thus be used to describe the hyperbolic profile of a solid disk. With this relation, differential equation (1.28) can be integrated by means of the linear combination of two mutually independent hypergeometric functions (see Chap. 7).

linearly ($a = 1$). With the exception of $a = 0$, all disks thus defined must be annular in order to prevent singularity at the axis. For $a > 0$, in fact, the thickness at the disk axis, where two sides of the profile or its extension converge, is zero; for $a < 0$, thickness becomes infinite. Figure 4.1b shows the curves for profiles tapering towards the outer radius of a disk with $\beta = 0.1$, for several negative values of the exponent a between $a = 0$ (constant thickness disk) and $a = -\infty$. For $a = -1$, the disk profile is that of an equilateral hyperbola for any value of β.

The stress and strain state in a hyperbolic disk subjected to centrifugal load and thermal load is analysed starting from differential equation (1.28). For the stresses resulting from thermal load, only the two functions of temperature variation with radius expressed by relations (3.1) and (3.2) will be considered below.

To solve the problem in the first case, the following general equation for the hyperbolic disk subjected to centrifugal load and thermal load varying according to function (3.1) is obtained by introducing the expressions for T and h, given respectively by (3.1) and (4.1), in (1.28) along with their first derivatives, and neglecting the term in T_0 (which is equivalent to evaluating the radial displacements by respect to the configuration at temperature T_0 assumed as a reference, i.e., as the initial assembly temperature).

$$\frac{d^2u}{dr^2} + \frac{(a+1)}{r} \cdot \frac{du}{dr} + (\nu \cdot a - 1) \cdot \frac{u}{r^2} - (1+\nu) \cdot \alpha \cdot (n+a) \cdot k \cdot r^{n-1} + \left(1-\nu^2\right) \cdot \frac{\gamma \cdot \omega^2 \cdot r}{E} = 0\,. \tag{4.4}$$

In this equation, the contributions of thermal load and of centrifugal load are given respectively by the next to last and the last addend in the first member.

In the second case, by introducing the expressions for T and h given respectively by (3.2) and (4.1) in (1.28) and neglecting the term in k_0 appearing in (3.2), whereby

$$T = \sum_{i=1}^{n} k_i \cdot r^i \tag{4.5}$$

$$\frac{dT}{dr} = \sum_{i=1}^{n} i \cdot k_i \cdot r^{i-1}\,, \tag{4.6}$$

we arrive at the following general equation for the hyperbolic disk subjected to centrifugal load and thermal load varying according to function (3.2):

$$\frac{d^2u}{dr^2} + \frac{(a+1)}{r} \cdot \frac{du}{dr} + (\nu \cdot a - 1) \cdot \frac{u}{r^2} - (1+\nu) \cdot \alpha \cdot \sum_{i=1}^{n} (i+a) \cdot k_i \cdot r^{i-1} + \left(1-\nu^2\right) \cdot \frac{\gamma \cdot \omega^2 \cdot r}{E} = 0. \tag{4.7}$$

Given the foregoing assumptions, with differential equation (4.7), where the contributions of thermal load and of centrifugal load are given respectively by the next to last and the last addend in the first member, radial displacements again refer to the configuration at temperature k_0 assumed as a reference, i.e., as the initial assembly temperature.

4.1 Annular Disk, Subjected to Centrifugal Load

We will now consider a hyperbolic disk subjected only to centrifugal load. In this case, the solving differential equation is (4.4) or (4.7), with the term for thermal load omitted from both. To integrate this equation, we set

$$u = z - \bar{C} \cdot r^3 \tag{4.8}$$

with $z = z(r)$ e $\bar{C} = const$, and we determine the latter by establishing the condition that, substituting relation (4.8) together with its first and second derivatives with respect to r in (4.4), the sum of all terms that do not contain the variable z is zero. With a few passages, we find:

$$\bar{C} = \frac{(1 - v^2) \cdot \gamma \cdot \omega^2}{E \cdot [8 + (3 + v) \cdot a]} . \tag{4.9}$$

Accordingly, relation (4.8) becomes:

$$u = z - \frac{(1 - v^2) \cdot \gamma \cdot \omega^2}{E \cdot [8 + (3 + v) \cdot a]} \cdot r^3 . \tag{4.10}$$

With the condition established above, the differential equation (4.4) from which the temperature term is omitted, becomes homogenous and takes the following form:

$$\frac{d^2 z}{dr^2} + \frac{(a + 1)}{r} \cdot \frac{dz}{dr} + (v \cdot a - 1) \cdot \frac{z}{r^2} = 0 . \tag{4.11}$$

This equation can be integrated by setting

$$z = C \cdot r^{\psi} \tag{4.12}$$

with C and ψ constant. By substituting relation (4.12) together with its first and second derivatives in (4.11), we arrive at the following second degree algebraic equation:

$$\psi^2 + a \cdot \psi + (\nu \cdot a - 1) = 0, \tag{4.13}$$

whose roots $p > 0$ and $q < 0$ are given respectively by the relations:

$$p = -\frac{a}{2} + \sqrt{1 - \nu \cdot a + \frac{a^2}{4}} \quad ; \qquad q = -\frac{a}{2} - \sqrt{1 - \nu \cdot a + \frac{a^2}{4}}. \tag{4.14}$$

The solution of the associated homogeneous equation can thus be written in the form

$$z = C_1 \cdot r^p + C_2 \cdot r^q, \tag{4.15}$$

where C_1 and C_2 are integration constants to be determined by imposing boundary conditions, while the general integral of differential equation (4.4), bearing relation (4.10) in mind, is given by:

$$u = C_1 \cdot r^p + C_2 \cdot r^q - \frac{(1 - \nu^2) \cdot \gamma \cdot \omega^2}{E \cdot [8 + (3 + \nu) \cdot a]} \cdot r^3. \tag{4.16}$$

Deriving this latter equation respect to r yields:

$$\frac{du}{dr} = p \cdot C_1 \cdot r^{p-1} + q \cdot C_2 \cdot r^{q-1} - \frac{3 \cdot (1 - \nu^2) \cdot \gamma \cdot \omega^2}{E \cdot [8 + (3 + \nu) \cdot a]} \cdot r^2. \tag{4.17}$$

Subsequently, substituting relations (4.16) and (4.17) in (1.27) from which temperature terms are omitted gives the following equations for σ_r and σ_t as a function of radius r:

$$\begin{cases} \sigma_r = \dfrac{E}{1 - \nu^2} \cdot \Big[(p + \nu) \cdot C_1 \cdot r^{p-1} + (q + \nu) \cdot C_2 \cdot r^{q-1} \\ \qquad - \dfrac{(3 + \nu) \cdot (1 - \nu^2) \cdot \gamma \cdot \omega^2}{E \cdot [8 + (3 + \nu) \cdot a]} \cdot r^2\Big] \\ \sigma_t = \dfrac{E}{1 - \nu^2} \cdot \Big[(1 + \nu \cdot p) \cdot C_1 \cdot r^{p-1} + (1 + \nu \cdot q) \cdot C_2 \cdot r^{q-1} \\ \qquad - \dfrac{(1 + 3\nu) \cdot (1 - \nu^2) \cdot \gamma \cdot \omega^2}{E \cdot [8 + (3 + \nu) \cdot a]} \cdot r^2\Big]. \end{cases} \tag{4.18}$$

Finally, by multiplying and dividing the three terms in square brackets in the above relations by r_e^{p-1} , r_e^{q-1} and r_e^2 respectively, remembering that between p, q and ν the following relations exist

$$p = -\frac{1 + \nu \cdot q}{\nu + q}; \quad q = -\frac{1 + \nu \cdot p}{\nu + p}, \tag{4.19}$$

and introducing the dimensionless variable ρ and the notations

$$\begin{aligned} A &= \frac{E}{1 - \nu^2} \cdot (p + \nu) \cdot C_1 \cdot r_e^{p-1}; \quad B = \frac{E}{1 - \nu^2} \cdot (q + \nu) \cdot C_2 \cdot r_e^{q-1} \\ C &= \frac{(3 + \nu)}{[8 + (3 + \nu) \cdot a]} \cdot \sigma_0; \quad D = \frac{(1 + 3\nu)}{[8 + (3 + \nu) \cdot a]} \cdot \sigma_0, \end{aligned} \tag{4.20}$$

with $\sigma_0 = \gamma \cdot \omega^2 \cdot r_e^{\,2}$, we arrive at the following relations expressing radial and hoop stresses as a function of ρ:

$$\begin{cases} \sigma_r = A \cdot \rho^{p-1} + B \cdot \rho^{q-1} - C \cdot \rho^2 \\ \sigma_t = -q \cdot A \cdot \rho^{p-1} - p \cdot B \cdot \rho^{q-1} - D \cdot \rho^2. \end{cases} \tag{4.21}$$

On the other hand, by introducing relations (4.21) in the second of (1.25), from which the temperature term is omitted and as usual bearing in mind that the second of (1.14) gives $u = r \cdot \varepsilon_t$, we obtain the following relation expressing radial displacement $u = u(\rho)$:

$$u = \frac{r_e}{E} \cdot \rho \cdot \left[-(q + \nu) \cdot A \cdot \rho^{p-1} - (p + \nu) \cdot B \cdot \rho^{q-1} - (D - \nu \cdot C) \cdot \rho^2\right]. \tag{4.22}$$

Relations (4.21) and (4.22) describe the distribution of radial and hoop stresses and radial displacement versus ρ in a hyperbolic annular disk subjected to centrifugal load. In these relations, C and D are known terms, while A and B are the integration constants to be determined by imposing boundary conditions. For a non-rotating disk, relations (4.21) and (4.22) will obviously not have the terms in ρ^2.

4.2 Non-Rotating Annular Disk, Loaded at the Outer Radius

In this case, the boundary conditions to be imposed to calculate the integration constants A and B appearing in relations (4.21) and (4.22), from which the terms in ρ^2 are omitted, are:

$$\begin{cases} \sigma_r = 0 & \text{for} \quad \rho = \beta \\ \sigma_r = \sigma_{re} & \text{for} \quad \rho = 1\,. \end{cases} \tag{4.23}$$

Accordingly, the system obtained from the first of relations (4.21), where $C = 0$, gives:

$$A = -\frac{\beta^{q-1}}{\beta^{p-1} - \beta^{q-1}} \cdot \sigma_{re}; \quad B = \frac{\beta^{p-1}}{\beta^{p-1} - \beta^{q-1}} \cdot \sigma_{re}. \tag{4.24}$$

Substituting the constants thus found in relations (4.21) and (4.22), where $C = D = 0$, gives the following expressions for σ_r, σ_t and u which provide a univocal solution of the problem:

$$\begin{cases} \sigma_r = \dfrac{\rho^{q-1} \cdot \beta^{p-1} - \rho^{p-1} \cdot \beta^{q-1}}{\beta^{p-1} - \beta^{q-1}} \cdot \sigma_{re} \\ \sigma_t = \dfrac{q \cdot \rho^{p-1} \cdot \beta^{q-1} - p \cdot \rho^{q-1} \cdot \beta^{p-1}}{\beta^{p-1} - \beta^{q-1}} \cdot \sigma_{re} \\ u = \dfrac{\left[(q+\nu) \cdot \rho^{p} \cdot \beta^{q-1} - (p+\nu) \cdot \rho^{q} \cdot \beta^{p-1}\right]}{\beta^{p-1} - \beta^{q-1}} \cdot \left(\dfrac{r_e}{E} \cdot \sigma_{re}\right). \end{cases} \tag{4.25}$$

4.3 Non-Rotating Annular Disk, Loaded at the Inner Radius

In this case, the boundary conditions to be imposed to calculate the integration constants A and B are:

$$\begin{cases} \sigma_r = \sigma_{ri} & \text{for} \quad \rho = \beta \\ \sigma_r = 0 & \text{for} \quad \rho = 1\,. \end{cases} \tag{4.26}$$

Accordingly, the system obtained from the first of relations (4.21), where $C = 0$, gives:

$$A = \frac{\sigma_{ri}}{\beta^{p-1} - \beta^{q-1}}; \quad B = -A = -\frac{\sigma_{ri}}{\beta^{p-1} - \beta^{q-1}}. \tag{4.27}$$

Substituting the constants thus found in relations (4.21) and (4.22), where $C = D = 0$, gives the following expressions for σ_r, σ_t and u:

$$\begin{cases} \sigma_r = \dfrac{\rho^{p-1} - \rho^{q-1}}{\beta^{p-1} - \beta^{q-1}} \cdot \sigma_{ri} \\ \sigma_t = \dfrac{-q \cdot \rho^{p-1} + p \cdot \rho^{q-1}}{\beta^{p-1} - \beta^{q-1}} \cdot \sigma_{ri} \\ u = \dfrac{\left[-(q+\nu) \cdot \rho^{p-1} + (p+\nu) \cdot \rho^{q-1}\right]}{\beta^{p-1} - \beta^{q-1}} \cdot \rho \cdot \left(\dfrac{r_e}{E} \cdot \sigma_{ri}\right). \end{cases} \tag{4.28}$$

4.4 Rotating Annular Disk

For an annular disk only rotating, i.e., not subjected to loads acting on the inner and outer edges, the boundary conditions to be imposed are:

$$\begin{cases} \sigma_r = 0 & \text{for} \quad \rho = 1 \\ \sigma_r = 0 & \text{for} \quad \rho = \beta. \end{cases} \tag{4.29}$$

Accordingly, the system obtained from the first of relations (4.21) gives:

$$A = C \cdot \frac{-\beta^{q-1} + \beta^2}{\beta^{p-1} - \beta^{q-1}} \quad B = C \cdot \frac{\beta^{p-1} + \beta^2}{\beta^{p-1} - \beta^{q-1}}. \tag{4.30}$$

Substituting the constants thus found in relations (4.21) and (4.22) and bearing in mind the third and fourth relations (4.20) gives the following expressions for σ_r, σ_t and u:

$$\begin{cases} \sigma_r = \sigma_0 \cdot \frac{(3+v)}{[8+(3+v)\cdot a]} \cdot \frac{\left[\rho^{p-1}\cdot\left(\beta^2-\beta^{q-1}\right)+\rho^{q-1}\cdot\left(\beta^{p-1}-\beta^2\right)-\rho^2\cdot\left(\beta^{p-1}-\beta^{q-1}\right)\right]}{\left(\beta^{p-1}-\beta^{q-1}\right)} \\ \sigma_t = \sigma_0 \cdot \frac{\left[-q\cdot(3+v)\cdot\rho^{p-1}\cdot\left(\beta^2-\beta^{q-1}\right)-p\cdot(3+v)\cdot\rho^{q-1}\cdot\left(\beta^{p-1}-\beta^2\right)-(1+3v)\cdot\rho^2\cdot\left(\beta^{p-1}-\beta^{q-1}\right)\right]}{[8+(3+v)\cdot a]\cdot\left(\beta^{p-1}-\beta^{q-1}\right)} \\ u = \frac{r_e}{E}\sigma_0 \cdot \frac{\left[-(q+v)\cdot(3+v)\cdot\left(\beta^2-\beta^{q-1}\right)\cdot\rho^p-(p+v)\cdot(3+v)\cdot\left(\beta^{p-1}-\beta^2\right)\cdot\rho^q-\left(1-v^2\right)\cdot\rho^3\cdot\left(\beta^{p-1}-\beta^{q-1}\right)\right]}{[8+(3+v)\cdot a]\cdot\left(\beta^{p-1}-\beta^{q-1}\right)}. \end{cases} \tag{4.31}$$

4.4.1 Example

A hyperbolic annular steel disk having r_i, r_e, h_i and h_e equal to 25, 200, 100 and 12.5 mm respectively is subjected only to centrifugal load. We will plot the diagrams representing radial and hoop stresses versus radius, and calculate the values of the latter which are of design interest. Radial displacement at the inner radius will also be calculated. The stress state in this disk will then be compared with that in a disk of constant thickness $h_i = h_e = 100$ mm, having the same inner and outer radii.

As $\beta = r_i/r_e = 0.125$, relation (4.3) written for $\rho = 1$ yields $a = -1$. For the steel disk ($v = 0.3$), relations (4.14) give the following values of roots p and q: $p = 1.75$ and $q = -0.75$. Given that $\beta^2 = 0.016$, $\beta^{p-1} = \beta^{0.75} = 0.210$ and $\beta^{q-1} = \beta^{-1.75} = 38.005$, the first two relations (4.31) yield:

$$\sigma_r = 0.70 \cdot \frac{\left(-38.04 \cdot \rho^{0.75} + 0.19 \cdot \rho^{-1.75} + 37.84 \cdot \rho^2\right)}{(-37.79)} \cdot \sigma_0$$

$$\sigma_t = \frac{\left(-94.15 \cdot \rho^{0.75} - 1.12 \cdot \rho^{-1.75} + 71.91 \cdot \rho^2\right)}{(-177.87)} \cdot \sigma_0 .$$

In particular, we have:

$$(\sigma_r)_{\rho=\beta} = (\sigma_r)_{\rho=1} = 0.$$
$$(\sigma_t)_{\rho=\beta} = 0.345 \cdot \sigma_0;\ (\sigma_t)_{\rho=1} = 0.132 \cdot \sigma_0.$$

At radius $\rho = 0.5$ we obtain:

$$(\sigma_r)_{\rho=0.5} = 0.231 \cdot \sigma_0;\ (\sigma_t)_{\rho=0.5} = 0.235 \cdot \sigma_0.$$

Profiles of the hyperbolic disk (solid-line curve) and a constant thickness disk (dashed-line curve) having the same h_i are shown in Fig. 4.2, whereas the solid-line curve in Fig. 4.2b represents the radial and hoop stresses in dimensionless form, i.e., relative to σ_0, versus radius in the hyperbolic disk in question. As can be seen from the latter figure, the absolute maximum stress value is here again that reached by hoop stress at the inner edge.

Rather than using the third (4.31) to calculate radial displacement at the inner radius, it is simpler to employ the second (1.14) together with the second (1.25), omitting the temperature term from the latter. We thus have:

$$(u)_{\rho=\beta} = \frac{\beta \cdot r_e}{E} \cdot \left[(\sigma_t)_{\rho=\beta} - \nu \cdot (\sigma_r)_{\rho=\beta}\right] = \frac{\beta \cdot r_e}{E} \cdot 0.345 \cdot \sigma_0.$$

Figure 4.2b also shows (dashed-line curve) the radial and hoop stresses, again in dimensionless form, versus radius in the constant thickness disk having $h_i = 100$ mm used for purposes of comparison. All other conditions remaining equal, this comparison between the stress states in the two disks clearly shows that the hyperbolic profile makes a more rational use of the disk material's strength properties. Passing from the constant profile disk to the hyperbolic disk having the same r_i, r_e, and h_i reduces hoop stress at the inner radius from $0.825 \cdot \sigma_0$ to $0.345 \cdot \sigma_0$, thus lowering maximum stress by more than 58 %.

A comparison of the radial displacement at the inner radius of the two disk types also shows the advantages of the hyperbolic disk. For the constant thickness disk, this radial displacement is:

$$(u)_{\rho=\beta} = \frac{\beta \cdot r_e}{E} \cdot \left[(\sigma_t)_{\rho=\beta} - \nu \cdot (\sigma_r)_{\rho=\beta}\right] = \frac{\beta \cdot r_e}{E} \cdot 0.825 \cdot \sigma_0.$$

Consequently, all other conditions remaining equal, the radial displacement at the inner radius for the hyperbolic disk is also over 58 % lower than that for the constant thickness disk.

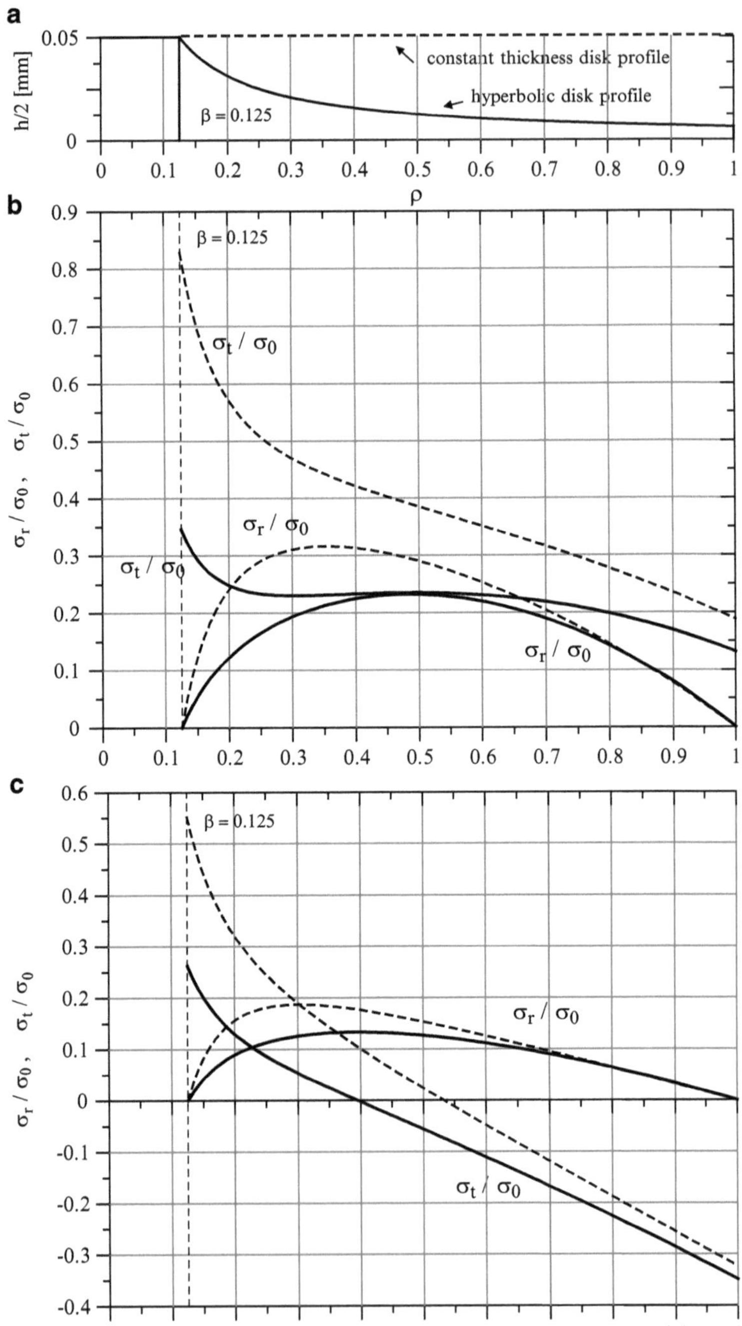

Fig. 4.2 (**a**) Hyperbolic disk and constant thickness disk with the same h_i; (**b**) σ_r/σ_0 e σ_t/σ_0 versus dimensionless radius ρ in the hyperbolic disk and the constant thickness disk with the same h_i, both subjected only to centrifugal load; (**c**) σ_r/σ_0 e σ_t/σ_0 versus dimensionless radius ρ in the hyperbolic disk and the constant thickness disk with the same h_i, both subjected to centripetal heat flow

It can thus be concluded that, *ceteris paribus*, the hyperbolic disk is intrinsically more robust than the corresponding constant thickness disk from the standpoint of both the stress state and the strain state.

4.5 Annular Disk, Subjected to Thermal Load

We will first consider a hyperbolic disk subjected only to thermal load characterized by a non-zero temperature gradient along the radius expressed by the relation $dT/dr = k \cdot n \cdot r^{n-1}$, deriving from the function $T = T(r)$ expressed by (3.1). In this case, the solving differential equation is (4.4), with the term for centrifugal load omitted. To integrate this equation, we set:

$$u = z - \overline{\overline{C}} \cdot r^{n+1} \tag{4.32}$$

with $z = z(r)$ and $\overline{\overline{C}} = const$, and, as before, we determine the latter by establishing the condition that, substituting relation (4.32) together with its first and second derivatives with respect to r in (4.4), the sum of all terms that do not contain the variable z is zero. With a few passages, we find:

$$\overline{\overline{C}} = -\frac{(1+\nu) \cdot \alpha \cdot k \cdot (n+a)}{[n \cdot (n+2) + a \cdot (n+1+\nu)]}. \tag{4.33}$$

Accordingly, relation (4.32) becomes:

$$u = z + \frac{(1+\nu) \cdot \alpha \cdot k \cdot (n+a)}{[n \cdot (n+2) + a \cdot (n+1+\nu)]} \cdot r^{n+1}. \tag{4.34}$$

With the condition established above, differential equation (4.4) yields the homogeneous equation (4.11), whose solution is given by relation (4.15), as was described earlier. The general integral of differential equation (4.4) is thus written in the following form:

$$u = C_1 \cdot r^p + C_2 \cdot r^q + \frac{(1+\nu) \cdot \alpha \cdot k \cdot (n+a)}{[n(n+2) + a(n+1+\nu)]} \cdot r^{n+1}. \tag{4.35}$$

Deriving this latter equation respect to r yields:

$$\frac{du}{dr} = p \cdot C_1 \cdot r^{p-1} + q \cdot C_2 \cdot r^{q-1} + \frac{(1+\nu) \cdot \alpha \cdot k \cdot (n+a) \cdot (n+1)}{[n \cdot (n+2) + a \cdot (n+1+\nu)]} \cdot r^n. \tag{4.36}$$

Subsequently, substituting relations (4.35) and (4.36) in (1.27), where it is established that $\alpha T = \alpha \cdot k \cdot r^n$, given that, as indicated several times, T_0 makes no contribution to stresses, yields the following equations for σ_r and σ_t as a function of radius r:

$$
\begin{cases}
\sigma_r = \dfrac{E}{1-\nu^2} \cdot \left[(p+\nu) \cdot C_1 \cdot r^{p-1} + (q+\nu) \cdot C_2 \cdot r^{q-1} \right. \\
\qquad \left. - \dfrac{(1-\nu^2) \cdot \alpha \cdot k \cdot n}{[n \cdot (n+2) + a \cdot (n+1+\nu)]} \cdot r^n \right] \\
\sigma_t = \dfrac{E}{1-\nu^2} \cdot \left[(1+\nu \cdot p) \cdot C_1 \cdot r^{p-1} + (1+\nu \cdot q) \cdot C_2 \cdot r^{q-1} \right. \\
\qquad \left. - \dfrac{(1-\nu^2) \cdot \alpha \cdot k \cdot n \cdot (n+1+a)}{[n \cdot (n+2) + a \cdot (n+1+\nu)]} \cdot r^n \right].
\end{cases} \tag{4.37}
$$

Finally, by multiplying and dividing the three terms in square brackets in the above relations by ${r_e}^{p-1}$, ${r_e}^{q-1}$ and ${r_e}^n$ respectively, remembering relations (4.19) which link p, q and ν, and introducing the dimensionless variable ρ and the first two notations (4.20), as well as the additional notations

$$
C' = \frac{E \cdot \alpha \cdot k \cdot n \cdot r_e^n}{[n \cdot (n+2) + a \cdot (n+1+\nu)]}; \quad D' = \frac{E \cdot \alpha \cdot k \cdot n \cdot (n+1+a) \cdot r_e^n}{[n \cdot (n+2) + a \cdot (n+1+\nu)]}, \tag{4.38}
$$

we arrive at the following relations, which express radial and hoop stresses as a function of ρ:

$$
\begin{cases}
\sigma_r = A \cdot \rho^{p-1} + B \cdot \rho^{q-1} - C' \cdot \rho^n \\
\sigma_t = -q \cdot A \cdot \rho^{p-1} - p \cdot B \cdot \rho^{q-1} - D' \cdot \rho^n.
\end{cases} \tag{4.39}
$$

On the other hand, by introducing relations (4.39) in the second of (1.25) and bearing in mind that the second of (1.14) gives $u = r \cdot \varepsilon_t$, we obtain the following relation expressing radial displacement $u = u(\rho)$ by respect to the configuration at reference temperature $T = T_0$, which here again is in general assumed to be the assembly temperature T_a:

$$
u = \frac{r_e}{E} \cdot \rho \cdot \left[-(q+\nu) \cdot A \cdot \rho^{p-1} - (p+\nu) \cdot B \cdot \rho^{q-1} - \left(D' - \nu \cdot C' - E \cdot \alpha \cdot k \cdot r_e^n \right) \cdot \rho^n \right]. \tag{4.40}
$$

Once again, it is obvious that if the assembly temperature were to differ from the reference temperature T_0, the further term $\alpha \cdot E \cdot (T_a - T_0)$ would appear within the square brackets in expression (4.20). Relations (4.39) and (4.40) describe the distribution of radial and hoop stresses and radial displacement in a disk subjected to a non-zero temperature gradient along the radius. Here again, as the temperature variation along the radius is known, C' and D' represent known terms for a given material, while A and B are the integration constants to be determined by imposing boundary conditions.

For the hyperbolic disk subjected only to thermal load, the boundary conditions to be imposed are:

$$\begin{cases} \sigma_r = 0 & \text{for} \quad \rho = 1 \\ \sigma_r = 0 & \text{for} \quad \rho = \beta \, . \end{cases} \tag{4.41}$$

Accordingly, the system obtained from the first of relations (4.39) gives:

$$A = C' \cdot \frac{\left(-\beta^{q-1} + \beta^n\right)}{\left(\beta^{p-1} - \beta^{q-1}\right)}; \qquad B = C' \cdot \frac{\left(\beta^{p-1} - \beta^n\right)}{\left(\beta^{p-1} - \beta^{q-1}\right)}, \tag{4.42}$$

with C' given by the first (4.38). Substituting the constants thus found in relations (4.39) and (4.40) gives the following expressions for σ_r, σ_t and u which provide a univocal solution of the problem:

$$\begin{cases} \sigma_r = \dfrac{E \cdot \alpha \cdot k \cdot n \cdot r_e^n}{[n \cdot (n+2) + a \cdot (n+1+v)]} \cdot \dfrac{1}{\left(\beta^{p-1} - \beta^{q-1}\right)} \\ \quad \times \left[\left(\beta^n - \beta^{q-1}\right) \cdot \rho^{p-1} + \left(\beta^{p-1} - \beta^n\right) \cdot \rho^{q-1} - \left(\beta^{p-1} - \beta^{q-1}\right) \cdot \rho^n\right] \\ \sigma_t = \dfrac{E \cdot \alpha \cdot k \cdot n \cdot r_e^n}{[n \cdot (n+2) + a \cdot (n+1+v)]} \cdot \dfrac{1}{\left(\beta^{p-1} - \beta^{q-1}\right)} \\ \quad \times \left[-q \cdot \left(\beta^n - \beta^{q-1}\right) \cdot \rho^{p-1} - p \cdot \left(\beta^{p-1} - \beta^n\right) \cdot \rho^{q-1} + \right. \\ \qquad \left. -(n+1+a) \cdot \left(\beta^{p-1} - \beta^{q-1}\right) \cdot \rho^n\right] \\ u = \dfrac{r_e}{E} \cdot \rho \cdot \dfrac{E \cdot \alpha \cdot k \cdot n \cdot r_e^n}{[n \cdot (n+2) + a \cdot (n+1+v)]} \cdot \dfrac{1}{\left(\beta^{p-1} - \beta^{q-1}\right)} \\ \quad \times \left[-(q+v) \cdot \left(\beta^n - \beta^{q-1}\right) \cdot \rho^{p-1} - (p+v) \cdot \left(\beta^{p-1} - \beta^n\right) \cdot \rho^{q-1} + \right. \\ \quad \left. - \left(\beta^{p-1} - \beta^{q-1}\right) \cdot \left(n+1+a-v - \dfrac{[n \cdot (n+2) + a \cdot (n+1+v)]}{n}\right) \cdot \rho^n\right]. \end{cases} \tag{4.43}$$

We will now consider a hyperbolic disk subjected only to thermal load characterized by a non-zero temperature gradient along the radius expressed by relation (4.5), deriving from the function $T = T(r)$ expressed by (3.2). In this case, the solving differential equation is (4.7), with the term for centrifugal load omitted.

To integrate this equation, we set:

$$u = z + F \cdot \sum_{i=1}^{n} \cdot r^{i+1}, \tag{4.44}$$

with $z = z(r)$ and $F = const$ and, as in the previous case, we determine the latter by establishing the condition that, substituting relation (4.44) together with its first and second derivatives with respect to r in (4.7), the sum of all terms that do not contain the variable z is zero. With a few passages, we find:

$$F = \frac{(1+v)\cdot\alpha\cdot\sum_{i=1}^{n} k_i\cdot(i+a)}{\left[\sum_{i=1}^{n}(a+1+i)\cdot(i+1)+(v\cdot a-1)\right]}. \tag{4.45}$$

Accordingly, relation (4.44) becomes:

$$u = z + \frac{(1+v)\cdot\alpha\cdot\sum_{i=1}^{n} k_i\cdot(i+a)}{\left[\sum_{i=1}^{n}(a+1+i)\cdot(i+1)+(v\cdot a-1)\right]}\cdot\sum_{i=1}^{n} r^{i+1}. \tag{4.46}$$

Note that relation (4.46) is more general than relation (4.34), which is a particular case for $i = n$. With the condition established above, (4.7) again yields the homogeneous equation (4.11), whose solution is described above. The general integral of differential equation (4.7) is thus written in the form:

$$\begin{aligned} u &= C_1\cdot r^p + C_2\cdot r^q + \frac{(1+v)\cdot\alpha\cdot\sum_{i=1}^{n} k_i\cdot(i+a)}{\left[\sum_{i=1}^{n}(a+1+i)\cdot(i+1)+(v\cdot a-1)\right]}\cdot\sum_{i=1}^{n} r^{i+1} = \\ &= C_1\cdot r^p + C_2\cdot r^q + F\cdot\sum_{i=1}^{n} r^{i+1}. \end{aligned} \tag{4.47}$$

Deriving this latter equation respect to r yields:

$$\frac{du}{dr} = p\cdot C_1\cdot r^{p-1} + q\cdot C_2\cdot r^{q-1} + F\cdot\sum_{i=1}^{n}(i+1)\cdot r^i. \tag{4.48}$$

with F expressed by relation (4.45). Subsequently, substituting relations (4.47) and (4.48) in (1.27), where for product $\alpha\cdot T$ the ineffective term appearing in the function of temperature variation with the radius $T = T(r)$ is omitted, gives the following equations for σ_r and σ_t as a function of radius r:

$$\begin{cases} \sigma_r = A \cdot r^{p-1} + B \cdot r^{q-1} + \dfrac{E}{1-\nu^2} \\ \quad \times \left[F \cdot \sum_{i=1}^{n} (i+1+\nu) \cdot r^i - (1+\nu) \cdot \alpha \cdot \sum_{i=1}^{n} k_i \cdot r^i \right] \\ \sigma_t = -q \cdot A \cdot r^{p-1} - p \cdot B \cdot r^{q-1} + \dfrac{E}{1-\nu^2} \\ \quad \times \left[F \cdot \sum_{i=1}^{n} [1+\nu \cdot (i+1)] \cdot r^i - (1+\nu) \cdot \alpha \cdot \sum_{i=1}^{n} k_i \cdot r^i \right], \end{cases} \tag{4.49}$$

where we set:

$$A = \frac{E}{1-\nu^2} \cdot (p+\nu) \cdot C_1 ; \qquad B = \frac{E}{1-\nu^2} \cdot (q+\nu) \cdot C_2 . \tag{4.50}$$

Relations (4.49) can also be rewritten in the following more compact form:

$$\begin{cases} \sigma_r = A \cdot r^{p-1} + B \cdot r^{q-1} + C'_r \\ \sigma_t = -q \cdot A \cdot r^{p-1} - p \cdot B \cdot r^{q-1} + D'_r , \end{cases} \tag{4.51}$$

where C'_r and D'_r, given respectively by the relations:

$$\begin{aligned} C'_r &= \frac{E}{1-\nu^2} \cdot \left[F \cdot \sum_{i=1}^{n} (i+1+\nu) \cdot r^i - (1+\nu) \cdot \alpha \sum_{i=1}^{n} k_i \cdot r^i \right] \\ D'_r &= \frac{E}{1-\nu^2} \cdot \left[F \cdot \sum_{i=1}^{n} [1+\nu(i+1)] \cdot r^i - (1+\nu) \cdot \alpha \sum_{i=1}^{n} k_i \cdot r^i \right], \end{aligned} \tag{4.52}$$

are functions of radius r and, like F, depend on the material's elastic and thermophysical properties as well as on the constants appearing in the thickness and temperature variation functions.

By introducing relations (4.49) in the second (1.25) and bearing in mind that the second (1.14) gives $u = r \cdot \varepsilon_t$, we obtain the following relation expressing radial displacement $u = u(r)$ for the configuration at reference temperature $T = T_0$, which here again is in general assumed to be the assembly temperature T_a:

$$u = -\frac{r}{E} \cdot \left[(q+\nu) \cdot A \cdot r^{p-1} + (p+\nu) \cdot B \cdot r^{q-1} - F \cdot E \cdot \sum_{i=1}^{n} r^i \right]. \tag{4.53}$$

Once again, if the assembly temperature were to differ from the reference temperature T_0, the further term $\alpha \cdot E \cdot (T_a - T_0)$ would appear within the square brackets in expression (4.53). Relations (4.49) and (4.53) describe the distribution of radial and hoop stresses and radial displacement in a disk subjected to a non-zero temperature gradient along the radius. Here again, as the function of temperature variation along the radius and the exponent a characterizing the disk profile are known, C'_r and D'_r are functions that can be readily calculated for an assigned material, while A and B are the integration constants to be determined by imposing boundary conditions. For the hyperbolic disk subject to a non-zero temperature gradient along the radius expressed by (4.5), these constants are still given by relations (4.41). Accordingly, the system obtained from the first of relations (4.51) gives:

$$A = \frac{C'_{re} \cdot r_i^{q-1} - C'_{ri} \cdot r_e^{q-1}}{r_i^{p-1} \cdot r_e^{q-1} - r_i^{q-1} \cdot r_e^{p-1}} \quad ; \quad B = -\frac{C'_{re} \cdot r_i^{p-1} - C'_{ri} \cdot r_e^{p-1}}{r_i^{p-1} \cdot r_e^{q-1} - r_i^{q-1} \cdot r_e^{p-1}}, \tag{4.54}$$

where C'_{re} and D'_{re} are the values of C'_r at the outer radius and inner radius respectively.

Substituting the constants thus found in relations (4.51) and (4.53) gives the following expressions for σ_r, σ_t and u which provide a univocal solution of the problem:

$$\begin{cases} \sigma_r = \dfrac{C'_{re} \cdot r_i^{q-1} - C'_{ri} \cdot r_e^{q-1}}{r_i^{p-1} \cdot r_e^{q-1} - r_i^{q-1} \cdot r_e^{p-1}} \cdot r^{p-1} - \dfrac{C'_{re} \cdot r_i^{p-1} - C'_{ri} \cdot r_e^{p-1}}{r_i^{p-1} \cdot r_e^{q-1} - r_i^{q-1} \cdot r_e^{p-1}} \cdot r^{q-1} + \\ \quad + \dfrac{E}{1-\nu^2} \cdot \left[F \cdot \sum_{i=1}^{n} (i+1+\nu) \cdot r^i - (1+\nu) \cdot \alpha \sum_{i=1}^{n} k_i \cdot r^i \right] \\ \sigma_t = -q \cdot \dfrac{C'_{re} \cdot r_i^{q-1} - C'_{ri} \cdot r_e^{q-1}}{r_i^{p-1} \cdot r_e^{q-1} - r_i^{q-1} \cdot r_e^{p-1}} \cdot r^{p-1} + p \cdot \dfrac{C'_{re} \cdot r_i^{p-1} - C'_{ri} \cdot r_e^{p-1}}{r_i^{p-1} \cdot r_e^{q-1} - r_i^{q-1} \cdot r_e^{p-1}} \cdot r^{q-1} + \\ \quad + \dfrac{E}{1-\nu^2} \cdot \left[F \cdot \sum_{i=1}^{n} [1+\nu(i+1)] \cdot r^i - (1+\nu) \cdot \alpha \sum_{i=1}^{n} k_i \cdot r^i \right] \\ u = -\dfrac{r}{E} \cdot \left[(q+\nu) \cdot \dfrac{C'_{re} \cdot r_i^{q-1} - C'_{ri} \cdot r_e^{q-1}}{r_i^{p-1} \cdot r_e^{q-1} - r_i^{q-1} \cdot r_e^{p-1}} \cdot r^{p-1} \right. \\ \quad -(p+\nu) \cdot \dfrac{C'_{re} \cdot r_i^{p-1} - C'_{ri} \cdot r_e^{p-1}}{r_i^{p-1} \cdot r_e^{q-1} - r_i^{q-1} \cdot r_e^{p-1}} \cdot r^{q-1} + \\ \quad \left. - \dfrac{(1+\nu) \cdot \alpha \cdot \sum_{i=1}^{n} k_i \cdot (i+a)}{\left[\sum_{i=1}^{n} (a+1+i) \cdot (i+1) + (\nu \cdot a - 1) \right]} \cdot E \cdot \sum_{i=1}^{n} r^i \right]. \end{cases} \tag{4.55}$$

4.5.1 Example

We will now consider the same hyperbolic annular disk discussed in the example in Sect. 4.4.1, subjected to a linear temperature distribution as in the example in, with $T_e = 120$ °C and $T_i = T_0 = 20$ °C. We will plot the diagrams representing radial and hoop stresses versus radius, and calculate the values of the latter which are of design interest. Radial displacement at the inner radius will also be calculated.

From the problem data, we thus have: $\alpha = 12{\cdot}10^{-6}$ °C^{-1}; $\nu = 0.3$; $E = 210$ GPa; $\beta = 0.125$; $p = 1.75$; $q = -0.75$; $a = -1$; $r_e = 200$ mm; $\beta^{p-1} = \beta^{0.75} = 0.210$ and $\beta^{q-1} = \beta^{-1.75} = 38.005$. Bearing in mind that linear temperature distribution along the radius can be expressed by means of relation $T = T_i + [(T_e - T_i)/(r_e - r_i)]{\cdot}(r - r_i) = T_i + [(T_e - T_i)/(1 - \beta)]{\cdot}(\rho - \beta)$, it follows that the value of parameter k appearing in the relation $T = T_0 + kr$ is 114.29 °C/m.

Using the first two relations (4.43) which express the stress field due to heat flow in the disk in question, where $n = 1$, we obtain the distributions of σ_r and σ_t shown in dimensionless form in Fig. 4.2c, i.e., relative to $E\alpha k r_e/[3 + a(2+\nu)]$. As can be seen from the figure, the absolute maximum stress value is that reached by hoop stress at the inner edge.

Rather than using the third (4.43) to calculate radial displacement at the inner radius, it is again simpler to employ the second (1.14) together with the second (1.25); we thus obtain:

$$(u)_{\rho=\beta} = \frac{\beta \cdot r_e}{E} \cdot \left[(\sigma_t)_{\rho=\beta} - \nu \cdot (\sigma_r)_{\rho=\beta}\right] = \frac{\beta \cdot r_e}{E} \cdot 0.825 \cdot \sigma_0$$

.

Chapter 5
Disk of Uniform Strength

5.1 Profile Definition

In general, a disk of uniform strength is defined as one in which the radial and hoop stresses resulting from centrifugal loading are constant and equal to each other at all points. Consequently, the following relation applies throughout the disk:

$$\sigma_r = \sigma_t = \sigma = const, \tag{5.1}$$

where σ is the stress at which the disk material works, which is obviously equal to the permissible stress σ_{am} for that material. This configuration, where disk profile varies according to an exponential function, was first introduced by De Laval in the late 1800s.

For a disk consisting of isotropic material subjected only to centrifugal load (zero thermal gradient along the radius), the above relations satisfies compatibility equations (1.23) or (1.24). A disk with these properties is thus possible from the standpoint of compatibility of the stress field. However, as a turbine disk is also always subjected to thermal load, the condition of uniform strength given by relation (5.1) can never be ensured, as it would not respect the compatibility equation. The following treatment thus applies to a hypothetical disk subjected only to centrifugal load. Although the uniform strength profile can be considered as a starting point in designing actual turbine disks, it is better to start from the profiles that will be discussed in the following chapters, as they can ensure an optimal distribution of principal stresses even when thermal loading is involved.

Introducing relation (5.1) in the first equilibrium equation (1.10) yields the following relation, which is the solving equation for the uniform strength disk:

$$\sigma \cdot r \cdot \frac{dh}{dr} + \gamma \cdot \omega^2 \cdot r^2 \cdot h = 0. \tag{5.2}$$

V. Vullo and F. Vivio, *Rotors: Stress Analysis and Design*,
Mechanical Engineering Series, DOI 10.1007/978-88-470-2562-2_5,
© Springer-Verlag Italia 2013

This is a first order differential equation with separable variables which can be written in the following form:

$$\frac{dh}{h} = -\frac{\gamma \cdot \omega^2}{\sigma} \cdot r \cdot dr. \tag{5.3}$$

Integrated, this relation gives:

$$\ln h = -\frac{\gamma \cdot \omega^2}{2 \cdot \sigma} \cdot r^2 + \ln C, \tag{5.4}$$

where C is the integration constant, which can be determined by establishing that the disk thickness assumes an assigned value at a given radius. Passing from the logarithm to the function, relation (5.4) gives:

$$h = C \cdot e^{-\frac{\gamma \cdot \omega^2}{2 \cdot \sigma} \cdot r^2}. \tag{5.5}$$

Independently of constant C, the profile thus defined features an inflection point at the radius

$$r = \sqrt{\frac{\sigma}{\gamma \cdot \omega^2}}, \tag{5.6}$$

which can be found by equalling the second derivative of relation (5.5) to zero. For values of radius r lesser or greater than that given by this relation, the profile's concavity will face inward or outward respectively as viewed by an observer located on the mid-plane bisecting the disk's thickness.

On the basis of relation (5.5), and depending on whether we impose the condition

$$h = h_e \quad \text{for} \quad r = r_e \tag{5.7}$$

or the condition

$$h = h_0 \quad \text{for} \quad r = 0, \tag{5.8}$$

the integration constant C will be given by:

$$C = h_e \cdot e^{\frac{\gamma \cdot \omega^2}{2 \cdot \sigma} \cdot r_e^2} \tag{5.9}$$

or

$$C = h_0. \tag{5.10}$$

Consequently, the variation function for thickness h with radius r can be expressed respectively with the following two equivalent relations:

$$h = h_e \cdot e^{\frac{\gamma \cdot \omega^2}{2 \cdot \sigma}\left(r_e^2 - r^2\right)}. \tag{5.11}$$

or

$$h = h_0 \cdot e^{-\frac{\gamma \cdot \omega^2}{2 \cdot \sigma} \cdot r^2}. \tag{5.12}$$

Finally, after collecting the term r_e^2 in the exponent in relation (5.11), multiplying and dividing the exponent appearing in relation (5.12) by r_e^2 and introducing the usual reference stress $\sigma_0 = \gamma \cdot \omega^2 \cdot r_e^2$ as well as the dimensionless variable ρ, relations (5.11) and (5.12) can be rewritten in the following form:

$$h = h_e \cdot e^{\frac{\sigma_0}{2 \cdot \sigma}\left(1 - \rho^2\right)}. \tag{5.13}$$

$$h = h_0 \cdot e^{-\frac{\sigma_0}{2 \cdot \sigma} \cdot \rho^2}. \tag{5.14}$$

Both relations (5.12) and (5.14) define a solid disk with thickness h_0 at the axis and thickness h decreasing rapidly from the axis outwards according to a one-parameter exponential function[1] given by the ratio $\sigma_0/2\sigma$; the first relation describes this disk in terms of r, and the second in terms of ρ. From both relations (5.12) and (5.14), we can conclude that the outer radius is not defined, and could thus be regarded as virtually infinite (Fig. 5.1a). If the disk were to be cut off at a given outer radius r_e (as is in any case necessary for design and construction reasons, as the geometrical dimensions of the interface on which the blades act are generally fixed), the condition (5.1) would not be respected on the cylindrical surface of radius $\rho = 1$, since we would necessarily have $\sigma_r = 0$ on this surface.

To respect the above condition, the disk must feature a peripheral crown ring whose dimensions are such as to develop the radial stress σ at the disk interface. Usually, in fact, the outer blades are not sufficient in order to have $\sigma_r = \sigma_t = \sigma$ at radius $r = r_e$; in a disk featuring an array of blades spaced at equal angles on the

[1] Eraslan and Orçan [15] considered a profile with thickness varying according to a two-parameter exponential function, defined by the relation $h = h_0 \cdot e^{-n \cdot \rho^k}$ where, with the meaning of the other symbols remaining the same, n and k are geometric parameters controlling thickness at the outer edge relative to that at the axis and the profile shape respectively. This relation makes it possible to describe solid and annular disks with concave, convex and inflection point profiles, but not conical disks; with this relation, differential equation (1.28) can be integrated by means of the linear combination of two mutually independent hypergeometric functions. From the application standpoint, it should be noted that an optimization process for the two geometric parameters n and k leads, as would be expected, to the De Laval uniform strength disk. Consequently, this relation is of purely mathematical interest.

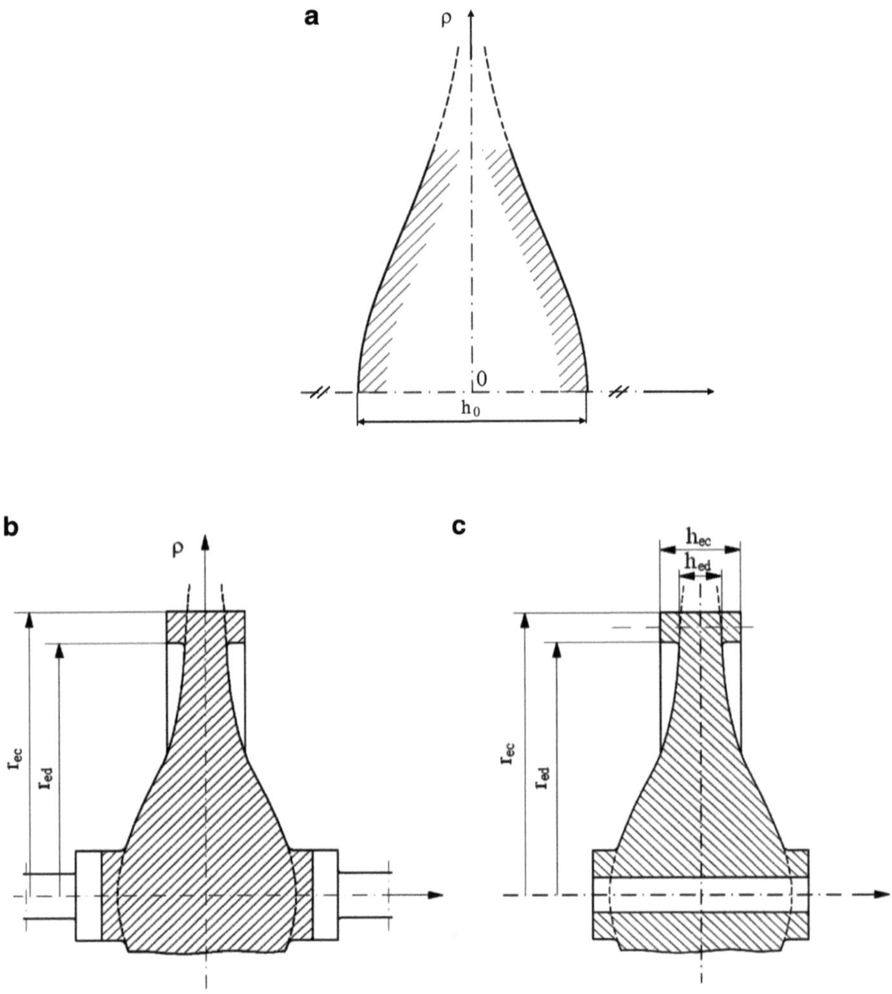

Fig. 5.1 (**a**) Uniform strength disk, unlimited at outer radius; (**b**) solid uniform strength disk with lateral flanges for connection to two flanges at the ends of a multi-pieces shaft; (**c**) uniform strength disk with annular hub to be shrunk onto a one-piece shaft

outer radius having total mass m_p and centre of gravity at a distance r_p from the rotational axis, this condition would be respected only if mass m_p were:

$$m_p = \frac{\pi \cdot \gamma \cdot r_e^3 \cdot h_0 \cdot e^{-\sigma_0/2\sigma}}{(\sigma_0/2\sigma) \cdot r_p}. \tag{5.15}$$

This value of m_p is determined by establishing that the total centrifugal force $F_c = m_p \cdot \omega^2 \cdot r_p$ to which the blades are subjected is evenly distributed on the outer periphery of the disk, having a surface area of $2\pi \cdot r_e \cdot h_e$. However, as relation (5.15) is not generally satisfied in a bladed disk, it is necessary to provide a crown ring of appropriate dimensions.

Similarly, if the disk were to feature a central hole, condition (5.1) would not be respected on the cylindrical surface of radius r_i, as here again radial stress would be zero. For this condition to be respected, the disk would have to be solid and integral with the shaft, or would have to feature a solid or annular central portion of constant thickness and blended with the uniform-strength portion, In this latter case, the central portion is a hub to be designed in such a way as to generate a tensile stress equal to σ at the interface with the variable-thickness portion. Connection to the shaft is generally accomplished by means of bolts or screws which secure the disk to flanges on the shaft.

5.2 Technical Solutions

In turbines, disks with profile as described by relation (5.14) occasionally have no central hole and are secured at both sides to two flanged piece of a multi-pieces shafts by means of threaded connections (Fig. 5.1b). More frequently, however, disks with a central hole are used, and are shrunk onto a one-piece shaft (Fig. 5.1c). In this case, the central portion of the disk is configured to form a long hub whose profile is very far from that of the uniform strength disk defined by relations (5.13) and (5.14). Figure 5.2a shows the design solution actually used in both the medium/low-pressure section and the convergent double-flow low-pressure section, which are mounted on the same shaft, of a steam turbine for a thermonuclear power plant. As can be seen, the solid disks also serve as the drive shaft, as they are torsionally and flexurally connected to each other by means of welded joints on the mating faces of the two consecutive disks' crown rings. On a larger scale, Fig. 5.2b shows a design solution similar to that represented in Fig. 5.1a. In other designs by the same manufacturer (see Fig. 5.3a, illustrating the four-flow low-pressure section of a thermonuclear power plant steam turbine), the disks are torsionally coupled, not at the frontal surfaces of the crown rings carrying seats for the blade attachments on their periphery, but at the frontal surfaces of two dummy crown rings at a smaller radius. In any case, turbine disks always feature a crown ring housing the blades on their outer periphery (Fig. 5.1b, c).

Though the hub and crown ring do not have profiles complying with the thickness variation functions characterizing the solid disk of uniform strength, if we bear the centrifugal forces involved and the deformability of the individual parts (hub, disk and crown ring) in mind, we can nevertheless determine the dimensions of these parts in such a way that the stresses σ that satisfy condition (5.1) are generated at the limit cross sections of the disk that is at outer and inner radii of the uniform strength disk. Figures 5.3a, b show two design solutions actually uses for, respectively, a steam turbine and a gas turbine.

With reference to relations (5.13) and (5.14), Fig. 5.4 shows ratio h/h_e (Fig. 5.4a) and ratio h/h_0 (Fig. 5.4b) versus ρ, for different values of parameter $\sigma_0/2\sigma$, as can be seen, bell shapes are heavily influenced by ratio $\sigma_0/2\sigma$, with maximum thickness at the axis. Pronounced thickness gradients do not exist only where the values of ratio

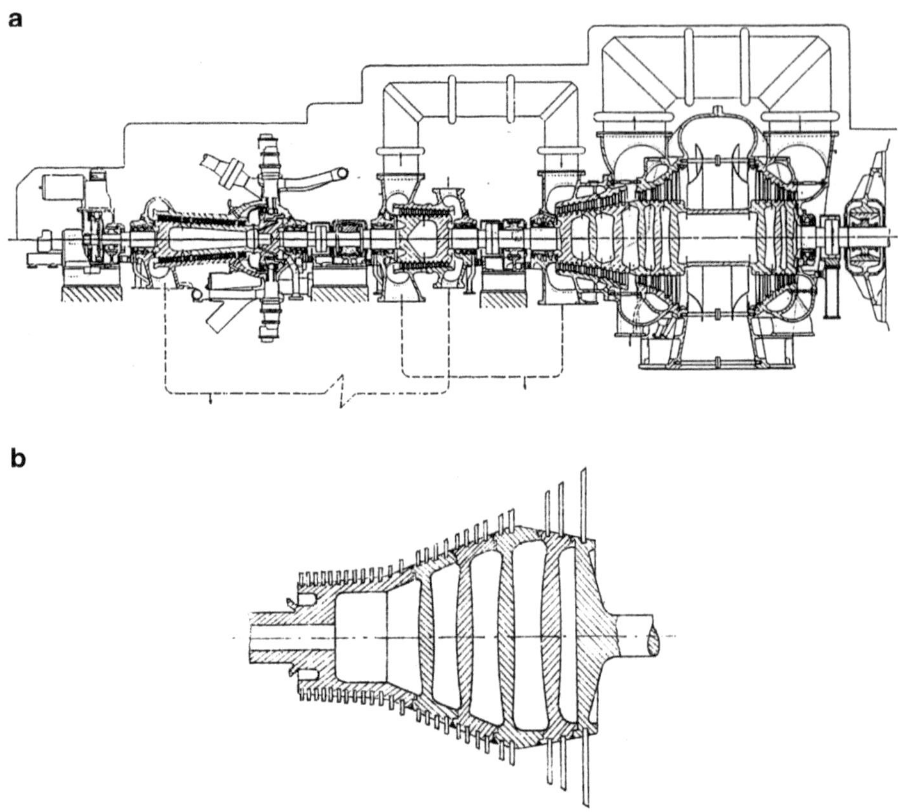

Fig. 5.2 (**a**) Uniform strength disks in the medium/low-pressure and double-flow low-pressure sections of a Brown Boveri steam turbine for a thermonuclear power plant, with disks connected at the crown rings; (**b**) detail of a similar steam turbine design

$\sigma_0/2\sigma$ are not particularly high. If ratio h_0/h_e exceeds 3, and $\sigma_0/2\sigma$ is thus higher than 1.1, the disk shape may be impossible to actually produce, or may be extremely expensive both from the manufacturing standpoint and because of its axial dimensions. As a result, the use of uniform strength disks is now limited to values of $\sigma_0/2\sigma$ below 1.1 (though certain design applications involve values of $\sigma_0/2\sigma$ up to (1.6–1.8), but always below 2) and, consequently, to relatively low peripheral velocities $\omega \cdot r_e$ or high strength materials, i.e., materials with high permissible stress σ.

In view of the foregoing considerations, the limit often given in the literature (Giovannozzi, [29]) for the ratio $\sigma_0/2\sigma = 3.22$, which applies to disks having a thickness at the outer radius equal to 1/25 of the thickness at the axis, must be regarded as quite far from that imposed by current manufacturing and functional requirements. It should also be specified that, when designing the disk, thickness at the axis must not exceed $(0.25–0.30) \cdot r_e$.

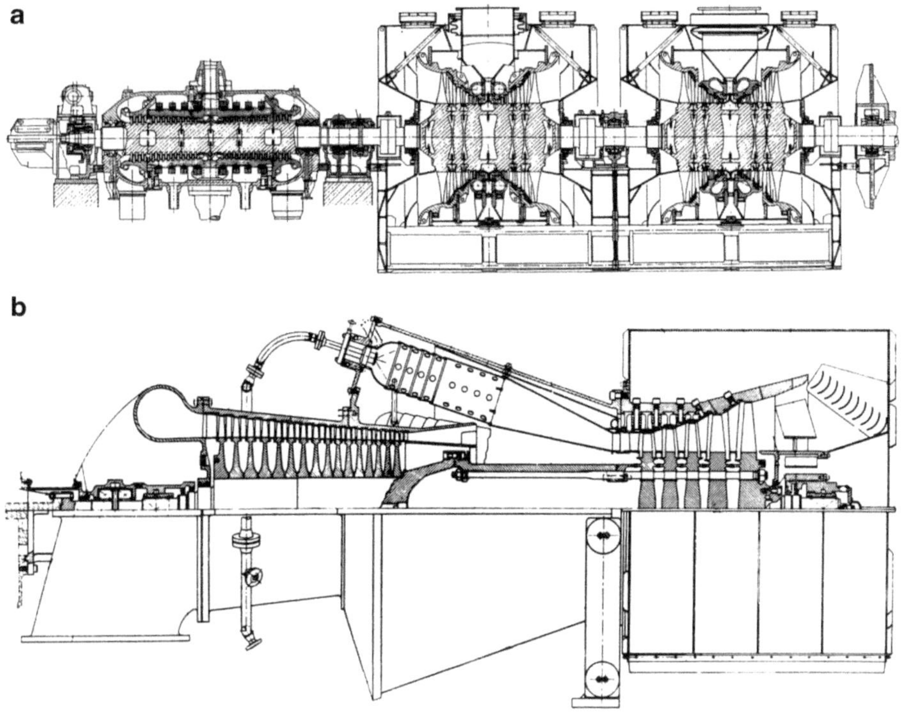

Fig. 5.3 (**a**) Uniform strength disks in the four flow low-pressure section of a Brown Boveri steam turbine for a thermonuclear power plant, with disks connected at dummy crown rings at a smaller radius than that of the bladed crown rings; (**b**) longitudinal section of a Westinghouse-Fiat gas turbine

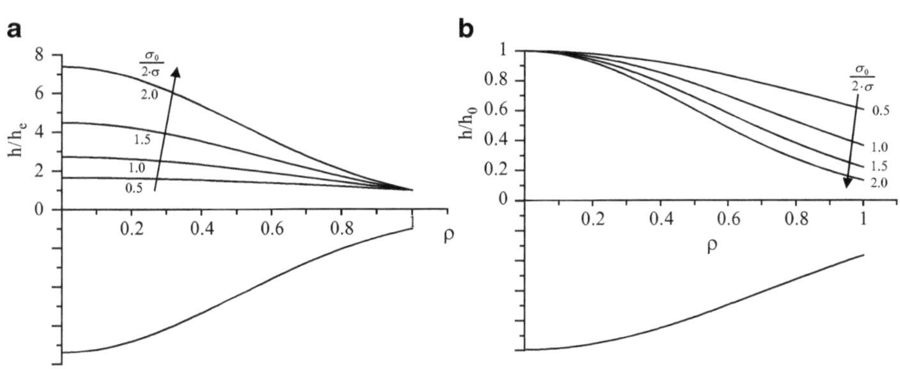

Fig. 5.4 (**a**) Distributions of ratio h/h_e; (**b**) ratio h/h_0 versus radius ρ for different values of parameter $\sigma_0/2\sigma$ in uniform strength disks

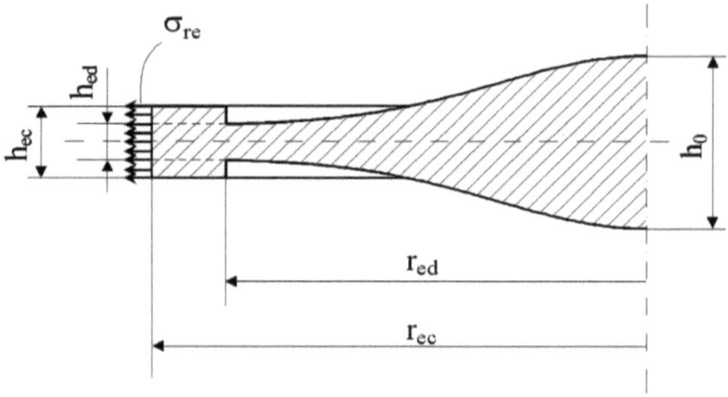

Fig. 5.5 Geometry of a uniform strength disk with a crown ring of uniform thickness

5.3 Crown Ring Design

To determine (Fig. 5.1b, c) the radial thickness ($r_{ec} - r_{ed}$) of the crown ring which, together with the blades at its outer radius, can guarantee condition (5.1) at the interface with the disk, we will consider a uniform strength disk with no central hole and with a crown ring of constant thickness, featuring the geometry shown in Fig. 5.5. Let $\beta = r_{ed}/r_{ec}$ and $\delta = h_{ec}/h_{ed}$; the crown ring's outer radius r_{ec} and its axial thickness h_{ec} are generally known, as they are determined beforehand when designing the blades and the associated seats in the crown ring.

For a rough calculation, we can consider the crown ring to be a disk of constant thickness stressed at the outer radius r_{ec} by a radial stress σ_{re} due to the blades, and at the inner radius r_{ed} by a radial stress $\sigma_{ri} = \sigma/\delta$ due to the disk. Stress σ_{ri} is justified by the fact that, for there to be equilibrium, the product ($\sigma_r{\cdot}h$) must remain constant at the disk/ring interface. At this interface, in any case, there is a discontinuity in function $h = h(r)$ if $\delta \neq 1$ or in its first derivative if $\delta = 1$.

Assuming a plane stress state in this area is thus a very rough approximation. With this interface regarded as part of the crown ring, there would be a redistribution of stress σ_{ri} for $\delta \neq 1$ in areas that are not in fact loaded. Nevertheless, this approximation is indispensable if we wish to use the relations of the mono-dimensional thin disk theory, and in any case leads to results that are acceptable from the design standpoint, given that, according to Saint Venant's principle, if a system of forces acting on a small portion of the surface of an elastic body is replaced by another statically equivalent system of forces acting on the same portion of the surface, the redistribution of loading produces substantial changes in the stresses only at the local level, but has negligible effects on the stress state at distances which are large by comparison with the linear dimensions of the surface to which the equivalent system of loads was applied.

For the structure to satisfy compatibility conditions, the radial displacement of the outer edge of the disk must be equal to the radial displacement of the inner edge of the crown ring. Bearing in mind the second geometric relation (1.14) and the second (1.25) from which the temperature term is omitted, and considering condition (5.1), the disk's radial displacement at radius r_{ed} will be:

$$u = r_{ed} \cdot \varepsilon_t = \frac{r_{ed}}{E} \cdot (\sigma_t - \nu \cdot \sigma_r) = \frac{r_{ed}}{E} \cdot \sigma \cdot (1 - \nu). \tag{5.16}$$

By applying the principle of superposition, the crown ring's radial displacement at the same radius r_{ed} can be calculated as the sum of the displacements resulting from centrifugal load, stress at the inner radius, and stress at the outer radius. Consequently, the third relations of (2.34), (2.16) and (2.19) written for $\rho = \beta$ yield:

$$u = \frac{r_{ed}}{E} \cdot \beta \cdot \left\{ \frac{\sigma_0}{4} \cdot \left[(3 + \nu) + \beta^2 \cdot (1 - \nu) \right] - \frac{\sigma}{\delta} \cdot \left[\frac{\beta^2 \cdot (1 - \nu) + (1 + \nu)}{1 - \beta^2} \right] + \frac{2 \cdot \sigma_{re}}{1 - \beta^2} \right\}, \tag{5.17}$$

where $\sigma_0 = \gamma \cdot \omega^2 \cdot r_{ec}^2$.

Equating (5.16) and (5.17) and solving for β, we obtain:

$$\beta = \sqrt{\frac{2 \cdot \sigma}{\sigma_0 \cdot \delta} \cdot \left[(\delta - 1) + 2 \cdot \sqrt{\frac{(\sigma_0 \cdot \delta^2 / 2\sigma) \cdot [(\sigma_0 / 2\sigma) - 1 + \nu]}{(1 - \nu^2)} + \frac{(\delta - 1)^2}{4} + \frac{\sigma_0 \cdot \delta^2 \cdot \sigma_{re}}{2\sigma^2 \cdot (1 - \nu)}} \right] - \frac{1 + \nu}{1 - \nu}} \tag{5.18}$$

Equation (5.18) can be used to determine the crown ring's radial thickness once all other parameters have been determined. The ratio of crown ring thickness h_{ec} to thickness h_0 at the centre of the disk can be readily determined by means of relation (5.14), thus leading to the following relation:

$$h_{ec} / h_0 = \delta \cdot e^{-\sigma_0 / 2\sigma}. \tag{5.19}$$

In design calculations, crown ring outer radius r_{ec} and axial length h_{ec} are imposed by manufacturing reasons, and rotor angular velocity ω is assigned. Once ratio δ and permissible stress σ are established as design choices (the latter on the basis of the material to be used), (5.18) can be used to obtain ratio β and, consequently, outer radius r_{ed} of the uniform strength disk. Disk profile is then determined by means of (5.13).

The disk profile can also be constructed via a graphic procedure, using the Rowe diagram [16] shown in Fig. 5.6 and taken from several technical manuals. To this end, once r_{ed} has been determined, we introduce the peripheral load per unit of tangential length $P = h_{ed} \cdot \sigma$ and use the diagram to find, for a selected number of radii ρ, the ratio $(h \cdot \sigma)/P$ for an assigned σ_0/σ. The thicknesses corresponding to these radii are obtained with the relation:

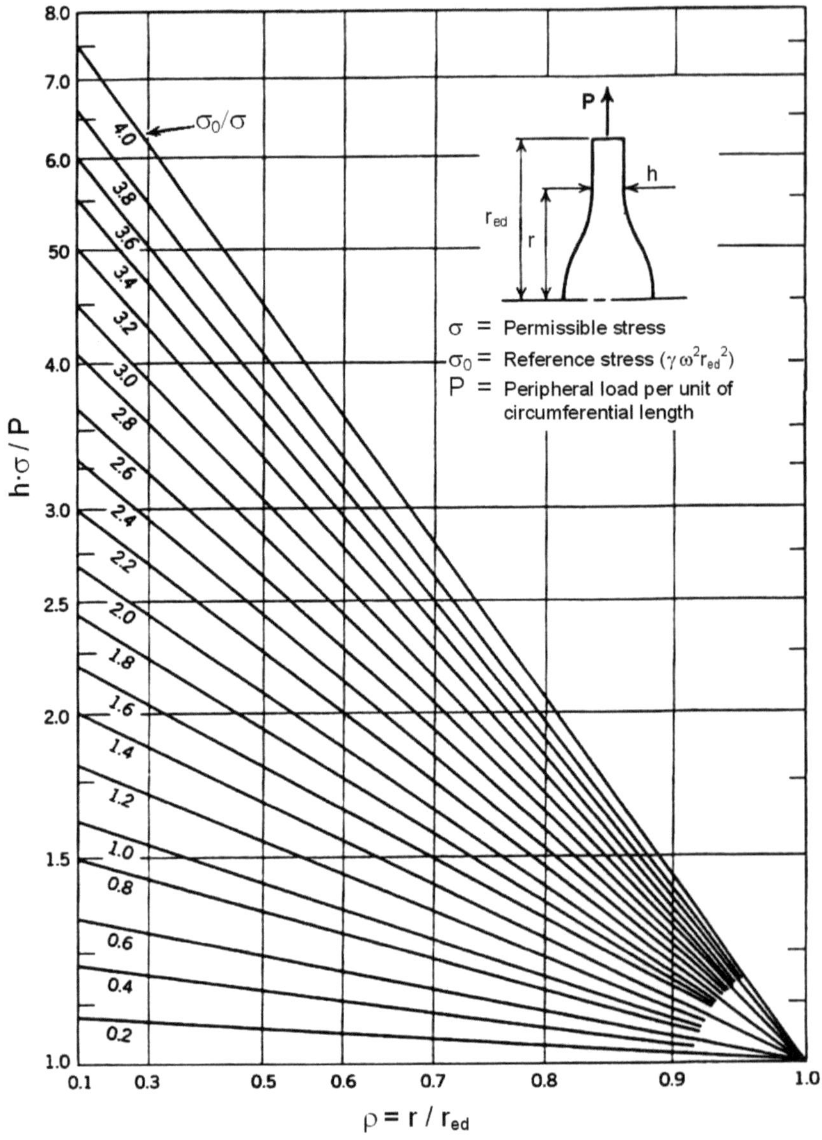

Fig. 5.6 Rowe diagram for constructing the profile of a uniform strength disk via a graphic procedure

$$h = \left(\frac{h \cdot \sigma}{P}\right) \cdot \left(\frac{P}{\sigma}\right). \tag{5.20}$$

Though modern computing power might appear to have made this diagram obsolete, it has been included here because it clarifies the influence of the various parameters on disk profile at a glance.

To conclude our discussion of the uniform strength disk, it should be borne in mind that, from the manufacturing standpoint, the large number of design constraints, some of which are dictated by the process cycle, can at times make it impracticable to employ disks which are of uniform strength in the strict sense. As a result, it may be preferable to forego the undeniable advantages they offer in terms of stress and strain states, and opt for hyperbolic disks (or, as we will see in Chaps. 6 and 7, conical disks and non-linearly variable thickness disks) which approximate them within certain limits (for instance, a uniform strength disk is characterized by an inflection point, which does not exist in hyperbolic, conical or non-linearly variable thickness disks).

5.4 Example

We will now consider a uniform strength steel disk of the type shown in Fig. 5.1c, whose geometry at the centre is defined by magnitudes $r_e = r_{ed} = 0.250$ m, $r_i = 0.050$ m, $h_e = h_{ed} = 0.015$ m and $h_0 = 0.075$ m, and rotating at angular velocity $\omega = 750$ rad/s ($n \approx 7{,}160$ rpm). We will determine the stress in the disk in question and its profile. We will then consider an annular hyperbolic disk, also consisting of steel, and also having at the central part $r_e = 0.250$ m, $r_i = 0.050$ m, $h_e = 0.015$ m and h_i corresponding to those of the uniform strength disk at radius $r_i = 0.050$ m, comparing the stress states at radius r_i of the two disks, as well as the radial displacements at the same radius.

Equating relations (5.13) and (5.14), we obtain:

$$\frac{h_e}{h_0} = \frac{e^{-\frac{\sigma_0}{2\cdot\sigma}\cdot\rho^2}}{e^{\frac{\sigma_0}{2\cdot\sigma}\cdot\left(1-\rho^2\right)}} = e^{-\frac{\sigma_0}{2\cdot\sigma}}.$$

whereby

$$\frac{0.015}{0.075} = \frac{1}{5} = e^{-\frac{\sigma_0}{2\cdot\sigma}}, \quad \text{then } \ln\frac{1}{5} = -\frac{\sigma_0}{2\cdot\sigma}$$

and hence

$$\sigma = \frac{\sigma_0}{3.219} = 0.310\cdot\sigma_0 = 85.19 \text{ MPa}$$

given that $\sigma_0 = \gamma\cdot\omega^2\cdot r_e^2 = 7.8\cdot 10^3\cdot 750^2\cdot 0.25^2 = 274.22$ MPa.

Using relation (5.14), we can then conclude that the profile of the uniform strength disk in question is defined by the following relation

$$h = h_0\cdot e^{-\frac{\sigma_0}{2\cdot\sigma}\cdot\rho^2} = 0.075\cdot e^{-1.609\cdot\rho^2}$$

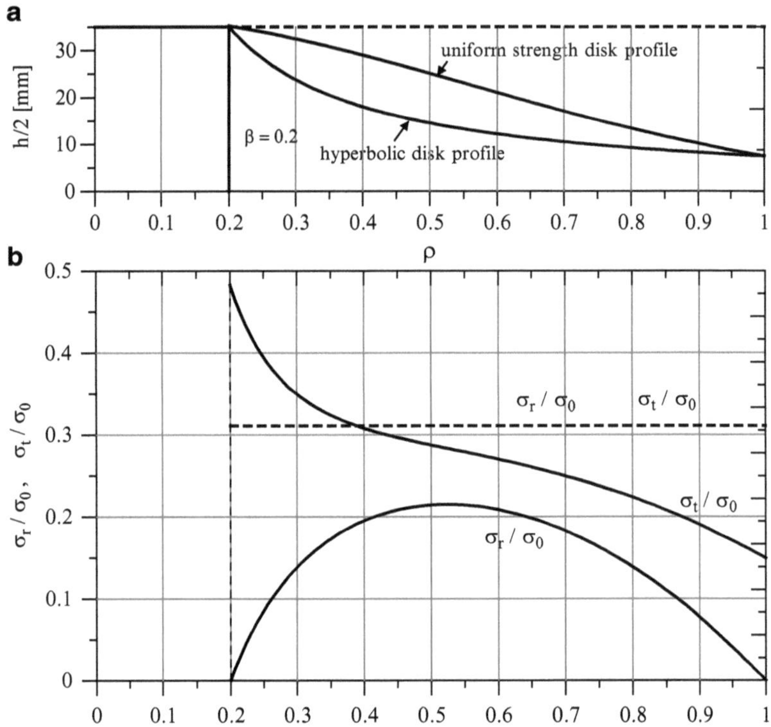

Fig. 5.7 (**a**) Profiles of the uniform strength disk and of the hyperbolic disk with the same h_i; (**b**) dimensionless principal stresses σ_r/σ_0 and σ_t/σ_0 versus dimensionless radius ρ in the uniform strength disk and in the hyperbolic disk with the same h_i, both subjected only to centrifugal load

as a function of dimensionless variable ρ, or as a function of variable r by the relation

$$h = h_0 \cdot e^{-\dfrac{\gamma \cdot \omega^2 \cdot r^2}{2 \cdot \sigma}} = 0.075 \cdot e^{-25.751 \cdot r^2}.$$

From both of these relations, it can be concluded that at radius $r_i = 0.050$ m, i.e., for $\beta = r_i/r_e = 0.2$, thickness h_i of the uniform strength disk is 0.070 m.

The hyperbolic disk to be compared with that considered here is thus defined by the following magnitudes: $r_e = 0.250$ m, $r_i = 0.050$ m, $h_e = 0.015$ m and $h_i = 0.070$ m.

As $\beta = 0.2$, (4.3) written for $\rho = 1$ gives $a = -0.957$. For the steel disk ($\nu = 0.3$), relations (4.14) yield the following values of roots p and q: $p = 1.710$, $q = -0.753$. As $\beta^2 = 0.04$, $\beta^{p-1} = \beta^{0.710} = 0.32$ and $\beta^{q-1} = \beta^{-1.753} = 16.80$, the first two relations (4.31) give:

$$\sigma_r = 0.68 \cdot \frac{\left(-16.76 \cdot \rho^{0.71} + 0.28 \cdot \rho^{-1.75} + 16.48 \cdot \rho^2\right)}{(-16.48)} \cdot \sigma_0$$

$$\sigma_t = \frac{\left(-41.65 \cdot \rho^{0.71} - 1.58 \cdot \rho^{-1.75} + 31.31 \cdot \rho^2\right)}{(-79.79)} \cdot \sigma_0.$$

In particular, we obtain:

$$(\sigma_r)_{\rho=\beta} = (\sigma_r)_{\rho=1} = 0; \quad (\sigma_t)_{\rho=\beta} = 0.482 \cdot \sigma_0; \quad (\sigma_t)_{\rho=1} = 0.149 \cdot \sigma_0.$$

In comparing the stress states of the two disks, we thus see that the hoop stress at the inner radius of the hyperbolic disk is over 55% higher than that in the uniform strength disk (Fig. 5.7).

The radial displacement at the inner radius of the uniform strength disk is:

$$(u)_{\rho=\beta} = \frac{r_e \cdot \beta}{E} \cdot \left[(\sigma_t)_{\rho=\beta} - \nu \cdot (\sigma_r)_{\rho=\beta}\right] = \frac{r_e \cdot \beta}{E} \cdot \sigma \cdot (1 - \nu) = \frac{r_e \cdot \beta}{E} \cdot 0.217 \cdot \sigma_0,$$

while that at the inner radius of the corresponding hyperbolic disk is:

$$(u)_{\rho=\beta} = \frac{r_e \cdot \beta}{E} \cdot \left[(\sigma_t)_{\rho=\beta} - \nu \cdot (\sigma_r)_{\rho=\beta}\right] = \frac{r_e \cdot \beta}{E} \cdot 0.482 \cdot \sigma_0.$$

From the standpoint of displacements as well as that of stresses, then, there can be no doubt that using a uniform strength disk is more advantageous, all other conditions remaining equal, than employing a comparable hyperbolic disk.

Chapter 6
Conical Disk

6.1 Conical Profile Geometry and General Differential Equations

Turbine disks frequently employ so-called conical profiles (also referred to as biconical), i.e., profiles whose thickness varies with the radius according to a linear function $h = h(r)$. This is because these profiles are easier, and thus less costly, to produce than the corresponding hyperbolic disks, uniform strength disks or disks whose thickness varies according to a power of a linear function such as (7.12), which will be discussed below in Chap. 7. As was mentioned in Chap. 4, (4.1) which defines Stodola's hyperbolic profile can also be used to describe the profile of an annular disk which diverges according to a linear function from the inner to the outer radius, or in other words, the diverging conical annular disk. In this case, (4.1), (4.2) and (4.3) will have $a = 1$ and $C = h_e/r_e = h_i/r_i$, and thus $h_i/h_e = r_i/r_e = \beta$.

The elastic stress and strain of solid or annular rotating converging conical disks, symmetric with respect to the mid-plane, was first analysed by E. Honegger [36], who also introduced the following function of thickness variation

$$h = h_0 \cdot (1 - t), \tag{6.1}$$

deriving from the position

$$h = a + b \cdot r = a \cdot \left(1 + \frac{b}{a} \cdot r\right) = a \cdot (1 - m \cdot r) = a \cdot (1 - t), \tag{6.2}$$

with $a = h_0$ as the disk thickness at the axis, b negative when the disk is tapered towards the outer radius, $m = -(b/a)$, $t = -(b/a)\cdot r = r/R$ and $R = -(a/b)$ distance from the rotational axis of the apex V where the two symmetrical sides of the profile intersect. As discussed in Chap. 7, in the relation (6.1) will be introduced an exponent k that generalizes, so that profiles whose thickness varies nonlinearly according to a power of a linear function can be analysed.

V. Vullo and F. Vivio, *Rotors: Stress Analysis and Design*,
Mechanical Engineering Series, DOI 10.1007/978-88-470-2562-2_6,

© Springer-Verlag Italia 2013

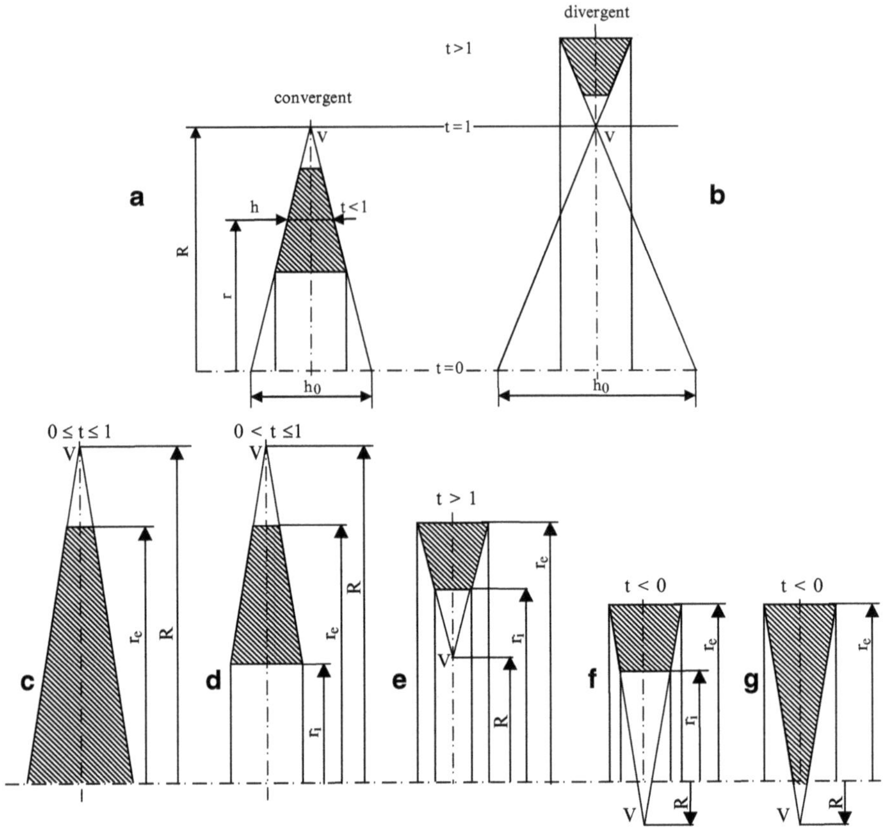

Fig. 6.1 Disks with conical profile: (**a**) convergent profile; (**b**) divergent profile; (**c**) convergent solid disk; (**d**) convergent annular disk; (**e**) divergent annular disk; (**f**) divergent annular disk; (**g**) divergent solid disk

Subsequently, R. Giovannozzi [26] used the same function (6.1) to extend Honegger's analysis, applying it to the same converging conical disk subjected to thermal load and featuring a fictitious variation of the density along its radius (this fictitious variation in density exists when the disk has radial blades on its lateral surfaces, which increase the disk's mass without appreciably contributing to its load capacity), as well as to the diverging conical disk, whether solid or annular. According to Giovannozzi's generalization as shown Fig. 6.1a, b, h_0 in relation (6.1) is either the actual or virtual thickness at the disk axis, and $t = r/R$ is the dimensionless radius, i.e., the ratio of generic radius r and the distance R from the rotational axis to the apex where the two symmetrical sides of the profile or its extension intersect.

The theoretical approach proposed by Honegger and used by Giovannozzi is based on the formulation originated by Meissner [53], which transformed the homogeneous second order differential equation governing the stress and strain

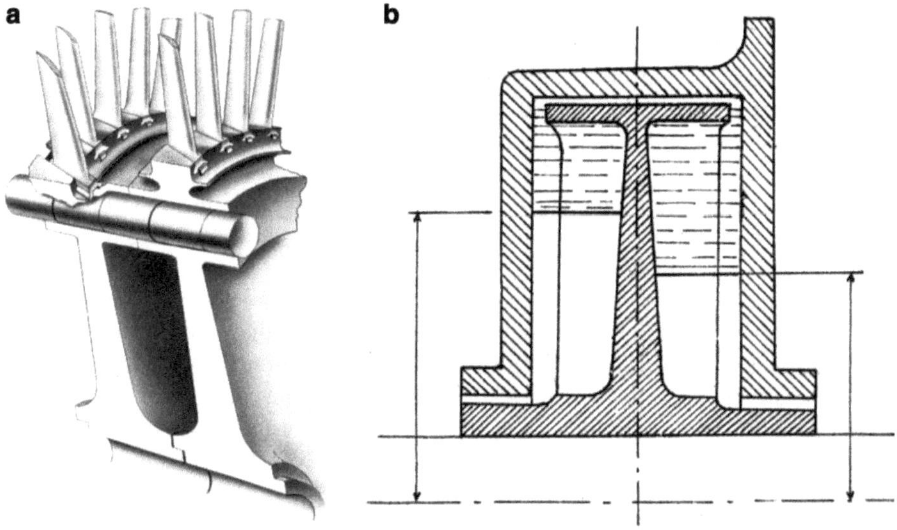

Fig. 6.2 (**a**) Conical disks in an axial compressor; (**b**) conical disk with radial ribs in a hydraulic seal for a steam turbine

states of conical shells of revolution under axisymmetric loads, and whose thickness varies according to a linear function (increasing or decreasing from the axis to the periphery), into the Gauss-type equation. Afterwards, Honegger [35] further developed the subject and supplied the exact analytical solution of the Gauss-type equation. Several years later Honegger [36] extended this approach to rotating tapered disks as analysed here.

Based on the considerations set out in the previous chapters, it can be readily seen that a rotor configured entirely as a divergent conical disk (Fig. 6.1e, f, g) is of little technical use because of its unfavourable mass distribution and, consequently, its higher rotational inertia. In this context, an analysis of such a rotor would be entirely divorced from the realm of actual engineering applications, and would thus be of almost exclusively mathematical interest. In reality, however, many turbine disks show linear diverging tapering where the disk is coupled to the crown ring. However, this area, which may show high thickness gradients, is very small in size and, if the apex V converges on the axis, can be dealt with by means of a closed formula, as discussed in Chap. 4, according to the hyperbolic profile disk with $a = 1$. Figure 6.2 shows two typical applications of conical disks.

Despite these limitations on the use of the divergent conical disk, our intention here is to present the most general possible treatment. Accordingly, we will continue to use (6.1) since, as Giovannozzi notes, it can describe the profile shapes of annular and solid conical disks converging from the inner to the outer radius or diverging from the inner to the outer radius (Fig. 6.1).

We will thus have $t < 1$ for the solid or annular converging disk, $t > 1$ for the annular diverging disk and $t < 0$ (i.e., $R < 0$) for the solid or annular diverging disk. More specifically: $0 \leq t \leq 1$, for the solid or annular converging disk, clearly with a physical singularity for $t = 1$, with zero thickness at the apex V ($r = R$), and a pointed-tip disk (Fig. 6.1c, d); $t > 1$, for the annular diverging disk, showing two singularity points, a physical one for $t = 1$ (which implies a null coupling surface with the shaft at radius $r_i = R$) and a numerical one for $t = \infty$, i.e., for $R = 0$, due to the fact that the point of convergence of the profile side extensions lies on the axis (Fig. 6.1e); $t < 0$, for the solid or annular diverging disk, either with its mass distributed near its rotation axis, and for the very slightly tapering solid or annular diverging disk, where the intersection of the profile side extensions lies beyond the axis (Fig. 6.1f, g, with $R < 0$).

To deduce the differential equation that governs the general displacement field for any configuration of these conical disk profiles, the relations for T and h, given respectively by (3.3) and (6.1), are introduced in (1.28) along with their first derivatives with respect to t, taking into account that $d/dr = (1/R)\cdot d/dt$, $d^2/dr^2 = (1/R^2)\cdot d^2/dt^2$, $dh/dt = -h_0$,$(1/h)\cdot(dh/dt) = -1/(1-t)$ and $dr = R\cdot dt$. We thus obtain the following non-homogeneous differential equation of second order in the variable u:

$$\frac{d^2u}{dt^2} + \left(\frac{1}{t} - \frac{1}{1-t}\right)\cdot\frac{du}{dt} - \left(\frac{1}{t} + \frac{\nu}{1-t}\right)\cdot\frac{u}{t} - (1+\nu)\cdot\alpha\cdot R\cdot\left(\frac{dT}{dt} - \frac{T}{1-t}\right) + \frac{(1-\nu^2)\cdot\gamma\cdot\omega^2\cdot R^3\cdot t}{E} = 0. \quad (6.3)$$

This equation describes the displacement field of the linear tapered disk, having constant density and subjected to centrifugal load and thermal field according to the generic function for temperature variation along the radius $T = T(r)$. As will be seen in Chap. 7, differential equation (6.3) can be deduced from (7.3), taking $n = k = 1$ and passing from the variable ρ to variable t.

If either rotation or thermal loads are present alone, the above equation yields the following relations:

$$\frac{d^2u}{dt^2} + \left(\frac{1}{t} - \frac{1}{1-t}\right)\cdot\frac{du}{dt} - \left(\frac{1}{t} + \frac{\nu}{1-t}\right)\cdot\frac{u}{t} + \frac{(1-\nu^2)\cdot\gamma\cdot\omega^2\cdot R^3\cdot t}{E} = 0. \quad (6.4)$$

$$\frac{d^2u}{dt^2} + \left(\frac{1}{t} - \frac{1}{1-t}\right)\cdot\frac{du}{dt} - \left(\frac{1}{t} + \frac{\nu}{1-t}\right)\cdot\frac{u}{t} - (1+\nu)\cdot\alpha\cdot R\cdot\left(\frac{dT}{dt} - \frac{T}{1-t}\right) = 0. \quad (6.5)$$

General solutions of the above three non-homogeneous differential equations are obtained by adding the solution of the following common associated homogeneous equation to particular solutions of the same complete equations:

$$\frac{d^2u}{dt^2} + \left(\frac{1}{t} - \frac{1}{1-t}\right)\cdot\frac{du}{dt} - \left(\frac{1}{t} + \frac{\nu}{1-t}\right)\cdot\frac{u}{t} = 0, \quad (6.6)$$

Equation (6.6) is the Gauss hypergeometric differential equation, featuring coefficients that are analytical functions of the independent variable t and showing three singular points ($t = 0; t = 1; t = \infty$). It can be considered as a Fuchs class or totally Fuchsian equation, because it satisfies the Fuchs conditions in all its singular points, including the point at infinity [75].

6.2 Rotating Conical Disk Having Constant Density

As indicated above, the differential equation governing the displacement field of a rotating conical disk is (6.4), whose solution is the sum of its associated homogeneous equation (6.6), and of a particular integral of the complete equation.

6.2.1 Particular Integral and Corresponding Stress and Strain State

To obtain the particular integral of differential equation (6.4), it is best to start from:

$$h \cdot r \cdot \frac{d^2u}{dr^2} + \left(h + \frac{dh}{dr} \cdot r\right) \cdot \frac{du}{dr} + \left(v \cdot \frac{dh}{dr} - \frac{h}{r}\right) \cdot u = -\frac{(1 - v^2) \cdot \gamma \cdot \omega^2 \cdot r^2 \cdot h}{E}, \tag{6.7}$$

which is obtained directly from (1.28), eliminating the term related to the thermal field and premultiplying by $h \cdot r$.

By introducing the function of thickness variation in terms of the variable r, written in the form given by the first equality appearing in (6.2), and its first derivative with respect to variable r and introducing the notation

$$C = \frac{(1 - v^2)}{E} \cdot \gamma \cdot \omega^2, \tag{6.8}$$

we obtain:

$$(a + b \cdot r) \cdot r \cdot \frac{d^2u}{dr^2} + (a + 2 \cdot b \cdot r) \cdot \frac{du}{dr} + \left(v \cdot b - \frac{a + b \cdot r}{r}\right) \cdot u = -C \cdot a \cdot r^2 - C \cdot b \cdot r^3. \tag{6.9}$$

To obtain the particular integral u_p of (6.9), we first put

$$u_p = c_1 \cdot r + c_2 \cdot r^2 + c_3 \cdot r^3. \tag{6.10}$$

where c_1, c_2 and c_3 are constants that can be determined by substituting relation (6.10), along with its first and second derivatives with respect to variable r, in (6.9), and by equalling to zero the coefficients of various powers in the variable r. We thus obtain:

$$c_1 = C \cdot R^2 \cdot \frac{3 \cdot (3+\nu)}{(11+\nu)\cdot(5+\nu)\cdot(1+\nu)}; \quad c_2 = C \cdot R \cdot \frac{3+\nu}{(11+\nu)\cdot(5+\nu)}; \quad c_3 = -\frac{C}{11+\nu} \tag{6.11}$$

Finally, substituting these constants in relation (6.10), taking notation (6.8) into account and changing from variable r to variable t, u_p is obtained as:

$$u_p = -\frac{(1-\nu^2)}{(11+\nu)} \cdot t \cdot \left[t^2 - \frac{3+\nu}{5+\nu} \cdot \left(t + \frac{3}{1+\nu}\right)\right] \cdot \gamma \cdot \frac{\omega^2 \cdot R^3}{E}. \tag{6.12}$$

Through relations (1.27) from which temperature terms are omitted, and where $r = R \cdot t$ and $du_p/dr = (1/R)\,(du_p/dt)$ are introduced along with the derivative du_p/dt obtained from (6.12), this particular integral (6.12) corresponds to the following stress state due to the centrifugal field alone:

$$\begin{cases} \sigma_r = -\gamma \cdot \omega^2 \cdot R^2 \cdot g_r = \sigma_0 \cdot g_r \\ \sigma_t = -\gamma \cdot \omega^2 \cdot R^2 \cdot g_t = \sigma_0 \cdot g_t, \end{cases} \tag{6.13}$$

where $\sigma_0 = \gamma \cdot \omega^2 \cdot R^2$ is the reference stress (note that the reference stress refers here to the radius R instead of, as usual, to the outer radius r_e of the disk) and g_r and g_t are dimensionless functions of t and ν, respectively from:

$$\begin{cases} g_r = -\left[\frac{3+\nu}{11+\nu} \cdot t^2 - \frac{(2+\nu)\cdot(3+\nu)}{(11+\nu)\cdot(5+\nu)} \cdot t - \frac{3\cdot(3+\nu)}{(11+\nu)\cdot(5+\nu)}\right] \\ g_t = -\left[\frac{1+3\nu}{11+\nu} \cdot t^2 - \frac{(1+2\nu)\cdot(3+\nu)}{(11+\nu)\cdot(5+\nu)} \cdot t - \frac{3\cdot(3+\nu)}{(11+\nu)\cdot(5+\nu)}\right]. \end{cases} \tag{6.14}$$

It can be concluded from expression (6.12) that the particular integral u_p, its associated stress state given by relations (6.13) and the resulting strain state obtained from (1.25) from which temperature terms are omitted after substituting relations (6.13) in them, largely depends on the disk's material, geometrical dimensions and rotation speed. For any given material, u_p is proportional to R^3 and to ω^2. It should be noted that the structure of expression (6.12) shows that the particular integral can be defined once and for all: in other words, once u_p is calculated for a specific conical disk geometry and for a specific angular velocity, it is possible to determine u_p for any other disk and any other angular velocity by means of a simple scale shift.

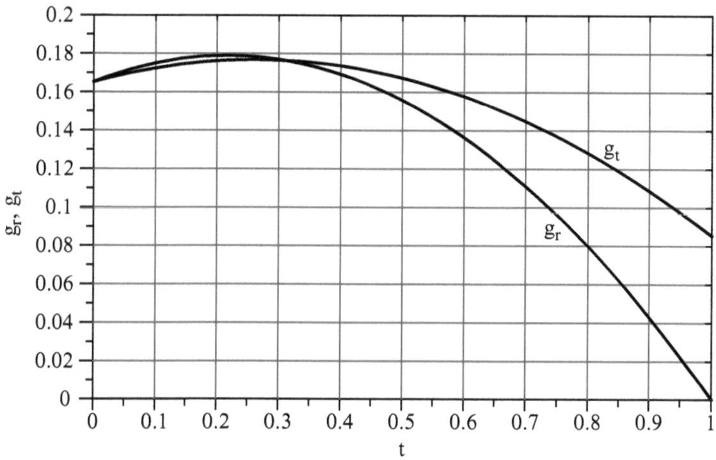

Fig. 6.3 Curves of g_r and g_t along dimensionless disk radius, in the interval $0 \leq t \leq 1$, for $\nu = 0.3$

Figure 6.3 shows functions g_r and g_t versus t, for steels ($\nu = 0.3$). Radial and hoop stress σ_r and σ_t, related to the integral u_p are proportional to these functions, through σ_0. Integral u_p, as we shall see later, has a precise physical meaning: it represents the radial displacement in a solid conical pointed-tip disk, i.e., extended up to its apex V where $r = R$ and $t = 1$ (Fig. 6.1c, d). It should be pointed out that g_r and g_t assume finite values throughout the interval $0 \leq t \leq 1$.

6.2.2 Solution of Homogeneous Differential Equation and Corresponding Stress and Strain State

To solve differential equation (6.6), which is the common homogeneous equation associated with the three non-homogeneous differential equations (6.3), (6.4) and (6.5), it is best to write (6.6) as:

$$\frac{d^2u}{dt^2} + \left(\frac{1}{t} + \frac{1}{t-1}\right) \cdot \frac{du}{dt} + \left[\frac{1}{t} + (\nu - 1)\right] \cdot \frac{u}{t \cdot (t-1)} = 0. \tag{6.15}$$

The above is a hypergeometric differential equation, showing three singular points ($t = 0$; $t = 1$; $t = \infty$) and corresponding to the general Papperitz form (see Tricomi [75]):

$$\frac{d^2u}{dt^2} + \left(\frac{1-\alpha-\alpha'}{t} + \frac{1-\gamma-\gamma'}{t-1}\right) \cdot \frac{du}{dt} + \left[-\frac{\alpha \cdot \alpha'}{t} + \frac{\gamma \cdot \gamma'}{t-1} + \beta \cdot \beta'\right] \cdot \frac{u}{t \cdot (t-1)} = 0, \tag{6.16}$$

where (α, α'), (β, β') and (γ, γ') are the three pairs of roots of the determinant equations for the three singular points, with

$$\alpha + \alpha' + \beta + \beta' + \gamma + \gamma' = 1. \tag{6.17}$$

This is because the sum of the roots of all determinant equations of a totally Fuchsian equation with $(i + 1)$ singular points must be $(i - 1)$. Thus, by comparing (6.16) with (6.15) and taking (6.17) into account, we can deduce the following relations:

$$\begin{gathered} 1 - \alpha - \alpha' = 1;\ 1 - \gamma - \gamma' = 1;\ \beta \cdot \beta' = v - 1; -\alpha \cdot \alpha' = 1; \\ \gamma \cdot \gamma' = 0;\ \beta + \beta' = 1. \end{gathered} \tag{6.18}$$

Here, we obtain:

$$\alpha = -1;\ \alpha' = 1;\ \gamma = \gamma' = 0;\ \beta = \frac{1}{2} + \xi;\ \beta' = \frac{1}{2} - \xi;\ \xi = \frac{1}{2} \cdot \sqrt{5 - 4 \cdot v}. \tag{6.19}$$

According to the related general theory, (6.16) allows for two independent integrals, each corresponding to one of the exponents α and α' of the variable t and therefore having the following expressions:

$$\begin{aligned} u_1 &= t^{\alpha'} \cdot (1 - t)^{\gamma} \cdot F(\alpha' + \beta + \gamma,\ \alpha' + \beta' + \gamma, 1 + \alpha' - \alpha, t) \\ u_{1'} &= t^{\alpha'} \cdot (1 - t)^{\gamma'} \cdot F(\alpha' + \beta + \gamma',\ \alpha' + \beta' + \gamma', 1 + \alpha' - \alpha, t) \\ u_2 &= t^{\alpha} \cdot (1 - t)^{\gamma} \cdot F(\alpha + \beta + \gamma,\ \alpha + \beta' + \gamma, 1 + \alpha - \alpha', t) \\ u_{2'} &= t^{\alpha} \cdot (1 - t)^{\gamma'} \cdot F(\alpha + \beta + \gamma',\ \alpha + \beta' + \gamma', 1 + \alpha - \alpha', t). \end{aligned} \tag{6.20}$$

The first two of these expressions are true for exponent α' and the remaining two are true for exponent α; we can thus shift between one equivalent independent integral and the other, either of which correspond to the exponent α' or to the exponent α, simply by replacing γ with γ' and vice versa. In the case presented here, by taking into account relations (6.19), equations (6.20) change into:

$$\begin{aligned} u_1 &= t \cdot F\left(\frac{3}{2} + \xi, \frac{3}{2} - \xi, 3, t\right) \\ u_{1'} &= t \cdot F\left(\frac{3}{2} + \xi, \frac{3}{2} - \xi, 3, t\right) \\ u_2 &= t^{-1} \cdot F\left(-\frac{1}{2} + \xi, -\frac{1}{2} - \xi, -1, t\right) \\ u_{2'} &= t^{-1} \cdot F\left(-\frac{1}{2} + \xi, -\frac{1}{2} - \xi, -1, t\right); \end{aligned} \tag{6.21}$$

therefore it is concluded that $u_1 = u_{1'}$ and $u_2 = u_{2'}$.

From a general point of view, it should be pointed out that when any of the numerical factors c, $(c - a - b)$ and $(a - b)$, with a, b and c defined below, are equal to an integer and when the difference between the greater and the smaller of exponents α' and α is an integer, or in other words when $\alpha' - \alpha = j$, j being a positive integer, the integrals corresponding to exponent $\alpha'$and those corresponding to exponent α are no longer independent of each other.

This is indeed the condition that obtains in the case in question, where $j = \alpha' - \alpha = 2$. Thus, only one of the four relations (6.21) is usable, either the first matching the second or the third matching the fourth. In fact, however, the latter two solutions are unusable, as the third factor in the hypergeometric series (the factor related to the integer numerator of the first addend, which in the differential equation (6.6) characterizes the coefficient du/dt), is a negative integer, in this specific case equal to -1. Accordingly, in the series expansion, all terms of the infinite series become infinite from the third term onwards, given that the denominator of each includes a null multiplier due to the increase by one unit of said factor, which equals -1.

In this context, it is necessary to select as the integral of (6.6) the one deriving from the first relation (6.15), which is here transcribed as:

$$u_1 = t^{\alpha'} \cdot (1-t)^{\gamma} \cdot F(a,b,c,t) = t \cdot F\left(\frac{3}{2} + \xi, \frac{3}{2} - \xi, 3, t\right) \tag{6.22}$$

where, for the sake of brevity, the following notation is used:

$$a = \alpha' + \beta + \gamma = \frac{3}{2} + \xi; \quad b = \alpha' + \beta' + \gamma = \frac{3}{2} - \xi; \quad c = 1 + \alpha' - \alpha = 3 \tag{6.23}$$

whereas $F(a, b, c, t)$ indicates the infinite hypergeometric series of the kind

$$F(a,b,c,t) = 1 + \frac{a \cdot b}{1! \cdot c} \cdot t + \frac{a \cdot (a+1) \cdot b \cdot (b+1)}{2! \cdot c \cdot (c+1)} \cdot t^2 + \\ + \ldots + \frac{a \cdot (a+1) \cdot \ldots \cdot (a+n-1) \cdot b \cdot (b+1) \cdot \ldots \cdot (b+n-1)}{n! \cdot c \cdot (c+1) \cdot \ldots \cdot (c+n-1)} \cdot t^n + \ldots . \tag{6.24}$$

The hypergeometric series $F(a, b, c, t)$ is divergent if the real part $\Re(c - a - b) \leq -1$, absolutely convergent if $\Re(c - a - b) > 0$ and, excluding the singularity point $t = 1$, conditionally convergent if $-1 < \Re(c - a - b) \leq 0$. Here, given that a, b and c are real coefficients and $(c - a - b) = 0$, this is a case of very slow, conditional convergence.

If we examine relations (6.23) in the light of relations (4.14) calculated for $a = 1$ (remember once again that the profile of the divergent annular conical disk can be described by the Stodola function that defines the profile of the hyperbolic disk, with $a = 1$), it follows that:

$$a = p + 2; \quad b = q + 2; \quad c = a + b = p + q + 4; \tag{6.25}$$

Consequently, the diverging conical annular disk becomes a particular case of a hyperbolic disk for $a = 1$ and as such can be treated, if it is of engineering interest, with a closed form formulation, there being no need for series expansions. This is true only if apex V is on the axis. In the general case where apex V lies off the axis, the procedure covered in this chapter must be used.

Given that the first derivative of the radial displacement $u = u(r)$ must be calculated to determine stress and strain states (in this regard, see relations (1.27) and (1.25)), it should be borne in mind that the first derivative with respect to variable t in series (6.24) is as follows:

$$\frac{d}{dt}F(a,b,c,t) = \frac{a \cdot b}{c} \cdot F(a+1,\ \mathrm{b}+1, c+1, t). \tag{6.26}$$

It is now necessary to find a second integral of differential equation (6.6) which is independent of the one resulting from relation (6.22). In this respect, it should be noted that the series appearing in the last two relations (6.21), featuring infinite terms from a certain value onwards, actually become finite when the interfering negative integer factors are freely increased by a small quantity ε, which, however, will need to be minimized by multiplying it by a suitable constant coefficient so as to prevent sensible terms from tending to infinity.

Without addressing this aspect of the problem (which is dealt with in specific texts on series and hypergeometric differential equations, such as [39]), we will limit ourselves here to pointing out that, upon establishing:

$$\begin{cases} u_1^* = & t^{\alpha'} \cdot (1-t)^{\gamma} \cdot F(a,b,c-\varepsilon,t) \\ u_2^* = & t^{\alpha'-j+\varepsilon} \cdot (1-t)^{\gamma} \cdot F(a-j+\varepsilon, b-j+\varepsilon, 2-c+\varepsilon, t) \\ m(\mathrm{j},\ \varepsilon) = & \prod_{\mathrm{i}=1}^{\mathrm{j}} \dfrac{i \cdot (\varepsilon - j + i)}{(a+\varepsilon-i) \cdot (b+\varepsilon-i)}, \end{cases} \tag{6.27}$$

a second independent integral of differential equation (6.6) is obtained, as:

$$u_2 = \lim_{\varepsilon \to 0} \frac{m \cdot u_2^* - u_1^*}{\varepsilon}. \tag{6.28}$$

Proceeding with calculation, the following expression of u_2 is obtained:

$$u_2 = \frac{j}{(a-1)\cdot(b-1)} \cdot t^{\alpha'-1} \cdot (1-t)^{\gamma} \cdot \left\{ 1 + \sum_{r=2}^{j} \left[\frac{1}{t^{r-1}} \cdot \prod_{\mathrm{s}=2}^{\mathrm{r}} \frac{(j-s+1)\cdot(1-s)}{(a-s)\cdot(b-s)} \right] \right\}$$
$$+ u_1 \cdot \ln t + (1-t)^{\gamma} \cdot t^{\alpha'} \cdot \sum_{i=1}^{\infty} C_i \cdot t^i \tag{6.29}$$

where

$$C_i = \frac{1}{i!}\left[\prod_{n=0}^{i-1}\frac{(a+n)\cdot(b+n)}{j+1+n}\right]\cdot\sum_{n=0}^{i-1}\left[\frac{1}{a+n}+\frac{1}{b+n}-\frac{1}{1+n}-\frac{1}{j+1+n}\right]. \tag{6.30}$$

Given that, in the case discussed here, $j=\alpha'-\alpha=2$, and taking relations (6.23) into consideration, relations (6.27), (6.29) and (6.30) are transformed respectively into:

$$\begin{cases} u_1^* = t\cdot F\left(\frac{3}{2}+\xi,\frac{3}{2}-\xi,3-\varepsilon,t\right) \\ u_2^* = t^{-1+\varepsilon}\cdot F\left(-\frac{1}{2}+\xi+\varepsilon,-\frac{1}{2}-\xi+\varepsilon,-1+\varepsilon,t\right) \\ m(2,\varepsilon)=\dfrac{\varepsilon\cdot(\varepsilon-1)}{\left(\frac{1}{2}+\xi+\varepsilon\right)\cdot\left(\frac{1}{2}-\xi+\varepsilon\right)\cdot\left(-\frac{1}{2}+\xi+\varepsilon\right)\cdot\left(-\frac{1}{2}-\xi+\varepsilon\right)} \end{cases} \tag{6.31}$$

$$u_2 = \frac{2}{\left(\frac{1}{2}+\xi\right)\cdot\left(\frac{1}{2}-\xi\right)}\cdot\left[1-\frac{t^{-1}}{\left(-\frac{1}{2}+\xi\right)\cdot\left(-\frac{1}{2}-\xi\right)}\right]+u_1\cdot\ln t+t\cdot\sum_{i=1}^{\infty}C_i\cdot t^i \tag{6.32}$$

$$C_i=\frac{\left(\frac{3}{2}+\xi\right)\cdot\left(\frac{3}{2}+\xi+1\right)\cdot\ldots\cdot\left(\frac{3}{2}+\xi+i-1\right)\cdot\left(\frac{3}{2}-\xi\right)\cdot\left(\frac{3}{2}-\xi+1\right)\cdot\ldots\cdot\left(\frac{3}{2}-\xi+i-1\right)}{i!\cdot 3\cdot 4\cdot\ldots\cdot(3+i-1)} \cdot\sum_{n=0}^{i-1}\left(\frac{1}{\frac{3}{2}+\xi+i}+\frac{1}{\frac{3}{2}-\xi+i}-\frac{1}{1+i}-\frac{1}{3+i}\right). \tag{6.33}$$

Thus, the second independent integral of the hypergeometric differential equation (6.6) is determined. Figure 6.4 shows the two independent integrals u_1 and u_2 versus t within the interval $0\le t\le 1$. From these curves, as well as from the structure of (6.22) and (6.32), and taking (6.33) into account, it can be seen that $u_1\to\infty$ for $t\to 1$, whereas $u_2\to\infty$ for $t\to 0$.

It should be noted that, on the real plane, u_1 is convergent within the interval $-\infty \le t\le 1$ $(-\infty\le r\le \mathrm{R})$, whereas u_2 is convergent within the interval $0\le t\le 1$ $(0\le r\le \mathrm{R})$. Within the interval $0\le t\le 1$ of interest to us here, both integrals u_1 and u_2 are convergent and, by contrast with the method adopted by Honegger and Giovannozzi (see Appendix B), are used directly in the following developments as envisaged by the authors [77].

When the two independent integrals u_1 and u_2 are used directly, the solution of the associated homogeneous equation (6.6) can be expressed as

$$u=C_1\cdot u_1+C_2\cdot u_2, \tag{6.34}$$

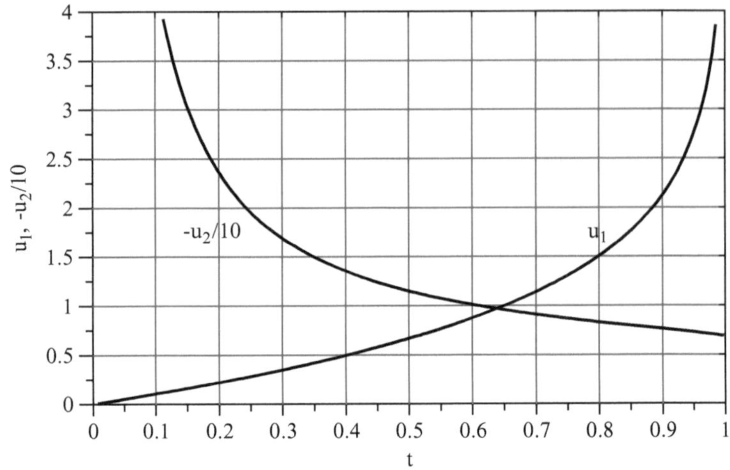

Fig. 6.4 Curves of u_1 and u_2 as a function of t within the interval $0 \leq t \leq 1$, for $\nu = 0.3$

where C_1 and C_2 are integration constants that can be evaluated from the boundary conditions.

Subsequently, the following relations are obtained by substituting relation (6.34) along with its first derivative with respect to t in relations (1.27) from which temperature terms are omitted, and by considering that $d/dr = (1/R) \cdot d/dt$. The following relations expressing σ_r and σ_t as a function of dimensionless radius t are thus obtained:

$$\begin{cases} \sigma_r = \dfrac{E}{R \cdot (1-\nu^2)} \cdot \left[C_1 \cdot \left(\dfrac{du_1}{dt} + \nu \cdot \dfrac{u_1}{t} \right) + C_2 \cdot \left(\dfrac{du_2}{dt} + \nu \cdot \dfrac{u_2}{t} \right) \right] \\ \sigma_t = \dfrac{E}{R \cdot (1-\nu^2)} \cdot \left[C_1 \cdot \left(\dfrac{u_1}{t} + \nu \cdot \dfrac{du_1}{dt} \right) + C_2 \cdot \left(\dfrac{u_2}{t} + \nu \cdot \dfrac{du_2}{dt} \right) \right]. \end{cases} \tag{6.35}$$

These relations can be expressed in the following compact form:

$$\begin{cases} \sigma_r = A \cdot a_r + B \cdot b_r \\ \sigma_t = A \cdot a_t + B \cdot b_t, \end{cases} \tag{6.36}$$

where

$$a_r = \frac{du_1}{dt} + \nu \cdot \frac{u_1}{t},\; b_r = \frac{du_2}{dt} + \nu \cdot \frac{u_2}{t},\; a_t = \frac{u_1}{t} + \nu \cdot \frac{du_1}{dt} \text{ and } b_t = \frac{u_2}{t} + \nu \cdot \frac{du_2}{dt} \tag{6.37}$$

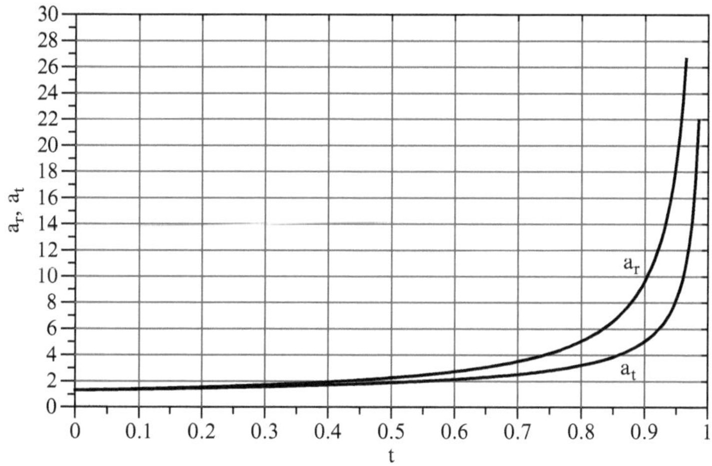

Fig. 6.5 Curves of a_r and a_t along dimensionless disk radius, within the interval $0 \leq t \leq 1$, for $\nu = 0.3$

are functions of t and ν, whereas

$$A = \frac{E \cdot C_1}{R \cdot (1 - \nu)} \text{ and } B = \frac{E \cdot C_2}{R \cdot (1 + \nu)} \tag{6.38}$$

are the dimensional integration constants, to be determined by imposing boundary conditions.

First derivatives du_1/dt and du_2/dt are obtained by taking relation (6.26) into consideration once functions u_1 and u_2 have been determined. Calculating functions a_r, b_r, a_t and b_t, each of which is related to different hypergeometric functions (u_1 or u_2), presents no difficulties.

Figures 6.5 and 6.6 show functions a_r, b_r, a_t and b_t versus dimensionless variable t, for steels ($\nu = 0.3$); the calculation interval is $0 < t < 1$, which is in fact that of the solid or annular converging conical disk of actual technical interest. It should be pointed out that a_r and a_t tend to infinity for $t \rightarrow 1$, given that $u_1 \rightarrow \infty$ at the outer edge of the pointed-tip disk ($r_e = R, t = 1$), whereas b_r and b_t tend to infinity for $t \rightarrow 0$.

The stress state given by relations (6.36) is caused by the distribution of surface loads to the outer radius and inner radius, in the case of an annular disk. The strain state related to the stress state given by (6.36) is then obtained from (1.25) from which temperature terms are omitted, after substituting relations (6.36) in them. Lastly, radial displacement is obtained by considering, as we know from (1.14), that $u = r \cdot \varepsilon_t$.

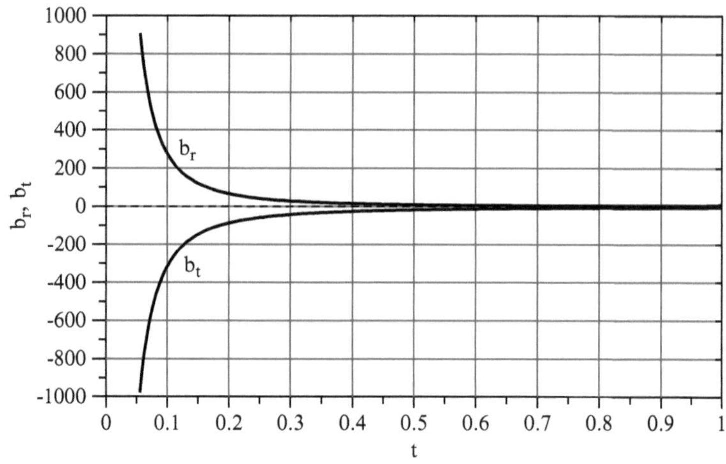

Fig. 6.6 Curves of b_r and b_t along dimensionless disk radius, within the interval $0 \leq t \leq 1$, for $\nu = 0.3$

6.2.3 General Integral and Corresponding Stress and Strain State

As was specified at the beginning of this chapter, the general solution of the non-homogeneous differential equation (6.4) governing the displacement field of the rotating disk without thermal load is the sum of the solution of its associated homogeneous equation (6.6), given by (6.34), and of the particular integral u_p, previously obtained and given by relation (6.12). Thus, radial displacement u at the dimensionless radius t will be obtained from the relation:

$$u = C_1 \cdot u_1 + C_2 \cdot u_2 + u_p. \tag{6.39}$$

By using the method of superposition in calculating total strains and stresses and by considering relations (6.36) as well as relations (6.13), where g_r and g_t are obtained from relations (6.14), we find that radial and hoop stresses in a rotating conical disk also subjected to surface force distribution at inner and outer radii are expressed by means of the following relations:

$$\begin{cases} \sigma_r &= A \cdot a_r + B \cdot b_r + \sigma_0 \cdot g_r \\ \sigma_t &= A \cdot a_t + B \cdot b_t + \sigma_0 \cdot g_t. \end{cases} \tag{6.40}$$

Here again, the strain distributions related to the stress distributions given by (6.40) are obtained from (1.25) from which temperature terms are omitted, after substituting relations (6.40) in them.

If the conical disk has no hub, the first of (6.40) can be used to determine the dimensional constants A and B by imposing boundary conditions at the inner and outer radii. The stress state in the disk is thus completely defined. For a solid disk, as stresses σ_r and σ_t must have a finite value for $t = 0$, while functions b_r and b_t tend to infinity for t tending to zero, we must necessarily have $B = 0$; consequently, relations (6.40) are simplified as the middle term of the second member vanishes. For details of the procedures, the reader is referred to the calculation examples given in Sect. 6.6.

6.3 Conical Disk Having Constant Density Subjected to Thermal Load

We will now consider a conical profile disk of constant density and subjected to thermal load only; this disk features a temperature gradient distribution along its radius expressed by relation (4.6) deriving from the function $T = T(r)$ given by (4.5). In this case, the solving differential equation (6.5) becomes:

$$\frac{d^2u}{dt^2}+\left(\frac{1}{t}-\frac{1}{1-t}\right)\cdot\frac{du}{dt}-\left(\frac{1}{t}+\frac{\nu}{1-t}\right)\cdot\frac{u}{t}-\alpha\cdot(1+\nu)\cdot R\cdot\left(\sum_{i=1}^{n} i\cdot k_i\cdot t^{i-1}-\frac{\sum_{i=1}^{n} k_i\cdot t^i}{1-t}\right)=0. \tag{6.41}$$

The general integral of this equation is the sum of the solution of the homogeneous equation, which is already known, and of a particular integral of the non-homogeneous equation. Using the superposition method, the latter can be obtained as the linear combination of contributions deriving from component n terms $k_i \cdot t^i$ (with $i = 1, 2, \ldots, n$) of the function $T = T(r)$; thus, by considering the *i-th* term of this function, differential equation (6.41) becomes:

$$\frac{d^2u}{dt^2}+\left(\frac{1}{t}-\frac{1}{1-t}\right)\cdot\frac{du}{dt}-\left(\frac{1}{t}+\frac{\nu}{1-t}\right)\cdot\frac{u}{t}+\alpha\cdot(1+\nu)\cdot R\cdot k_i\cdot\frac{(i+1)\cdot t^i-i\cdot t^{i-1}}{1-t}=0. \tag{6.42}$$

To obtain a particular integral of differential equation (6.42), we take:

$$u = a_{i+1} \cdot t^{i+1} + a_i \cdot t^i + a_{i-1} \cdot t^{i-1} + \ldots., \tag{6.43}$$

where $a_{i+1}, a_i, a_{i-1}, \ldots.$ are constants; by substituting this relation, along with its first and second derivatives, in (6.42) and by equalling to zero the coefficients of various powers in the variable t, the following relations are obtained:

$$\begin{aligned} a_{i+1} &= \alpha \cdot R \cdot k_i \cdot \frac{(i+1)\cdot(1+\nu)}{(i+1)\cdot(i+2)-(1-\nu)} \\ a_i &= \alpha \cdot R \cdot k_i \cdot \frac{i\cdot(1-\nu^2)}{[(i+1)\cdot(i+2)-(1-\nu)]\cdot[i\cdot(i+1)-(1-\nu)]} \end{aligned} \tag{6.44}$$

while for the subsequent coefficients a_{i-p}, with $p = 1, 2, \ldots$, we can derive the general relation

$$a_{i-p} = a_{i-p+1} \cdot \frac{(i-p) \cdot (i-p+2)}{(i-p+1) \cdot (i-p) - (1-v)}. \tag{6.45}$$

If we take $i = p$ in relation (6.45), we obtain $a_0 = 0$. On the basis of the same relation, all coefficients after a_0 related to all negative powers of t are thus zero. Therefore, given i as any positive integer, the polynomial (6.43), within the first $(i + 1)$ terms starting from the linear term $a_1 \cdot t$, always represents a particular integral of differential equation (6.42).

By substituting the series expansion (6.43) and its first derivative in relations (1.27), while considering relations (6.44) and (6.45) as well as the *i-th* term of the function $T = T\ (r)$, i.e., $T = k_i \cdot t^i$, we obtain the following expressions of the corresponding stresses σ_r and σ_t as polynomials of grade i in t, with a constant, non-zero term, which are valid for any value of i:

$$\begin{cases} \sigma_r = E \cdot \alpha \cdot k_i \cdot \left(A_i \cdot t^i + A_{i-1} \cdot t^{i-1} + \ldots. + A_2 \cdot t^2 + A_1 \cdot t + A_0\right) = E \cdot \alpha \cdot k_i \cdot \varphi_i(t) \\ \sigma_t = E \cdot \alpha \cdot k_i \cdot \left(B_i \cdot t^i + B_{i-1} \cdot t^{i-1} + \ldots. + B_2 \cdot t^2 + B_1 \cdot t + B_0\right) = E \cdot \alpha \cdot k_i \cdot \psi_i(t). \end{cases} \tag{6.46}$$

The coefficients A and B appearing in expressions (6.46) are obtained from the following relations:

$$\begin{aligned} A_i &= -\frac{i}{(i+1) \cdot (i+2) - (1-v)} \\ A_{i-1} &= \frac{i \cdot (1+v)}{[(i+1) \cdot (i+2) - (1-v)] \cdot [i \cdot (i+1) - (1-v)]} \\ B_i &= -\frac{i \cdot (i+2)}{(i+1) \cdot (i+2) - (1-v)} \\ B_{i-1} &= \frac{i \cdot (1+v \cdot i)}{[(i+1) \cdot (i+2) - (1-v)] \cdot [i \cdot (i+1) - (1-v)]}, \end{aligned} \tag{6.47}$$

while all the subsequent coefficients A_{i-1-p} and B_{i-1-p}, with $p = 1, 2, \ldots, (i - 1)$, are derived progressively through the relations:

$$\begin{aligned} A_{i-1-p} &= A_{i-p} \cdot \frac{(i-p+v)}{(i-p+1+v)} \cdot \frac{(i-p) \cdot (i-p+2)}{(i-p) \cdot (i-p+1) - (1-v)} \\ B_{i-1-p} &= B_{i-p} \cdot \frac{1+v \cdot (i-p)}{1+v \cdot (i-p+1)} \cdot \frac{(i-p) \cdot (i-p+2)}{(i-p) \cdot (i-p+1) - (1-v)}. \end{aligned} \tag{6.48}$$

These relations are obtained by means of simple proportions between the coefficients of the powers in t^{i-p-1} and t^{i-p}. Given that coefficients A_{i-1} and B_{i-1}

are known, as directly defined from the second and the fourth expressions (6.47) respectively, relations (6.48) can be used to progressively determine all other coefficients present in expressions (6.46). It can be readily seen that, for a specific value of i, $\sum_{i=0}^{n} A_i = 0$. These relations, in any case, clearly show that coefficients A and B appearing in polynomials $\varphi_i(t)$ and $\psi_i(t)$, are simply functions of i and v.

In general terms, for any value of i, the relations whereby functions $\varphi_i(t)$ and $\psi_i(t)$ appearing in expressions (6.46) can be calculated as follows:

$$\varphi_i(t) = -\frac{i \cdot t^i}{i^2+3i+1+v} + i \cdot \left[\prod_{j=2}^{i}(j^2-1)\right] \cdot \sum_{m=1}^{i-1} \frac{(m+1+v)\cdot t^m}{\left[\prod_{j=1}^{m}(j^2+2j)\right] \cdot \left[\prod_{j=m}^{i}(j^2+3j+1+v)\right]} + \frac{i \cdot \prod_{j=2}^{i}(j^2-1)}{\prod_{j=1}^{i}(j^2+3j+1+v)};$$

$$\psi_i(t) = -\frac{i \cdot (i+2) \cdot t^i}{1+3i+i^2+v} + i \cdot \left[\prod_{j=2}^{i}(j^2-1)\right] \cdot \sum_{m=1}^{i-1} \frac{[1+(m+1)\cdot v]\cdot t^m}{\left[\prod_{j=1}^{m}(j^2+2j)\right] \cdot \left[\prod_{j=m}^{i}(j^2+3j+1+v)\right]} + \frac{i \cdot \prod_{j=2}^{i}(j^2-1)}{\prod_{j=1}^{i}(j^2+3j+1+v)}. \tag{6.49}$$

These general relations derived by the authors [77] can be used to obtain several others of greater technical relevance for engineering calculations, viz.: $i = 1$ (linear variation of temperature with radius); $i = 2$ (temperature variation with radius according to a second-order function); $i = 3$ (temperature variation with radius according to a third-order function). Indeed, almost all distributions $T = T(r)$ found in commonly used disks can be satisfactorily approximated with a third-order polynomial.

The explicit relations whereby $\varphi_i(t)$ and $\psi_i(t)$ can be calculated for $i = 1, 2, 3$ are as follows:

(a) For $i = 1$, $T = k_1 \cdot t$ and, hence:

$$\varphi_1 = \frac{1-t}{5+v}; \quad \psi_1 = \frac{1-3t}{5+v}; \tag{6.50}$$

(b) For $i = 2$, $T = k_2 \cdot t^2$ and, hence:

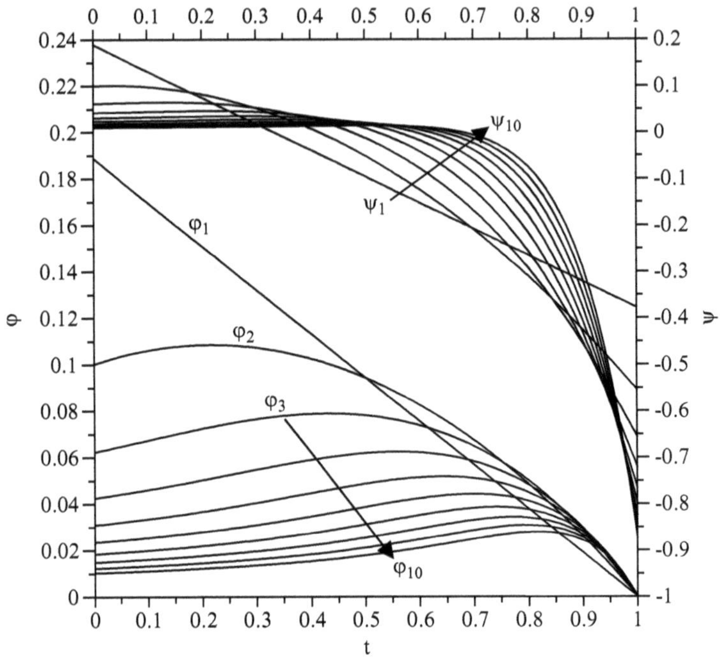

Fig. 6.7 Distribution of $\varphi_i(t)$ and $\psi_i(t)$ along dimensionless radius of disk, within the interval $0 \leq t \leq 1$, up to $i = 10$, for $v = 0.3$

$$\begin{aligned} \varphi_2 &= \frac{2}{11+v} \cdot \left(-t^2 + \frac{2+v}{5+v} \cdot t + \frac{3}{5+v} \right); \\ \psi_2 &= \frac{2}{11+v} \cdot \left(-4t^2 + \frac{1+2v}{5+v} \cdot t + \frac{3}{5+v} \right); \end{aligned} \tag{6.51}$$

(c) For $i = 3$, $T = k_3 \cdot t^3$ and, hence:

$$\begin{aligned} \varphi_3 &= 3 \cdot \frac{-(5+v)\cdot(11+v)\cdot t^3 + (5+v)\cdot(3+v)\cdot t^2 + 8\cdot(2+v)\cdot t + 24}{(5+v)\cdot(11+v)\cdot(19+v)}; \\ \psi_3 &= 3 \cdot \frac{-5\cdot(5+v)\cdot(11+v)\cdot t^3 + (5+v)\cdot(3v+1)\cdot t^2 + 8\cdot(1+2v)\cdot t + 24}{(5+v)\cdot(11+v)\cdot(19+v)}. \end{aligned} \tag{6.52}$$

Figure 6.7 shows functions $\varphi_i(t)$ and $\psi_i(t)$ up to $i = 10$ versus dimensionless variable t for steels ($v = 0.3$). It should be noted that the literature gives analytical relations for $\varphi_i(t)$ and $\psi_i(t)$ only for $i = 1$ and $i = 2$ ([26], [28]), as well as diagrams of said functions up to $i = 5$ [26].

On the basis of the foregoing considerations and relations (6.36), the stress state can be expressed as follows where a non-zero temperature gradient is present along the radius given by relation (4.6):

$$\begin{cases} \sigma_r = A \cdot a_r + B \cdot b_r + E \cdot \alpha \cdot (k_1 \cdot \varphi_1 + k_2 \cdot \varphi_2 + k_3 \cdot \varphi_3 + ...) \\ \quad = A \cdot a_r + B \cdot b_r + E \cdot \alpha \cdot \varphi(t) \\ \sigma_t = A \cdot a_t + B \cdot b_t + E \cdot \alpha \cdot (k_1 \cdot \psi_1 + k_2 \cdot \psi_2 + k_3 \cdot \psi_3 + ...) \\ \quad = A \cdot a_t + B \cdot b_t + E \cdot \alpha \cdot \psi(t), \end{cases} \tag{6.53}$$

with $\varphi_i = \varphi_i\,(t)$ and $\psi_i = \psi_i(t)$.

Here again, the strain state associated with the stress state given by expressions (6.53) is obtained from (1.25), after substituting (6.53) in them; radial displacement, on the other hand, is obtained from the relation $u = r \cdot \varepsilon_t$.

6.4 Rotating Disk Having Density Variation on Radius

We will now consider a rotating conical profile disk featuring a density variation along its radius as expressed by the polynomial function:

$$\gamma = \gamma_0 + \gamma_1 \cdot t + \gamma_2 \cdot t^2 + ... = \sum_{i=0}^{n} \gamma_i \cdot t^i \tag{6.54}$$

in whose series γ_0 is the density of the rotor material while $\gamma_1, \gamma_2, \ldots$ are constants, and $i = 1, 2, \ldots, n$. Function (6.54) is similar to (3.44), except that the dimensionless variable t takes the place of the radial coordinate r. Thus, by considering relation (6.54) and by introducing it in the differential equation (6.4), we obtain:

$$\frac{d^2u}{dt^2} + \left(\frac{1}{t} - \frac{1}{1-t}\right) \cdot \frac{du}{dt} - \left[\frac{1}{t^2} + \frac{\nu}{t \cdot (1-t)}\right] \cdot u + \left(1-\nu^2\right) \cdot \frac{\omega^2 \cdot R^3 \cdot t}{E} \cdot \sum_{i=0}^{n} \gamma_i \cdot t^i = 0. \tag{6.55}$$

In this case too, the general integral of (6.55) is the sum of the solution of the homogeneous equation, which is already known, and of a particular integral of the non-homogeneous equation. The latter can be obtained as the linear combination of contributions deriving from component n terms $\gamma_i \cdot t^i$ (with $i = 0, 1, 2, \ldots, n$) of the function $\gamma = \gamma(t)$; thus, by considering the *i-th* term of this function, (6.55) becomes:

$$\frac{d^2u}{dt^2} + \left(\frac{1}{t} - \frac{1}{1-t}\right) \cdot \frac{du}{dt} - \left[\frac{1}{t^2} + \frac{\nu}{t \cdot (1-t)}\right] \cdot u + \left(1-\nu^2\right) \cdot \frac{\omega^2 \cdot R^3}{E} \cdot \gamma_i \cdot t^{i+1} = 0. \tag{6.56}$$

This relation, with $i = 0$, can be used to calculate the displacement $u = u(t)$ related to density γ_0 of the conical profile disk's basic material; the same equation, with $i = 1$, 2, ..., n, makes it possible to determine the contributions related to the fictitious variation in density along the radius deriving from radial blades on lateral surfaces, which increase the disk's mass without appreciably contributing to its load capacity.

The particular integrals of differential equation (6.56) can be obtained using the same procedure as that adopted for (6.41) and (6.42) governing thermal loading. For the sake of brevity, the analytical developments will not be specified here. Through these developments, the stress state for a conical disk with a fictitious density variation along its radius is expressed as the sum of the partial contributions of the particular integrals, and thus in the following form:

$$\begin{cases} \sigma_r = \omega^2 \cdot R^2 \cdot (\gamma_0 \cdot g_r + \gamma_1 \cdot \eta_1 + \gamma_2 \cdot \eta_2 + \ldots) \\ \sigma_t = \omega^2 \cdot R^2 \cdot (\gamma_0 \cdot g_t + \gamma_1 \cdot \zeta_1 + \gamma_2 \cdot \zeta_2 + \ldots) \end{cases} \tag{6.57}$$

where $g_r, g_t, \eta_1, \zeta_1, \eta_2, \zeta_2, \ldots$ are functions of t and v.

Functions g_r and g_t related to the particular integral in equation (6.56) with $i = 0$ are obtained from relations (6.14).

In general terms, for any value of i, the relations whereby functions $\eta_i(t)$ and $\zeta_i(t)$ appearing in expressions (6.57) can be calculated as follows:

$$\begin{aligned} \eta_i(t) = & -\frac{(i+3+v)\cdot t^{i+2}}{i^2+7i+11+v} + (i+3+v)\cdot\sum_{m=1}^{i+1}\frac{(m+1+v)\cdot t^m \cdot \prod_{j=m+1}^{i+1}(j^2+2j)}{\prod_{j=m}^{i+2}(j^2+3j+1+v)} \\ & + \frac{(i+3+v)\cdot\prod_{j=1}^{i+1}(j^2+2j)}{\prod_{j=1}^{i+2}(j^2+3j+1+v)} \\ \zeta_i(t) = & -\frac{[1+(i+3)\cdot v]\cdot t^{i+2}}{i^2+7i+11+v} + (i+3+v)\cdot\sum_{m=1}^{i+1}\frac{[1+(m+1)\cdot v]\cdot t^m \cdot \prod_{j=m+1}^{i+1}(j^2+2j)}{\prod_{j=m}^{i+2}(j^2+3j+1+v)} \\ & + \frac{(i+3+v)\cdot\prod_{j=1}^{i+1}(j^2+2j)}{\prod_{j=1}^{i+2}(j^2+3j+1+v)}. \end{aligned} \tag{6.58}$$

These general relations, which were also derived by the authors [77], can be used to obtain several others of greater technical relevance for engineering calculations, viz.: $i = 1$, $i = 2$ and $i = 3$. Indeed, almost all functions $\gamma = \gamma\ (r)$ found in commonly used disks can be satisfactorily approximated with a third-order polynomial.

The explicit relations whereby $\eta_i(t)$ and $\zeta_i(t)$ can be calculated for $i = 1, 2$ are as follows (explicit expressions of $\eta_3(t)$ and $\zeta_3(t)$ are omitted because they are very long):

(a) For $i = 1, \gamma = \gamma_1 \cdot t$ and, hence:

$$\begin{aligned}\eta_1(t) &= -\frac{(4+v)}{(19+v)}\cdot t^3 + \frac{(4+v)\cdot(3+v)}{(19+v)\cdot(11+v)}\cdot t^2 + \frac{(4+v)\cdot[8\cdot(2+v)\cdot t+24]}{(19+v)\cdot(11+v)\cdot(5+v)};\\ \zeta_1(t) &= -\frac{(1+4v)}{(19+v)}\cdot t^3 + \frac{(4+v)\cdot(1+3v)}{(19+v)\cdot(11+v)}\cdot t^2 + \frac{(4+v)\cdot[8\cdot(1+2v)\cdot t+24]}{(19+v)\cdot(11+v)\cdot(5+v)};\end{aligned} \tag{6.59}$$

(b) For $i = 2, \gamma = \gamma_2 \cdot t^2$ and, hence:

$$\begin{aligned}\eta_2(t) &= -\frac{(5+v)}{(29+v)}\cdot t^4 + \frac{(5+v)\cdot(4+v)}{(29+v)\cdot(19+v)}\cdot t^3\\ &\quad + \frac{15\cdot(5+v)\cdot(3+v)\cdot t^2 + 120\cdot(2+v)\cdot t + 360}{(29+v)\cdot(19+v)\cdot(11+v)};\\ \zeta_2(t) &= -\frac{(5v+1)}{(29+v)}\cdot t^4 + \frac{(5+v)\cdot(4v+1)}{(29+v)\cdot(19+v)}\cdot t^3\\ &\quad + \frac{15\cdot(5+v)\cdot(3v+1)\cdot t^2 + 120\cdot(2v+1)\cdot t + 360}{(29+v)\cdot(19+v)\cdot(11+v)}.\end{aligned} \tag{6.60}$$

Figure 6.8 shows functions $\eta_i(t)$ and $\zeta_i(t)$ up to $i = 10$ versus dimensionless variable t for steels ($v = 0.3$). It should be noted that the literature gives analytical relations for $\eta_i(t)$ and $\zeta_i(t)$ only for $i = 1$ (Giovannozzi, [28]).

On the basis of (6.40) and (6.57), the stress state for a rotating conical disk featuring a fictitious density variation along its axis is obtained from the following relations:

$$\begin{cases}\sigma_r = A\cdot a_r + B\cdot b_r + \omega^2\cdot R^2\cdot(\gamma_0\cdot g_r + \gamma_1\cdot\eta_1 + \gamma_2\cdot\eta_2 + \gamma_3\cdot\eta_3 + \ldots)\\ \sigma_t = A\cdot a_t + B\cdot b_t + \omega^2\cdot R^2\cdot(\gamma_0\cdot g_t + \gamma_1\cdot\zeta_2 + \gamma_2\cdot\zeta_2 + \gamma_3\cdot\zeta_3 + \ldots).\end{cases} \tag{6.61}$$

Here again, the strain state associated with the above stress state is obtained from (1.25) from which temperature terms are omitted; radial displacement, on the other hand, is obtained from the relation $u = r \cdot \varepsilon_t$.

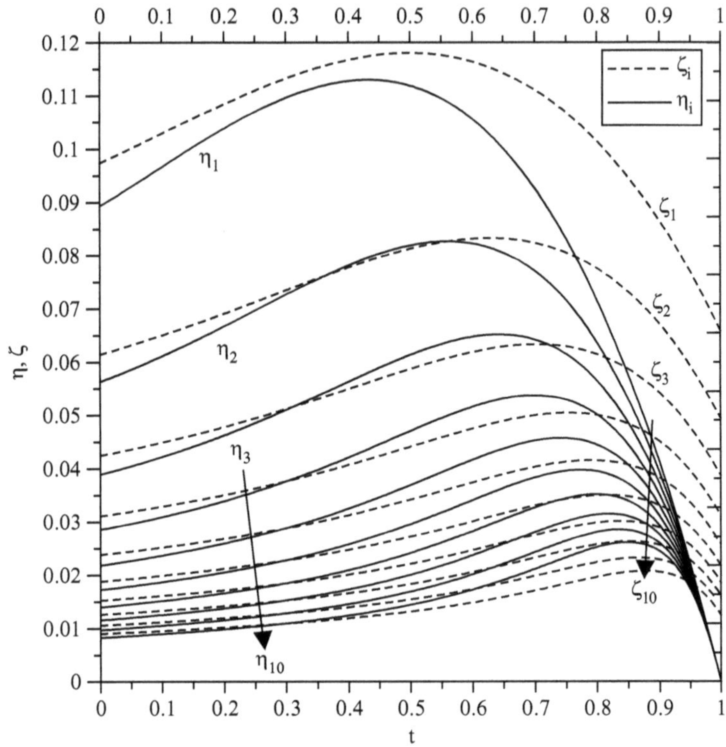

Fig. 6.8 Distribution of $\eta_i(t)$ and $\zeta_i(t)$ along dimensionless, radius of disk, within the interval $0 \leq t \leq 1$, up to $i = 10$, for $v = 0.3$

6.5 Conical Disk with Density Variation Subjected to Centrifugal and Thermal Loads

Lastly, we will consider a conical disc subjected to both centrifugal and thermal loads and featuring a density variation along the radius. It will be assumed that the functions $T = T(r)$ and $\gamma = \gamma(r)$ are the polynomials given by (4.5) and (6.54) respectively.

Since the principle of superposition applies in the linear elastic field, it is obvious that, taking (6.53) and (6.57) into account, the stress state will be given by the relations:

$$\begin{cases} \sigma_r = A \cdot a_r + B \cdot b_r + \omega^2 \cdot R^2 \cdot (\gamma_0 \cdot g_r + \gamma_1 \cdot \eta_1 + \gamma_2 \cdot \eta_2 + \gamma_3 \cdot \eta_3 + \ldots) \\ \qquad + E \cdot \alpha \cdot \varphi(t) \\ \sigma_t = A \cdot a_t + B \cdot b_t + \omega^2 \cdot R^2 \cdot (\gamma_0 \cdot g_t + \gamma_1 \cdot \zeta_2 + \gamma_2 \cdot \zeta_2 + \gamma_3 \cdot \zeta_3 + \ldots) \\ \qquad + E \cdot \alpha \cdot \psi(t), \end{cases} \tag{6.62}$$

while the associated strain state will be given by (1.25) after substituting the above relations in them. Radial displacement, on the other hand, will be derived from the usual relation $u = r \cdot \varepsilon_t$.

6.6 Examples of Conical Disks

This chapter provides a number of case studies of the analytical method laid out above, which is based on using the two independent integrals u_1 and u_2 of the homogeneous differential equation (6.6) directly throughout the whole interval $0 \leq t \leq 1$. These case studies will demonstrate:

- The advantages of this method over that used by Honegger and Giovannozzi, which is based on splitting this interval into two partial intervals, viz., $0 \leq t \leq 0.5$ and $0.5 < t \leq 1$, and on the use of four independent integrals of differential equation (6.6) for variable t: the first two, u_1 and u_2, being obtained by integrating (6.6) into variable t, and the second two, $\bar{u}_1$ and $\bar{u}_2$, being obtained by integrating (6.6) after changing variable $x = (1 - t)$ and then transformed into variable t (see Sect. 13.2 for further details);
- The advantages of this method over other methods (finite elements, boundary elements, etc.,) especially in the early stages of design, as it is more straightforward and versatile (nor should we forget that its advantages are even greater when the method is used to optimize geometrical shapes and their mechanical strength in the later stages of design).

The results for these case studies are compared with those obtained using finite element models with 4-node axially symmetric elements and three dofs per node. In general, the comparison indicates a perfect match between the analytical results and those from finite element analysis.

To avoid over-complication, however, the figures showing the results of theoretical analysis for each case study do not include the numerical results, with the exception of those for the most complex case, which can be regarded as subsuming the simpler cases. While the comparison of results achieved with theoretical analysis and with numerical models is shown for completeness, it is obvious that the analytical model has advantages over, and is more generally valid than, the finite element model, which is not only bound to a single case, but also presents problems in the neighbourhood of the axis of rotation when solid disks are modelled with axisymmetric elements.

6.6.1 Rotating Solid Conical Disk with Apex Singularity and Having Constant Density

This type of disk, although of little technical interest because of its pointed tip at outer radius $r = R$ (in reality, a few applications are to be found in centrifugal

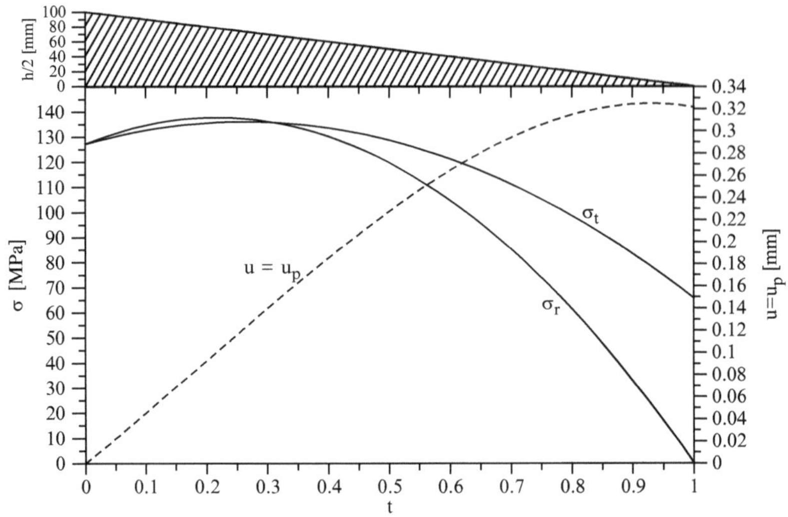

Fig. 6.9 Stress-distribution curves σ_r, σ_t and displacement-distribution $u = u_p$ in a solid conical disk with apex singularity and having constant density ($\nu = 0.3$)

machine disks, rotating cutters and centrifugal oil barrier devices), deserves attention inasmuch as it makes it possible to assign inherent physical significance to particular integral u_p, which directly represents radial displacement in the solid conical disk, extended to its apex R. Indeed, as was indicated earlier, given that u_1 tends to infinity at the outer edge of the disk ($r = R$, $t = 1$), whereas u_2 tends to infinity at the disk axis ($r = 0$, $t = 0$), the constants appearing in relation (6.39) must be zero in order to have finite values for displacement u. In other words, it is necessary that $C_1 = C_2 = 0$.

In this connection, it should be pointed out that relation (6.10), expressing u_p as a third-order polynomial in r, already fulfils the boundary conditions:

$$\begin{cases} \sigma_r = \sigma_t & \text{for } r = 0\,(t = 0) \\ \sigma_r = 0 & \text{for } r = R\,(t = 1). \end{cases} \tag{6.63}$$

Thus, it is sufficient to replace relation (6.10) and its first derivative with respect to r in (1.27) from which temperature terms are omitted, considering that, by means of the procedure outlined in Sect. 6.2.1, constants c_1, c_2 and c_3 are given by relations (6.11) shown therein.

Figure 6.9 shows stress-distribution curves σ_r and σ_t and displacement-distribution curve $u = u_p$ as functions of t in a steel solid conical disk ($\nu = 0.3$; $E = 204$ GPa, $\gamma = 7{,}800$ kg/m^3), extended to its apex radius R ($R = 1$ m), with thickness at axis $h_0 = 0.1$ m (note that this value does not affect calculation results) and rotating at angular velocity $\omega = 314$ rad/s. The stress-distributions show that σ_r

locally exceeds σ_t, though very slightly. Obviously, in this case there is no difference from the results obtained by Honegger and Giovannozzi.

6.6.2 *Rotating Annular Conical Disk with Apex Singularity and Having Constant Density*

In this case, also of little technical interest, given that u_1 tends to infinity at the disk apex ($r = R, t = 1$), constant C_1 appearing in relation (6.39) must be zero in order to obtain a finite value of radial displacement u at this radius. Moreover, given that the disk is annular, $t \neq 0$ at the inner radius, and there is no u_2-related singularity for $t = 0$, then $C_2 \neq 0$. The relation whereby the displacement field can be calculated in this case is (6.39) from which term u_1 is omitted. Here, the boundary conditions are as follows:

$$\begin{cases} \sigma_r = 0 \quad \text{for } r = R(t = 1). \\ \sigma_r = \sigma_{ri} \quad \text{for } r = r_i\left(t = t_i = \dfrac{r_i}{R}\right). \end{cases} \tag{6.64}$$

As was shown, the first boundary condition is already fulfilled by u_p and u_2, while integration constant C_2 is obtained from the second boundary condition. Supposing that $\sigma_{ri} = 0$, i.e., that the disk inner surface is not loaded, the value of constant C_2 is obtained by equalling to zero (for $r = r_i$) terms in square brackets appearing in the first of (1.27), from which temperature terms are omitted, and by introducing $u = C_2 \cdot u_2 + u_p$.

Figure 6.10 shows stress-distribution curves σ_r and σ_t and displacement-distribution curve u as functions of t in a steel annular disk ($v = 0.3$; $E = 204$ GPa, $\gamma = 7{,}800\,\text{kg/m}^3$), featuring inner radius $r_i = 0.1$ m and extended to radius R ($R = 1$ m), with thickness at axis $h_0 = 0.1$ m and rotating at angular velocity $\omega = 314$ rad/s.

6.6.3 *Rotating Solid Conical Disk with* $\mathbf{r_e < R}$ *and Having Constant Density*

In this type of disk, as u_2 tends to infinity at the disk axis ($r = 0, t = 0$), constant C_2 appearing in relation (6.39) must be zero in order to obtain a finite value of radial displacement u at this radius. Moreover, given that the disk is of the frustum type ($r_e < R$), $t < 1$ at the disk outer radius, and as there is no u_1-related singularity for $t = 1$, then $C_1 \neq 0$. The relation whereby the displacement field can be calculated in this case is (6.39) from which term u_2 is omitted. Here, the boundary conditions are as follows:

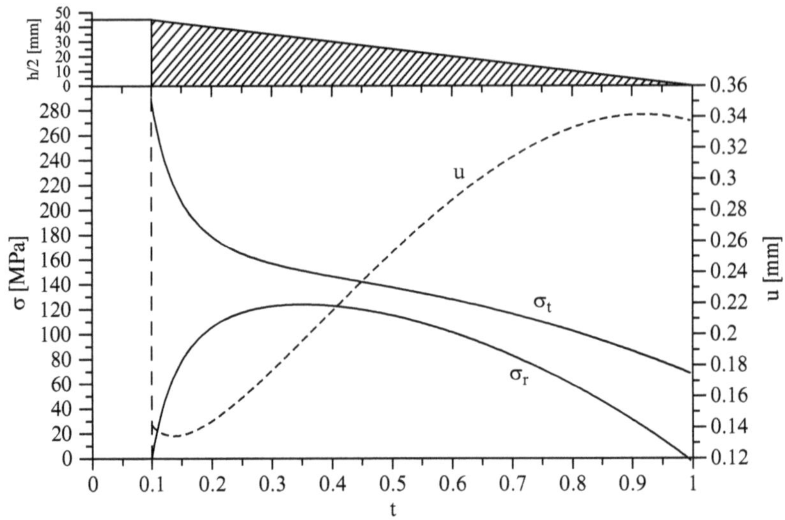

Fig. 6.10 Stress-distribution curves σ_r, σ_t and displacement u in an annular conical steel disk with apex singularity and having constant density ($\nu = 0.3$)

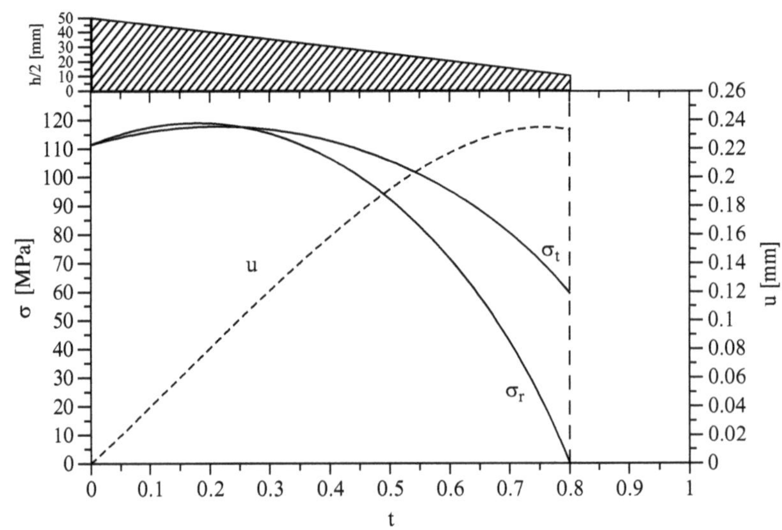

Fig. 6.11 Stress-distribution curves σ_r, σ_t and displacement-distribution u in a solid conical steel disk with $r_e < R$ and having constant density ($\nu = 0.3$)

$$\begin{cases} \sigma_r = \sigma_t \quad \text{for } r = 0(t = 0). \\ \sigma_r = \sigma_{re} \quad \text{for } r = r_e\left(t = t_e = \dfrac{r_e}{R}\right). \end{cases} \tag{6.65}$$

As was shown, the first boundary condition is already fulfilled by u_p and u_1, while integration constant C_1 is obtained from the second boundary condition.

Supposing that $\sigma_{re} = 0$, i.e., that the disk outer surface is not loaded, the value of constant C_1 is obtained by equalling to zero (for $r = r_e$) terms in square brackets appearing in the first of (1.27), from which temperature terms are omitted, and by introducing $u = C_1 \cdot u_1 + u_p$.

Figure 6.11 shows stress-distribution curves σ_r and σ_t and displacement-distribution curve u as functions of t in a steel solid disk ($\nu = 0.3$; $E = 204$ GPa, $\gamma = 7{,}800$ kg/m^3), extended to outer radius $r_e = 0.8$ m, with $R = 1$ m, thickness at axis $h_0 = 0.1$ m and rotating at angular velocity $\omega = 314$ rad/s.

6.6.4 *Rotating Annular Conical Disk Without Singularities ($r_i > 0$ and $r_e < R$) and Having Constant Density*

As the majority of disks used in current machinery applications feature a central hole of radius $r_i > 0$ and an outer radius $r_e < R$, problems deriving from two singularity points $t = 0$ and $t = 1$ do not apply. In this case, radial displacement u is obtained from (6.39) complete with all terms. As there are no singularity points, it is advisable to calculate the stress field by using relations (6.40), determining integration constants A and B by establishing:

$$\begin{cases} \sigma_r = \sigma_{ri} & \text{for } r = r_i \\ \sigma_r = \sigma_{re} & \text{for } r = r_e. \end{cases} \tag{6.66}$$

Constants C_1 and C_2 in (6.39) can be determined by using relations (6.38).

Figure 6.12 shows stress-distribution curves σ_r and σ_t and displacement-distribution curve u as functions of t in a steel annular conical disk with $r_i = 0.1$ m, $r_e = 0.8$ m, $R = 1$ m, $h_i = 0.09$ m, $h_e = 0.02$ m, rotating at angular velocity $\omega = 314$ rad/s and not loaded at either outer or inner surfaces ($\sigma_{ri} = 0$; $\sigma_{re} = 0$).

6.6.5 *Rotating Conical Disk with Hub and Crown Ring and Having Constant Density*

If, as often occurs in machinery applications (Fig. 6.13), the conical disk features a crown ring and a hub, both of which are of constant thickness, it is first necessary to determine the radial stresses $\sigma_{r,A}$ and $\sigma_{r,B}$ present in section A ($r = r_i$) and in section B ($r = r_e$) respectively, and which constitute two unknown hyperstatic values.

By imposing boundary conditions on the conical disk, constants A and B are first determined from the first of relations (6.40), and are expressed as functions of unknown values $\sigma_{r,A}$ and $\sigma_{r,B}$. Equations 6.40 are then used to obtain relations expressing σ_r and σ_t as functions of $\sigma_{r,A}$ and $\sigma_{r,B}$. Lastly, by imposing equality of radial displacements of the interface A, first regarded as a part of the conical disk

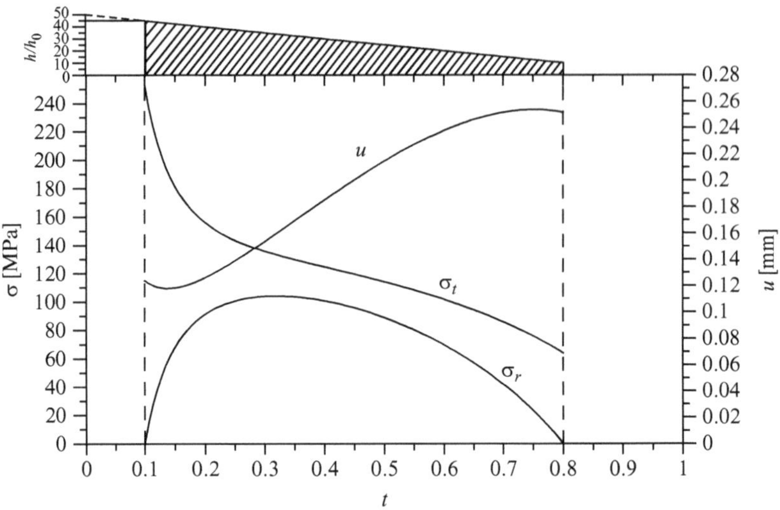

Fig. 6.12 Stress-distribution curves σ_r, σ_t and displacement-distribution u in an annular conical steel disk with $r_e < R$ and having constant density ($\nu = 0.3$)

and second as a part of the hub, and of the interface B, also regarded first as a part of the conical disk and second as a part of the crown ring, we determine unknown hyperstatic values $\sigma_{r,A}$ and $\sigma_{r,B}$, and, proceeding backwards, all other items (in this connection, see calculation example 3 in Sect. 2.5.3).

It should be noted that, by appropriately selecting the disk profile, along with crown ring and hub, stresses σ_r and σ_t can be made almost constant; in other words, a disk of uniform strength can be approximated within certain limits.

In the practical example in Fig. 6.13, which refers to a steel disk ($\nu = 0.3$), the crown ring is simulated by a radial stress distribution at B ($\sigma_{rB} = 120$ MPa), while the hub may assume three configurations which differ in mass. For each of these configurations, the corresponding interface stresses $\sigma_{r,A}$ are indicated, but not the geometry. Figure 6.13 shows stress-distribution curves σ_r and σ_t and displacement-distribution curve u, clarifying that the most advantageous stress distribution is obtained with the hub having greater mass.

6.6.6 Conical Disk Without Singularities and Subjected to Temperature Gradient

We will now examine a conical disk which, like that analysed in Sect. 6.6.4, has no singularity points. Here, however, the disk is not rotating and is subjected to thermal load only, with temperature varying according to function (4.5), with $i = 3$. In this case, the stress state is obtained from relations (6.53), while integration constants A and B are determined by imposing boundary conditions (6.66) of Sect. 6.6.4.

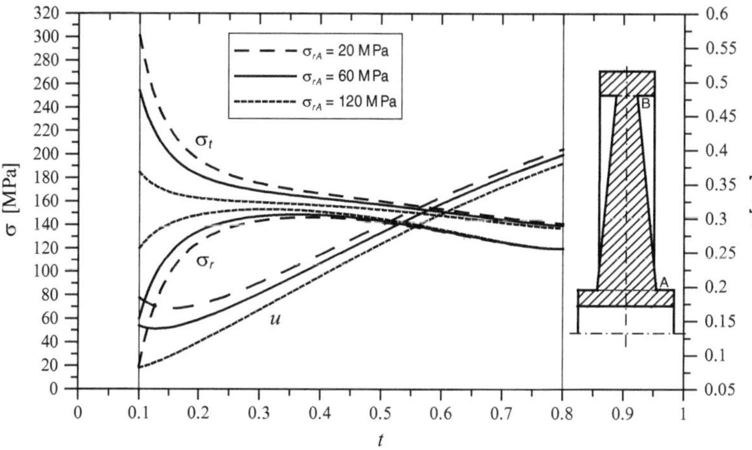

Fig. 6.13 Stress-distribution curves σ_r, σ_t and displacement-distribution u in a conical steel disk with hub and crown ring, for three different hub configurations, corresponding to three interface stress distributions $\sigma_{r,A}$ ($\nu = 0.3$)

Figure 6.14 shows stress-distribution curves σ_r and σ_t and displacement-distribution curve u as functions of t in an annular conical disk featuring the same geometry and made of the same material ($\alpha = 12{\cdot}10^{-6}{\cdot}°C^{-1}$) as the one examined in Sect. 6.6.4, not loaded at either inner or outer surface and subjected to a temperature gradient along its radius according to function $T = T_0 + k_1 \cdot r + k_2 \cdot r^2 + k_3 \cdot r^3$. The same figure shows the curve of the temperature distribution, T being dimensionless relative to reference temperature T_0, as well as coefficients of the third-order polynomial law.

6.6.7 *Conical Disk Without Singularities and Having Density Variation on Radius*

We will now examine a conical disk which, like that analysed in Sect. 6.6.4, has no singularity points and is not subjected to thermal load. Here, however, the disk is rotating and has variable density along its radius according to function (6.54), with $i = 3$. In this case, the stress state is obtained from relations (6.61), while integration constants A and B are determined by imposing boundary conditions (6.66).

Figure 6.15 shows stress-distribution curves σ_r and σ_t and displacement-distribution curve u as functions of t in an annular conical disk featuring the same geometry and made of the same material as the one examined in Sect. 6.6.4, rotating at angular velocity $\omega = 314$ rad/s, not loaded at either inner surface or outer surface, but featuring variable density along its radius according to function $\gamma = \gamma_0 + \gamma_1 \cdot t + \gamma_2 \cdot t^2 + \gamma_3 \cdot t^3$. The same figure shows the curve of the density variation, γ being dimensionless relative to the basic material density γ_0, as well as coefficients of the third-order polynomial function.

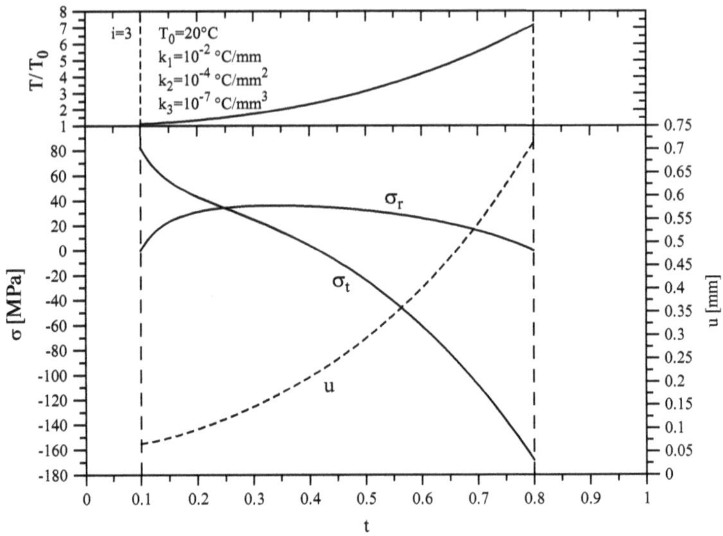

Fig. 6.14 Stress-distribution curves σ_r, σ_t and displacement-distribution u in an annular conical steel disk without singularities and subjected to temperature gradient ($\nu = 0.3$)

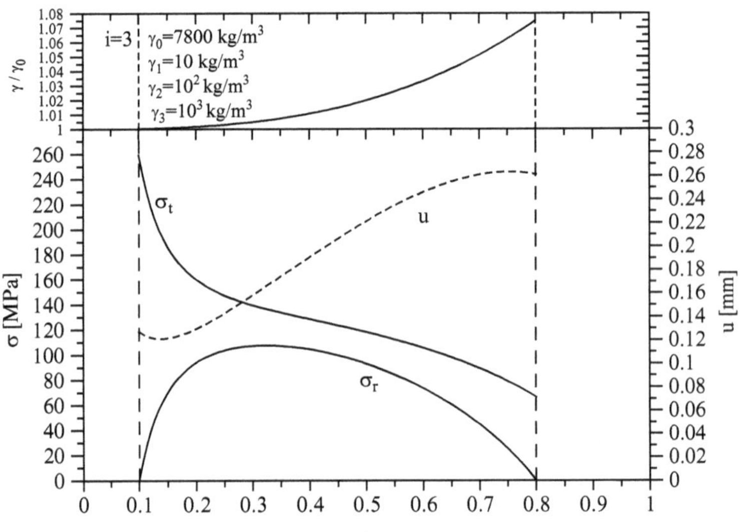

Fig. 6.15 Stress-distribution curves σ_r, σ_t and displacement-distribution u in an annular conical steel disk without singularities and having density variation on radius ($\nu = 0.3$)

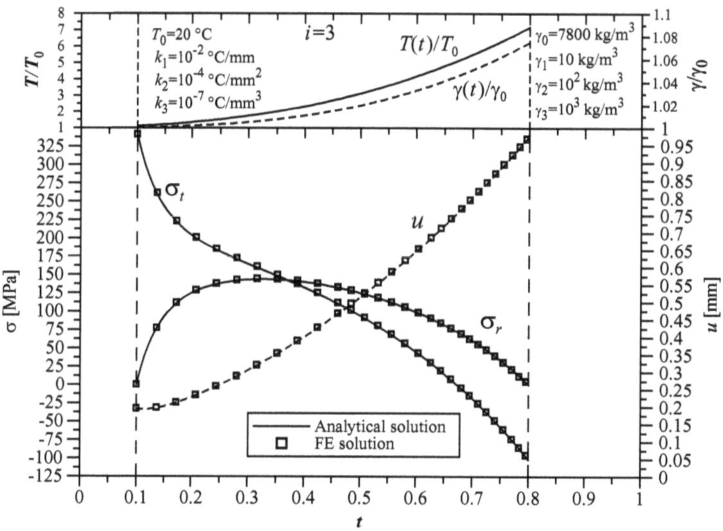

Fig. 6.16 Stress-distribution curves σ_r, σ_t and displacement-distribution u in an annular steel disk with density variation on radius and subjected to temperature gradient, and comparison with FE results ($\nu = 0.3$)

6.6.8 Rotating Conical Disk Without Singularities, with Density Variation on Radius and Subjected to Temperature Gradient

Here, the same disk featured in the previous two examples will be examined, but rotating and subjected to thermal load, with temperature varying according to function (4.5), with $i = 3$, and having variable density along its radius according to function (6.54), still with $i = 3$. In this case, the stress state is obtained from relations (6.62), while integration constants A and B are determined by imposing boundary conditions (6.66).

Figure 6.16 shows stress-distribution curves σ_r and σ_t and displacement-distribution curve u as functions of t in an annular conical disk featuring the same geometry and made of the same material as the one examined in the previous two examples, not loaded at either inner or outer surface, subject to a temperature gradient along its radius and featuring variable density along its radius. The functions of variation for T and γ are the cubical ones referred to in the previous two examples, and shown in the diagram in dimensionless form.

The figure also compares the results obtained by applying the analytical method proposed here and the numerical results obtained, all other conditions remaining unchanged, through FEM. Stress and displacement distributions clearly show that the results of the numerical model and those of the analytical model match perfectly. As mentioned above, the same verification was successfully carried out with the same procedure for all previous examples.

Chapter 7
Non-Linearly Variable Thickness Disks

7.1 General Differential Equations and Variable Thickness Profile Geometry

The topic discussed here goes well beyond those normally addressed in courses on mechanical design. However, it is advisable to provide complete coverage of the mathematical basics of the method for integrating the differential equations that govern the displacement field of rotating disks whose thickness varies according to a two-parameter linear function, and thus make it possible to determine their stress and strain states. With this approach, which is a generalization of the conical disk problem covered in the preceding Chapter (the conical disk can be obtained as a particular case of this non-linearly variable thickness disk), we can analyse disk profile configurations of undoubted design interest.

The theory presented here can be used to solve any problem involving a disk whose thickness varies according to relation (7.1), with developments entailing the same degree of complexity found in Chap. 6 for the conical disk. From the standpoint of the analytical developments and the complexity of the calculations involved, it is advisable to use this approach where strictly necessary (convergent and/or divergent solid disks with concave and/or convex profile, convergent annular disks with convex profile, divergent annular disks with concave profile) for which, from the purely technical point of view, *secundum non datur*!

When dealing with (Fig. 4.1a) annular disks with divergent conical profile, tapered disks with concave profile, or divergent disks with convex profile, it is advisable to use Stodola's relations (4.1) or (4.2), which characterize the hyperbolic disk. In fact, once the disk's radii r_i and r_e have been established together with the corresponding thicknesses h_i and h_e, it will almost always be possible – with the exceptions indicated in Chap. 4 – to find a value of exponent a appearing in relations (4.1) and (4.2), with which the profile geometry can be correctly

V. Vullo and F. Vivio, *Rotors: Stress Analysis and Design*,
Mechanical Engineering Series, DOI 10.1007/978-88-470-2562-2_7,
© Springer-Verlag Italia 2013

described. It will thus be possible to proceed much more simply on the basis of the hyperbolic disk theory, which as we have seen makes it possible to arrive at closed form solutions for the differential equations governing the displacement field.

The disks analysed here are characterized by a non-linearly variation of thickness with the radius expressed by a power of a two-parameter linear function according to relation (7.1), also used by other researchers (see, for example, Güven [32]). As will be discussed below, this relation can define the geometry of a fourfold infinity of profiles, including convergent and divergent conical profiles; the only profiles that cannot be simulated are those with an inflection point, such as uniform strength disks. This family of variable thickness disks was investigated in the elastic field by Eraslan and Argeşo [14] as part of a study addressing the limit angular velocity where, consequently, the rotor is stressed beyond the elastic limit.

Here it is assumed that the disk always has a plane of symmetry perpendicular to the axis, which bisects the thickness, and that the thickness variation function $h = h(r) = h(\rho) = h(t)$ has any one of the following equivalent expressions:

$$h = h_0 \cdot (1 - m \cdot r)^k = h_0 \cdot (1 - n \cdot \rho)^k = h_0 \cdot (1 - t)^k, \tag{7.1}$$

where: h_0 is the thickness at the axis of the disk (which is real for the solid disk and fictitious for the annular disk); r_e is the outer radius; R, for the profiles of interest to us here (see below) is the distance of the axis from apex V (the term apex is used here even if the disk does not feature a pointed tip, e.g., in convex profiles – Fig. 7.1d) where the two symmetrical sides of the profile intersect (Fig. 7.1a, c, d); m, $n = m{\cdot}r_e$ and k are constant geometrical parameters, r is the radius; $\rho = r/r_e$ and $t = r/R = n{\cdot}\rho$ are dimensionless radii relative to r_e and R respectively. The domains of variables ρ and t are $0 \le \rho \le 1$ and $0 \le t \le n$ for solid disks, and $\beta = r_i/r_e \le \rho \le 1$ and $r_i/R \le t \le n$ for annular disks respectively, r_i being the inner radius. The two domains coincide only for disks profiled in order to feature nil thickness at radius r_e (in this case $n = 1$).

By carefully selecting parameters k and m (and therefore n), the above functions can be used to describe a fourfold infinity of disk profiles. Indeed, with function $h = h(\rho)$ we have:

- For $k = 0$, with any value of n, and for $n = 0$, with any value of k, the disk shows constant thickness (see profiles shown with a horizontal dashed line in Fig. 7.1a–d);
- For $k = 1$ and $0 < n < 1$, the disk shows a converging conical profile (Fig. 7.1a), with no limitations for $0 < n{\cdot}\rho \le 1$ (with $n = 1$, the conical disk ends with a pointed tip at the apex, i.e., for $r_e = R$ and $\rho = 1$); also for $k = 1$ and $n > 1$, we have the limitation $0 \le \rho \le 1/n$ (if the disk outer radius r_e is a design datum, such profiles have no physical meaning); again for $k = 1$, but with $n < 0$, the disk shows a diverging conical profile (Fig. 7.1a), with no limitations;
- For $k < 0$, the disk profile is always concave ($d^2h/d\rho^2 > 0$), but with $n < 0$ ($dh/d\rho < 0$), the disk is always converging, whereas with $n > 0$ ($dh/d\rho > 0$) the

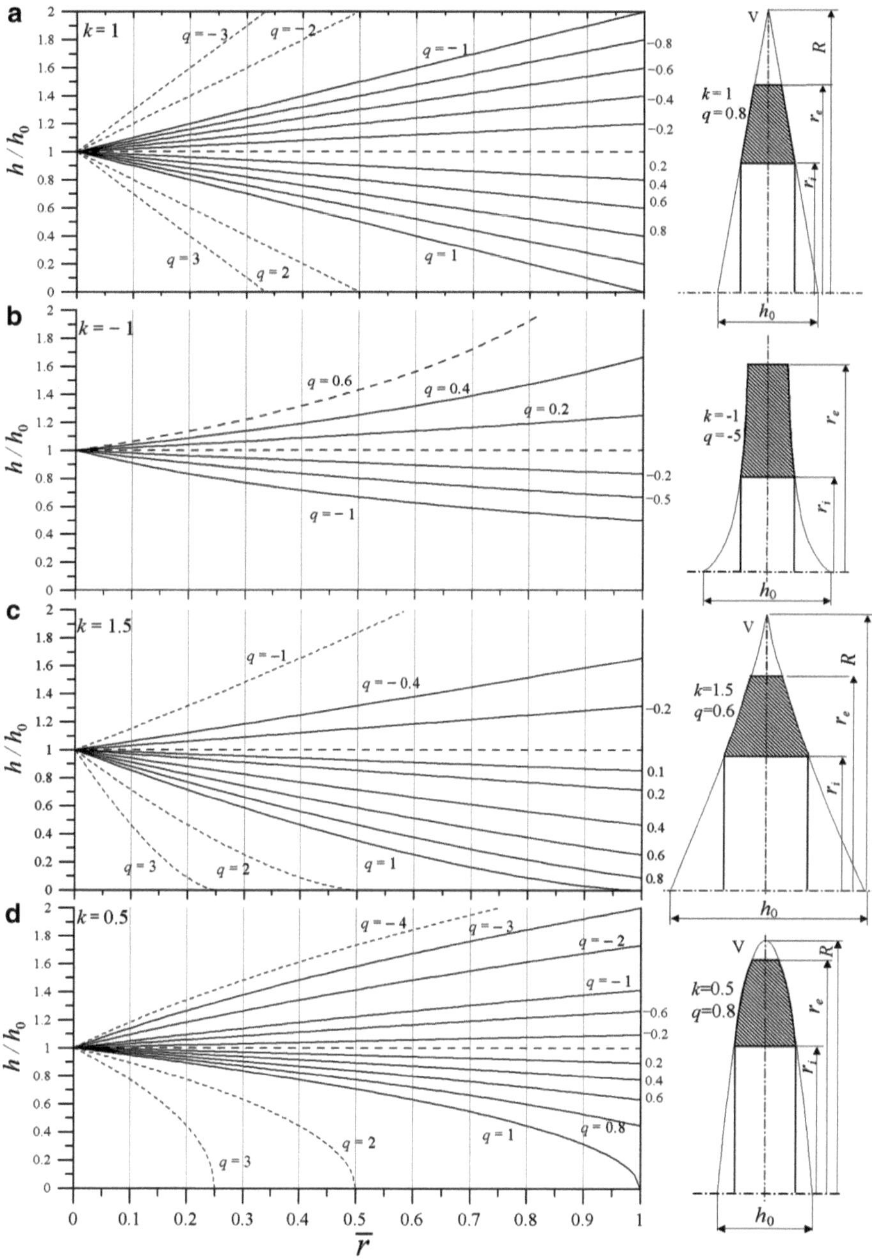

Fig. 7.1 Profiles of solid or annular non-linearly variable thickness disks: (**a**) converging and diverging conical disk ($k = 1$ and positive and negative n); (**b**) converging and diverging concave disk ($k < 0$ and positive and negative n); (**c**) converging and diverging concave disk ($k > 1$ and positive and negative n); (**d**) converging and diverging convex disk ($0 < k < 1$ and positive and negative n)

disk is always diverging, though with a limitation, for $n > 1$, $0 \le \rho \le 1/n$ (Fig. 7.1b shows several profiles for $k = -1$ and different values of n);

- For $k > 1$, the disk profile is always concave ($d^2h/d\rho^2 > 0$), but with $n < 0$ ($dh/d\rho > 0$) the disk is always diverging, whereas with $n > 0$ the disk is converging ($dh/d\rho < 0$), though with a limitation, for $n > 1$, $0 \le \rho \le 1/n$ (Fig. 7.1c shows several profiles for $k = 1.5$ and different values of n);
- For $0 < k < 1$, the disk profile is always convex ($d^2h/d\rho^2 < 0$), but with $n < 0$ ($dh/d\rho > 0$) the disk is always diverging, whereas with $n > 0$ the disk is converging ($dh/d\rho < 0$), though with a limitation, for $n > 1$, $0 \le \rho \le 1/n$ (Fig. 7.1d shows several profiles for $k = 0.5$ and different values of n);
- For $n = -1$ and $k = a$, we have the generalized Stodola disk whose profile, as indicated in Sect. 1.3 and in note 2 to Sect. 3.1, is defined by the relation $h = h_0 \cdot (1 + \rho)^a$ and which, unlike the Stodola profile in the strict sense as given by relations (4.2) or (4.3), does not show a singularity at the axis;
- For $n \to \infty$ e $k \cong a$, we have the true Stodola disk. Bearing in mind (7.1) written in terms of variable ρ together with (4.2) and establishing that the inner, mean and outer radii r_i, r_m and r_e coincide with the corresponding thicknesses h_i, h_m and h_e, we obtain three equalities from which the following relations can be obtained (Calderale et al., 2012):

$$\frac{\ln \beta}{\ln \rho_m} = \frac{\ln\left(\frac{1-n\beta}{1-n}\right)}{\ln\left(\frac{1-n\rho_m}{1-n}\right)}; \quad k = \frac{a \ln \beta}{\ln\left(\frac{1-n\rho_m}{1-n}\right)}; \quad h_0 = \frac{h_e}{(1-n)^k}. \tag{7.2}$$

Solved in succession, these relations make it possible to calculate n, k and h_0. Solved iteratively, the first relation (7.2) gives very high values for the modulus of parameter n, tending to infinity. Thus, the limit for $n \to \infty$ for the second member of the first relation (7.2) is $\ln \beta / \ln \rho_m$. Consequently, the second relation (7.2) yields $k \cong a$, as the limit for $n \to \infty$ of the denominator is $\ln \beta$. The third relation (7.2) then gives $h_0 = \infty$, confirming the singularity of the hyperbola at the disk's rotational axis.

The same conclusions are reached if thickness functions h = h(r) and h = h(t) are considered. As Manna [47] noted for only rotating disks, if we set $t = (r/R)^{2/q} = (n\rho)^{2/q}$ in the last relation (7.1), the family of disk profiles that can be analysed can be extended to include, as shown in Fig. 7.2, those with an inflection point, such as uniform strength disks, as well as those that have geometrical singularities at radius R (zero thickness or infinite thickness). The following treatment applies equally to the profile defined by a power of a linear function and a power of a nonlinear function. For the latter, in the relations written in terms of t, we will have $t = (r/R)^{2/q} = (n\rho)^{2/q}$ rather than $t = r/R = n\rho$, while in the relations written in terms of ρ, we will have $\rho = (1/n)t^{q/2}$ instead of $\rho = t/n$.

The terms converging (tapered is a synonym) and diverging are used here to express thickness variation from the axis (or the inner radius) to the outer

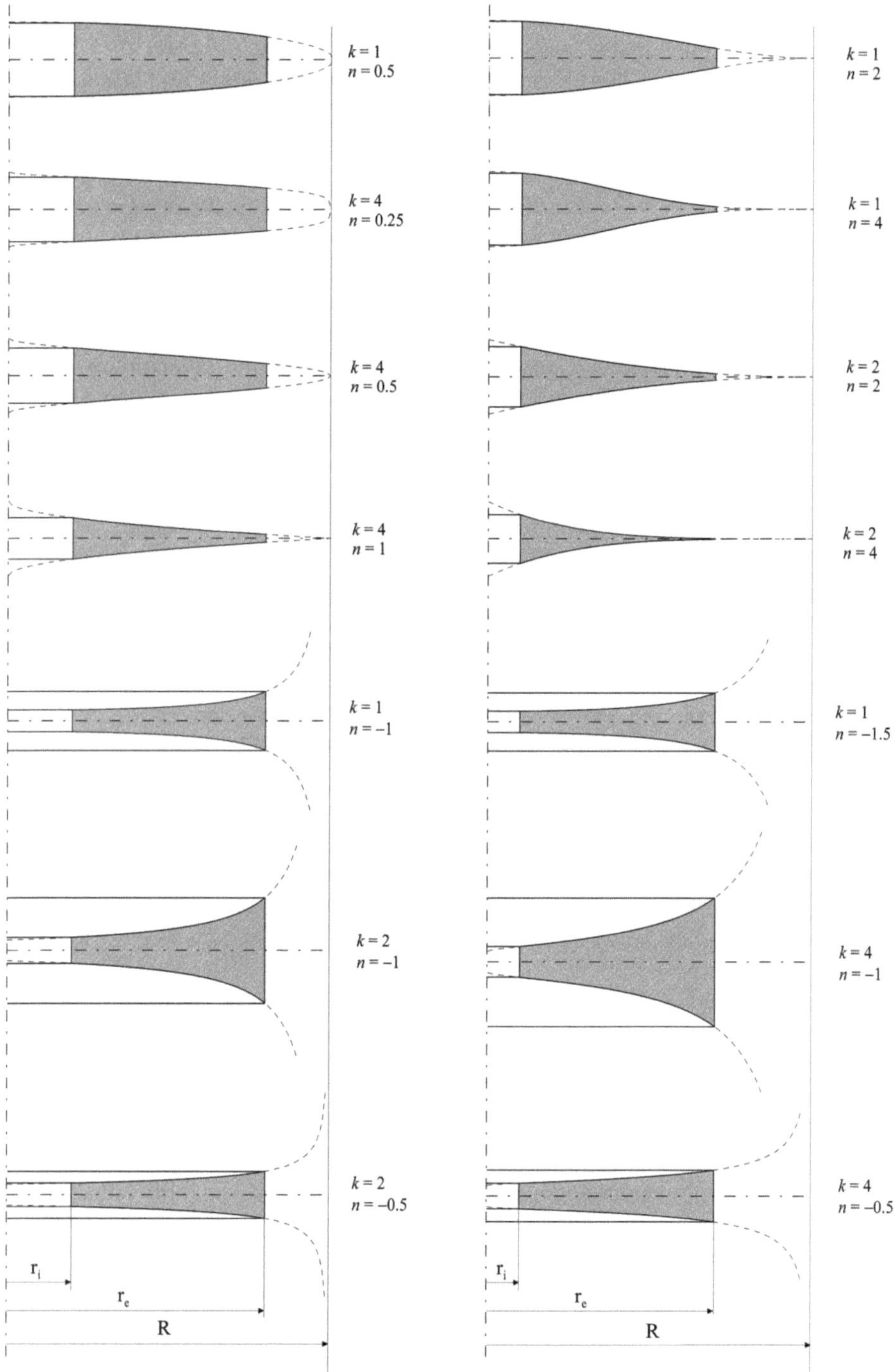

Fig. 7.2 Various profiles of solid or annular non-linear variable thickness disks according to (7.1) and with $t = (r/R)^{2/q} = (n\rho)^{2/q}$

radius. From the application standpoint, interest is limited to converging disks, given that diverging disks, with their unfavourable rotating mass distribution, involve high rotational inertia and equally unfavourable stress and strain states. For manufacturing and design reasons (and, above all, because of functional requirements for current turbine disks, where the axial dimensions of the surface coupling with the shaft must be limited in order to control vibration behaviour for any given distance between bearings), the ratio h_0/h_e must be lower than, or at the most equal to, 3 and thickness h_0 at the axis must not be higher than $(0.25 \div 0.30)r_e$. The outer radius r_e is also a design datum.

Accordingly, considering the values r_e, h_0 and h_e as design data, and selecting the exponent k so as to outline the actual disk profile as closely as possible, the value of n is univocally defined. In fact, $n = 1 - (h_e/h_0)^{1/k}$ at the outer radius, where $\rho = 1$ and $h = h_e$, given function $h = h(\rho)$. This shows that any higher value of n determines profiles of no physical meaning (see non-horizontal dashed profiles in Fig. 7.1). From an even more general viewpoint, by accepting slight variations for r_e, h_0 and h_e from design data, both parameters n and k may be varied, provided that the geometrical relationships between the quantities in relation (7.1) are respected.

As usual, dimensionless radius ρ will be taken into consideration in order to determine the disk size, given that r_e is a design datum. However, because the solution of the homogeneous differential equation entails hypergeometric series in terms of the variable $t = n{\cdot}\rho$, the more general dimensionless radius t will be considered for analytical development. Function $h = h(\rho)$ shows that the local value ρ_V of variable ρ at apex V (Fig. 7.1a, c, d) where, for converging disks, the two symmetrical surfaces of the profile meet, must necessarily be $\rho_V = R/r_e = 1/n$, given that at V the disk shows null thickness. Therefore, $R = r_e/n$ so that, shifting from variable ρ to variable t and considering that $n = m{\cdot}r_e$, then $m{\cdot}R = 1$, whereas the local value of t at radius r_e, where $\rho = 1$, will be $t = r_e/R = n$. Consequently, for converging profiles showing $0 \le h_e/h_0 \le 1$, domain definitions of n and t corresponding to $0 \le \rho \le 1$ are $0 \le n \le 1$ and $0 \le t \le n$ respectively. In this connection, it can be readily seen that, for $t = n$, then $\rho = 1$, whereas for $t = 1$, then $\rho = 1/n$ and $r = R$.

Figure 7.1b shows that a converging concave profile can be obtained from (7.1) by assuming negative values for both n and k. In this case, as will be seen later, of the two independent integrals of the associated homogeneous equation of the differential equation governing the displacement field, the first u_1 is defined in the interval $0 \le \rho \le 1/n$, whereas the second u_2 is defined in the interval $-1/n \le \rho \le 0$, so that neither is usable. Consequently, we will not consider disk profiles obtained from (7.1) showing negative values for n and k: values for which the radius R can only be defined conventionally. This, however, is not a limitation of the proposed procedure, given that, as shown in Fig. 7.2d, converging concave profiles can also be obtained with $0 < k < 1$ and $0 < n < 1$ respectively.

To deduce the differential equation that governs the general displacement field, whatever the configuration of these profiles of the non-linearly variable thickness disk, the expression of h given by (7.1) in terms of variable ρ, is introduced in (1.28) together with its first derivative with respect to ρ, taking into account that

$d/dr = (1/r_e) \cdot d/d\rho$, $d^2/dr^2 = (1/r_e^2) \cdot d^2/d\rho^2$, $dh/d\rho = -n \cdot h_0 \cdot k \cdot (1 - n \cdot \rho)^{k-1}$ and $(1/h) \cdot (dh/d\rho) = -n \cdot k/(1 - n \cdot \rho)$. We thus obtain the following general solving differential equation of the non-linearly variable thickness rotating disk according to relation (7.1), subjected to thermal load with the generic temperature variation function $T = T(r)$:

$$\frac{d^2u}{d\rho^2} + \left(\frac{1}{\rho} - \frac{n \cdot k}{1 - n \cdot \rho}\right) \cdot \frac{du}{d\rho} - \left[\frac{1}{\rho^2} + \frac{\nu \cdot n \cdot k}{\rho \cdot (1 - n \cdot \rho)}\right] \cdot u - (1 + \nu) \cdot \alpha \cdot r_e \cdot \left(\frac{dT}{d\rho} - \frac{n \cdot k \cdot T}{1 - n \cdot \rho}\right) + \frac{(1 - \nu^2) \cdot \gamma \cdot \omega^2 \cdot r_e^3 \cdot \rho}{E} = 0. \tag{7.3}$$

For cases in which the disk is subjected only to centrifugal load or only to thermal load, this equation specializes in the following relations:

$$\frac{d^2u}{d\rho^2} + \left(\frac{1}{\rho} - \frac{n \cdot k}{1 - n \cdot \rho}\right) \cdot \frac{du}{d\rho} - \left[\frac{1}{\rho^2} + \frac{\nu \cdot n \cdot k}{\rho \cdot (1 - n \cdot \rho)}\right] \cdot u + \frac{(1 - \nu^2) \cdot \gamma \cdot \omega^2 \cdot r_e^3 \cdot \rho}{E} = 0; \tag{7.4}$$

$$\frac{d^2u}{d\rho^2} + \left(\frac{1}{\rho} - \frac{n \cdot k}{1 - n \cdot \rho}\right) \cdot \frac{du}{d\rho} - \left[\frac{1}{\rho^2} + \frac{\nu \cdot n \cdot k}{\rho \cdot (1 - n \cdot \rho)}\right] \times u - (1 + \nu) \cdot \alpha \cdot r_e \cdot \left(\frac{dT}{d\rho} - \frac{n \cdot k \cdot T}{1 - n \cdot \rho}\right) = 0. \tag{7.5}$$

General solutions of the above three non-homogeneous differential equations are obtained by adding the solution of the common associated homogeneous equation to particular solutions of the same complete equations.

$$\frac{d^2u}{d\rho^2} + \left(\frac{1}{\rho} - \frac{n \cdot k}{1 - n \cdot \rho}\right) \cdot \frac{du}{d\rho} - \left[\frac{1}{\rho^2} + \frac{\nu \cdot n \cdot k}{\rho \cdot (1 - n \cdot \rho)}\right] \cdot u = 0. \tag{7.6}$$

Like (6.6), (7.6) is also a differential equation with coefficients that are analytical functions of the independent variable ρ; it can also be considered as a Fuchs class equation and, being characterized by three singular points ($\rho = 0$; $\rho = 1/n$, $\rho = \infty$), is the classic Gauss hypergeometric differential equation. Equation 7.6 makes it possible to treat both Stodola's hyperbolic disk and its generalization as particular cases. For the former, bearing in mind that the limits for $n \to \infty$ (here, we consider n in modulus) of ratios $-nk/(1 - n\rho)$ and $-\nu nk/[\rho(1 - n\rho)]$ appearing in (7.6) are k/ρ and $-\nu k/\rho^2$ respectively, and noting that $k \cong a$, it is clear that this equation reduces to the homogeneous equation associated with equations (4.4) and (4.7), as well as to (4.11). For the generalization of the hyperbolic disk, differential equation (7.6) must be rewritten in terms of $n = -1$ and $k = a$.

7.2 Rotating Disk Having Constant Density

As indicated above, the differential equation that governs the displacement field when the disk is subjected only to centrifugal load is (7.4), whose solution is the sum of the associated homogeneous equation, given by (7.6), and a particular integral of the complete equation.

7.2.1 *Particular Integral and Corresponding Stress and Strain State*

To obtain the particular integral of Eq. 7.4, it is best to start from (1.28) without thermal load. The following non-homogeneous differential equation is obtained by introducing the thickness function (7.1) in terms of the variable r in (1.28) together with its first derivative with respect to r, and introducing the notation (6.8):

$$\frac{d^2u}{dr^2} + \left(\frac{1}{r} - \frac{m \cdot k}{1 - m \cdot r}\right) \cdot \frac{du}{dr} - \left[\frac{1}{r^2} + \frac{\nu \cdot m \cdot k}{r \cdot (1 - m \cdot r)}\right] \cdot u = -C \cdot r. \qquad (7.7)$$

To obtain the particular integral u_p of differential equation (7.7), we put

$$u_p = c_1 \cdot r + c_2 \cdot r^2 + c_3 \cdot r^3, \qquad (7.8)$$

where c_1, c_2 and c_3 are constants that can be determined by substituting (7.8), along with its first and second derivatives with respect to variable r, in (7.7), and by equating to zero the coefficients of various powers in the variable r. We thus obtain:

$$c_1 = \frac{3 \cdot (3 + \nu) \cdot C}{m^2 \cdot (1 + \nu) \cdot [8 + k \cdot (3 + \nu)] \cdot [3 + k \cdot (2 + \nu)]};$$

$$c_2 = \frac{k \cdot (3 + \nu) \cdot C}{m \cdot [8 + k \cdot (3 + \nu)] \cdot [3 + k \cdot (2 + \nu)]}; \quad c_3 = -\frac{C}{8 + k \cdot (3 + \nu)}. \qquad (7.9)$$

By substituting (7.9) in (7.8), bearing in mind the notation (6.8), which expresses C and by changing from variable r to ρ, we obtain:

$$u_p = -\frac{(1 - \nu^2)}{[8 + k \cdot (3 + \nu)]} \cdot \rho \cdot \left\{\rho^2 - \frac{3 + \nu}{[3 + k \cdot (2 + \nu)]} \cdot \left[\frac{k \cdot \rho}{n} + \frac{3}{n^2 \cdot (1 + \nu)}\right]\right\} \cdot \gamma \cdot \frac{\omega^2 \cdot r_e^3}{E}. \qquad (7.10)$$

However, in the usual assumptions of small displacements and plane stress ($\sigma_z = 0$) made for disks (thickness small by respect to the outer radius r_e), radial and hoop stresses are given by relations (1.27). Consequently, the following stress state correlated to particular integral u_p is obtained by substituting (7.10) and its derivative $du_p/d\rho$, after changing from variable r to variable ρ, bearing in mind that $r = r_e \cdot \rho$ and $du_p/dr = (1/r_e) \cdot (du_p/d\rho)$ in relations (1.27), from which temperature terms are omitted:

$$\begin{cases} \sigma_r = -\gamma \cdot \omega^2 \cdot r_e^2 \cdot g_r = \sigma_0 \cdot g_r \\ \sigma_t = -\gamma \cdot \omega^2 \cdot r_e^2 \cdot g_t = \sigma_0 \cdot g_t, \end{cases} \tag{7.11}$$

where $\sigma_0 = \gamma \cdot \omega^2 \cdot r_e^2$ is the reference stress and g_r and g_t are dimensionless functions of ρ v, k and n, given respectively by:

$$\begin{cases} g_r = -\dfrac{1}{[8 + k \cdot (3 + v)]} \\ \quad \times \left[(3 + v) \cdot \rho^2 - \dfrac{(2 + v) \cdot (3 + v)}{[3 + k \cdot (2 + v)]} \cdot \dfrac{k}{n} \cdot \rho - \dfrac{3 \cdot (3 + v)}{[3 + k \cdot (2 + v)] \cdot n^2}\right] \\ g_t = -\dfrac{1}{[8 + k \cdot (3 + v)]} \\ \quad \times \left[(1 + 3v) \cdot \rho^2 - \dfrac{(1 + 2v) \cdot (3 + v)}{[3 + k \cdot (2 + v)]} \cdot \dfrac{k}{n} \cdot \rho - \dfrac{3 \cdot (3 + v)}{[3 + k \cdot (2 + v)] \cdot n^2}\right]. \end{cases} \tag{7.12}$$

It should be noted that relations (7.12), where t is substituted for ρ, are generalizations of (6.14), obtained for the conical disk, which is thus a specific case of the non-linearly variable-thickness disk. These relations can be used to obtain, for $n = k = 1$, relations for the converging conical disk, and for $n = -1$ and $k = 1$ relations for the diverging conical disk (Fig. 7.1a).

It can be concluded from (7.10) that the particular integral u_p, and thus its related stress state given by relations (7.11) and the resulting strain state given by (1.25), from which temperature terms are omitted (after substituting relations (7.11) in them, and bearing (7.12) in mind) largely depend on the disk's material, geometrical dimensions and rotation speed. For any given material, u_p is proportional to r_e^3 and to ω^2. It should be noted that, for a given disk of assigned geometry (n, k and r_e fixed) and for a specific angular velocity, the structure of (7.10) shows that the particular integral may be defined once and for all: in other words, once the disk geometry is established, u_p can be determined for any other angular velocity by means of a simple scale shift.

Figure 7.3 shows functions g_r and g_t versus ρ, for two steel disks ($v = 0.3$) with $h_0/h_e \cong 3$, one of which has a convex profile ($k = 0.5$ and $n = 0.89$), while the other has a concave profile ($k = 1.5$ and $n = 0.52$). Through σ_0, functions g_r and g_t are proportional to the radial and hoop stresses σ_r and σ_t related to particular integral u_p, which as will be demonstrated below has an inherent physical meaning:

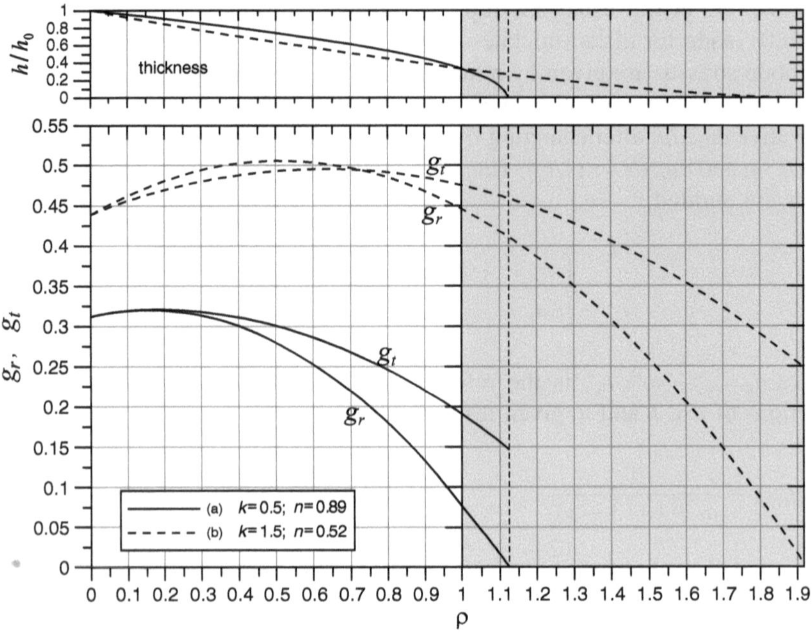

Fig. 7.3 Curves of g_r and g_t as a function of ρ within the interval $0 \leq \rho \leq 1/q$, for two steel disks ($v = 0.3$), with ratio $h_0/h_e \cong 3$ and with variable profile according to (7.1): (*a*) convex profile ($k = 0.5$ and $n = 0.89$); (*b*) concave profile ($k = 1.5$ and $n = 0.52$)

it represents radial displacement in a solid disk whose thickness varies according to a power of a linear function, with concave or convex profile, extended to its apex V where $r = R$ and $\rho = 1/n$. It should be noted that functions g_r and g_t assume finite values throughout the interval $0 \leq \rho \leq 1/n$ ($0 \leq t \leq 1$); only g_r becomes null at the right edge of the interval, i.e., for $\rho = 1/n$ ($t = 1$).

It should also be noted that relations (7.12) are only valid in the interval of the variable ρ showing physical meaning ($0 \leq \rho \leq 1$). To extend their validity from $\rho = 1$ to $\rho = 1/n$, to calculate functions g_r and g_t as well as related stresses, it would be necessary to reconsider the actual value of the outer radius which, in this case, shifts from r_e to R (this entails considering a scale factor equal to n^2). Clearly, the same consideration applies to calculating displacement u_p, as obtained by relation (7.10). For the sake of completeness, the curve of functions g_r and g_t in the interval $1 < \rho \leq 1/n$ is also shown in Figure 7.3, although said scale factor should be considered for calculation purposes. This interval is represented with a dashed line in the figure.

7.2.2 Solution of Homogeneous Differential Equation and Corresponding Stress and Strain State

To solve (7.6), which is the common homogeneous equation associated with the three complete differential equation (7.3), (7.4) and (7.5), it is best to introduce variable $t = n \cdot \rho$, whereby, as $d\rho = dt/n$ and $d\rho^2 = dt^2/n^2$, (7.6) becomes:

$$\frac{d^2u}{dt^2} + \left(\frac{1}{t} + \frac{k}{t-1}\right) \cdot \frac{du}{dt} + \left[\frac{1}{t} + (\nu \cdot k - 1)\right] \cdot \frac{u}{t \cdot (t-1)} = 0. \tag{7.13}$$

Like (6.15), this is a hypergeometric differential equation, showing three singular points ($t = 0$; $t = 1$; $t = \infty$) and corresponding to the general Papperitz form given by (6.16), again with:

$$\alpha + \alpha' + \beta + \beta' + \gamma + \gamma' = 1. \tag{7.14}$$

The following relations can be deduced from a comparison of (7.13) with (6.16), taking (7.14) into account:

$$\begin{gathered} 1 - \alpha - \alpha' = 1; \;\; 1 - \gamma - \gamma' = k; \;\; \beta \cdot \beta' = \nu \cdot k - 1; -\alpha \cdot \alpha' = 1; \\ \gamma \cdot \gamma' = 0; \;\; \beta + \beta' = k \end{gathered} \tag{7.15}$$

Consequently, for the case examined here we obtain:

$$\begin{gathered} \alpha = -1; \;\; \alpha' = 1; \;\; \gamma = 1 - k; \;\; \gamma' = 0; \;\; \beta = \frac{k}{2} + \xi; \;\; \beta' = \frac{k}{2} - \xi; \\ \text{with } \xi = \frac{1}{2} \cdot \sqrt{k^2 + 4 \cdot (1 - \nu \cdot k)}. \end{gathered} \tag{7.16}$$

According to the general theory mentioned in Sect. 6.2.2, (7.13), which is equivalent to (6.16), allows for two independent integrals, chosen from the four given by relations (6.20). However, in the case examined here, and taking (7.16) into account, relations (6.20) become:

$$\begin{aligned} u_1 &= t \cdot (1-t)^{1-k} \cdot F\left(2 - \frac{k}{2} + \xi, 2 - \frac{k}{2} - \xi, 3, t\right) \\ u_{1'} &= t \cdot F\left(1 + \frac{k}{2} + \xi, 1 + \frac{k}{2} - \xi, 3, t\right) \\ u_2 &= t^{-1} \cdot (1-t)^{1-k} \cdot F\left(-\frac{k}{2} + \xi, \, -\frac{k}{2} - \xi, \, -1, t\right) \\ u_{2'} &= t^{-1} \cdot F\left(-1 + \frac{k}{2} + \xi, \, -1 + \frac{k}{2} - \xi, \, -1, t\right). \end{aligned} \tag{7.17}$$

For convergence, the same considerations made for relations (6.21) also apply here.

In this context, it is necessary to select as the first integral of (7.13) the one deriving from the second relation (7.17), which is here designated u_1 and transcribed as:

$$u_1 = \quad t^{\alpha'} \cdot (1-t)^{\gamma} \cdot F(a,b,c,t) = t \cdot F\left(1 + \frac{k}{2} + \xi, 1 + \frac{k}{2} - \xi, 3, t\right) \tag{7.18}$$

where, for the sake of brevity, the following notation is used:

$$a = \alpha' + \beta + \gamma = 1 + \frac{k}{2} + \xi; \quad b = \alpha' + \beta' + \gamma = 1 + \frac{k}{2} - \xi; \quad c = 1 + \alpha' - \alpha = 3, \tag{7.19}$$

whereas $F(a, b, c, t)$ indicates the infinite hypergeometric series of the kind (6.24). This hypergeometric series slowly converges for $|\, t \,| = |\, n \cdot \rho \,| < 1$, which implies $0 < n < 1$. In other words, n is always lower than 1, as the actual disk profile mostly develops in the interval $0 \leq \rho \leq 1$. More specifically, the hypergeometric series $F(a, b, c, t)$ is divergent if the real part $\Re(c - a - b) \leq -1$, absolutely convergent if $\Re(c - a - b) > 0$ and, excluding the singularity point $t = 1$, conditionally convergent if $-1 < \Re(c - a - b) \leq 0$. Here, given that, a, b and c are real coefficients and $(c - a - b) = 1 - k$, for tapered disks, we have divergence for $k \geq 2$, absolute convergence for $0 < k < 1$ and conditional convergence for $1 \leq k < 2$. In the case of convergent disks, the range of variability of parameters n and k featured in (7.1) is thus limited, given that $0 < n < 1$, with $0 < k < 1$ for absolute convergence and $1 \leq n < 2$ for conditional convergence.

Given that the first derivative of the radial displacement $u = u(r)$ must be calculated in order to determine stress and strain states (in this regard, see (1.27) and (1.25)), it should be borne in mind that the first derivative with respect to variable t in series (6.24) is given by relation (6.26).

It is now necessary to find a second integral of differential equation (7.13) which is independent of that given by relation (7.18). In this respect, it should be noted that, here as earlier, the series appearing in the last two relations (7.17), featuring infinite terms from a certain value onwards, actually become finite when the interfering negative integer factors are freely increased by a small quantity ε, which, however, will need to be minimized by multiplying it by a suitable constant coefficient so as to prevent sensible terms from tending to infinity.

Therefore, given that relations (6.27), (6.28), (6.29) and (6.30) are still valid, developing the calculations with respect to the first and third of expressions (7.17) and with the necessary changes with respect to the conical disk (i.e., substituting notations (6.20) for the corresponding notations (6.23)), it can be shown that the following relations apply instead of relations (6.31):

$$\begin{cases} u_1^* = t \cdot F\left(1 + \frac{k}{2} + \xi, 1 + \frac{k}{2} - \xi, 3 - \varepsilon, t\right) \\ u_2^* = t^{-1+\varepsilon} \cdot F\left(-1 + \frac{k}{2} + \xi + \varepsilon, \, -1 + \frac{k}{2} - \xi + \varepsilon, \, -1 + \varepsilon, t\right) \\ \mathrm{m}(2,\varepsilon) = \dfrac{\varepsilon \cdot (\varepsilon - 1)}{\left(\frac{k}{2} + \xi + \varepsilon\right) \cdot \left(\frac{k}{2} - \xi + \varepsilon\right) \cdot \left(-1 + \frac{k}{2} + \xi + \varepsilon\right) \cdot \left(-1 + \frac{k}{2} - \xi + \varepsilon\right)} . \end{cases} \tag{7.20}$$

and that the second independent integral of (7.13) is given by relation:

$$u_2 = \frac{2}{\left(\frac{k}{2} + \xi\right) \cdot \left(\frac{k}{2} - \xi\right)} \cdot \left[1 - \frac{t^{-1}}{\left(-1 + \frac{k}{2} + \xi\right) \cdot \left(-1 + \frac{k}{2} - \xi\right)}\right] + u_1 \cdot \ln t + t \cdot \sum_{i=1}^{\infty} C_i \cdot t^i \tag{7.21}$$

with

$$C_i = \frac{\prod_{m=1}^{i} \left(1 + \frac{k}{2} + \xi + m - 1\right) \cdot \left(1 + \frac{k}{2} - \xi + m - 1\right)}{i! \cdot \prod_{m=1}^{i} (3 + m - 1)} \cdot \sum_{m=0}^{i-1} \left(\frac{1}{1 + \frac{k}{2} + \xi + m} + \frac{1}{1 + \frac{k}{2} - \xi + m} - \frac{1}{1 + m} - \frac{1}{3 + m}\right). \tag{7.22}$$

Thus, the second independent integral of the hypergeometric differential equation (7.13) is determined. Figure 7.4a, b show the two independent integrals u_1 and u_2 versus ρ within the interval $0 \leq \rho \leq 1/n$ for two convergent disks with profile according to (7.1), the first being a convex disk falling in the range of absolute convergence ($k = 0.5$ and $n = 0.89$), and the second a concave disk falling in the range of conditional convergence ($k = 1.5$ and $n = 0.52$). Here, too, it should be noted that, outside the interval where variable ρ has physical meaning, i.e., for $1 < \rho \leq 1/n$ (see Fig. 7.4a, b, to the right of the vertical dashed line, $\rho = 1$), the scale factor indicated in the previous section must be introduced for calculation purposes. In Fig. 7.4c, the curves of integrals u_1 and u_2 are shown for a concave converging disk for which both parameters n and k are negative ($k = -1$ and $n = -1$); in this case, integral u_2, by contrast with integral u_1, is not defined in

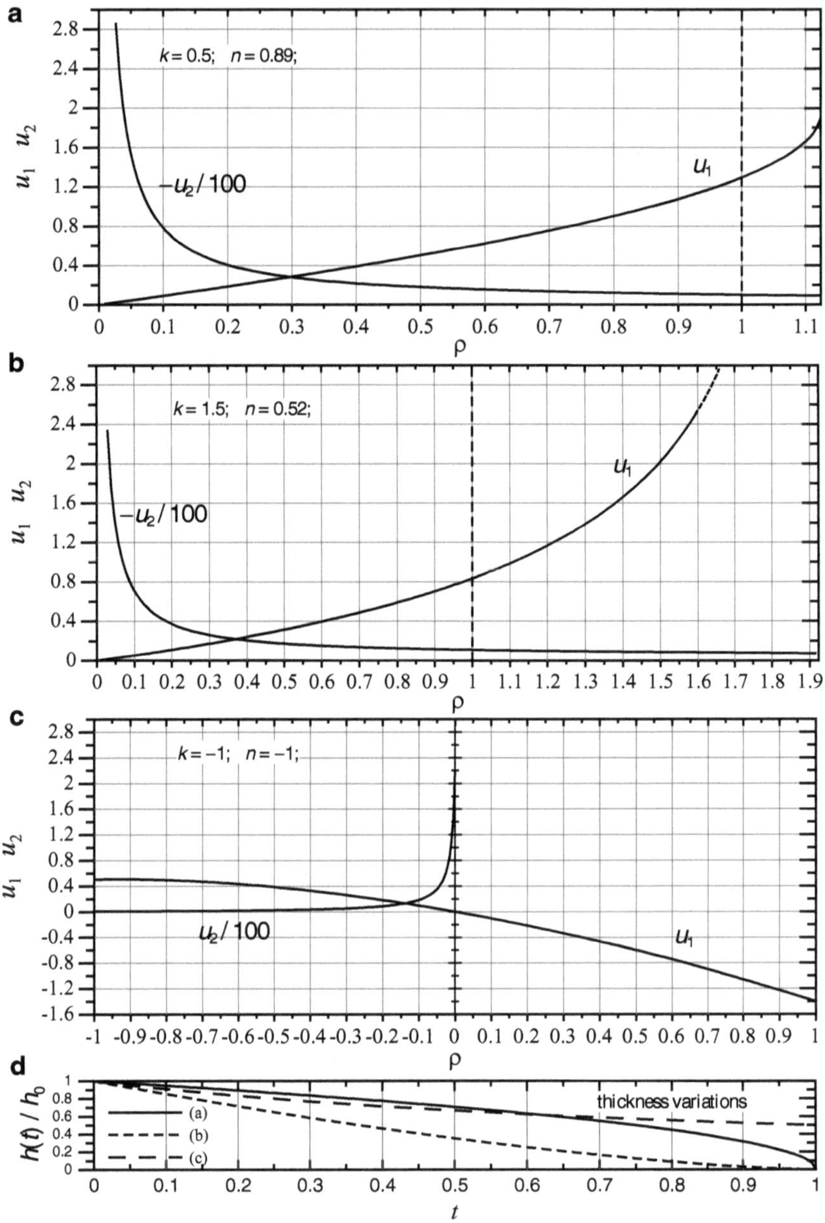

Fig. 7.4 Curves of u_1 and u_2 as a function of ρ, within the interval $0 \le \rho \le 1/n$, for three converging steel disks ($\nu = 0.3$) with profile varying according to (7.1): (**a**) convex profile ($k = 0.5$ and $n = 0.89$); (**b**) concave profile ($k = 1.5$ and $n = 0.52$); (**c**) concave profile ($k = -1$ and $n = -1$); (**d**) profiles of the three disk types

the interval $0 \leq \rho \leq 1/n$ considered here, so that it is not usable. However, this does not restrict the generality of this theoretical analysis, given that concave converging profiles (the same can be said for diverging disks, which are of no interest with regard to rotors) can be obtained with $0 < n < 1$ and $1 \leq k < 2$ (Fig. 7.4c), instead of with both n and k negative (Fig. 7.1b).

From Fig. 7.4a, b, as well as from the structure of relations (7.18) and (7.21), while considering (7.22), we can conclude that $u_1 \to \infty$ for $\rho \to 1/n$, whereas $u_2 \to \infty$ for $\rho \to 0$. It should also be noted that, if the profile at the apex V does not feature a pointed tip, as is the case for convex profiles (Fig. 7.1d), u_1 has a finite value for $\rho = 1/n$, i.e., there is no singularity.

Here, these integrals are used directly in the following developments, as was done for the conical disk (see Sect. 6.2.2). By using the integrals directly, the solution of the associated homogeneous equation (7.13) can be expressed by means of relation (6.34), where C_1 and C_2 are new integration constants that can be evaluated from the boundary conditions.

Subsequently, the following relations are obtained by replacing relation (6.34), written in terms of ρ, and its first derivative, in relations (1.27) from which temperature terms are omitted, and by considering that $d/dr = (1/r_e)d/d\rho$. These relations express σ_r and σ_t as a function of dimensionless radius ρ:

$$\begin{aligned}\sigma_r &= \frac{E}{r_e \cdot (1-\nu^2)} \cdot \left[C_1 \cdot \left(\frac{du_1}{d\rho} + \nu \cdot \frac{u_1}{\rho}\right) + C_2 \cdot \left(\frac{du_2}{d\rho} + \nu \cdot \frac{u_2}{\rho}\right)\right] \\ \sigma_t &= \frac{E}{r_e \cdot (1-\nu^2)} \cdot \left[C_1 \cdot \left(\frac{u_1}{\rho} + \nu \cdot \frac{du_1}{d\rho}\right) + C_2 \cdot \left(\frac{u_2}{\rho} + \nu \cdot \frac{du_2}{d\rho}\right)\right].\end{aligned} \tag{7.23}$$

These relations can be expressed in the compact form given by (6.36), where, however

$$a_r = \frac{du_1}{d\rho} + \nu \cdot \frac{u_1}{\rho},\ b_r = \frac{du_2}{d\rho} + \nu \cdot \frac{u_2}{\rho},\ a_t = \frac{u_1}{\rho} + \nu \cdot \frac{du_1}{d\rho} \text{ and } b_t = \frac{u_2}{\rho} + \nu \cdot \frac{du_2}{d\rho} \tag{7.24}$$

are functions of ρ, ν, k and n, whereas:

$$A = \frac{E \cdot C_1}{r_e \cdot (1-\nu)} \text{ e } B = \frac{E \cdot C_2}{r_e \cdot (1+\nu)} \tag{7.25}$$

are the dimensional integration constants, to be determined by imposing boundary conditions. Derivatives $du_1/d\rho$ and $du_2/d\rho$ are obtained by taking relation (6.26), written in terms of ρ, into consideration once functions u_1 and u_2 have been determined. Calculating functions a_r, b_r, a_t and b_t, each of which is related to different hypergeometric functions (u_1 or u_2), presents no difficulties.

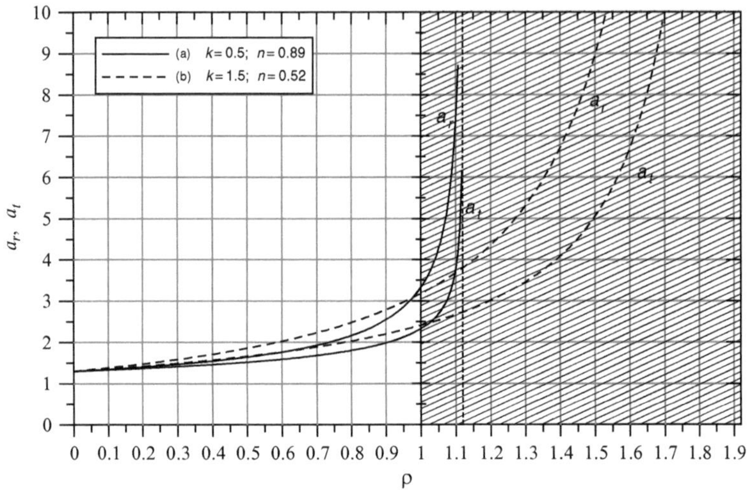

Fig. 7.5 Curves of a_r and a_t as a function of ρ within the interval $0 \leq \rho \leq 1/n$, for two steel disks ($\nu = 0.3$) with profile varying according to (7.1): (*a*) convex profile ($k = 0.5$ and $n = 0.89$); (*b*) concave profile ($k = 1.5$ and $n = 0.52$)

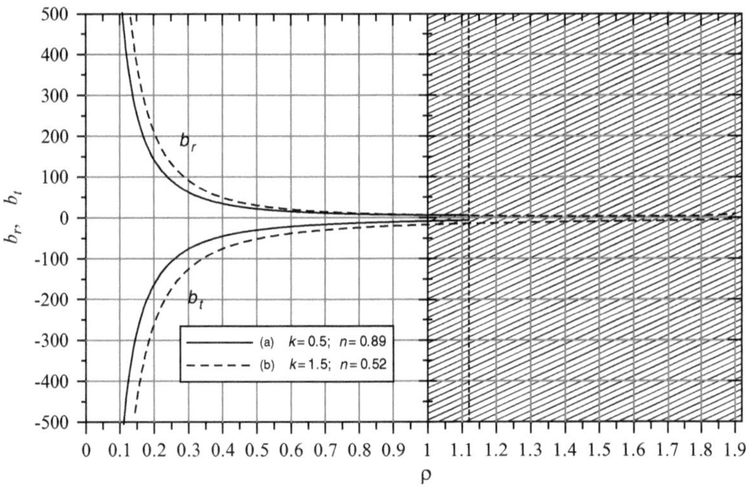

Fig. 7.6 Curves of b_r and b_t as a function of ρ within the interval $0 \leq \rho \leq 1/n$, for two steel disks ($\nu = 0.3$) with profile varying according to (7.1): (*a*) convex profile ($k = 0.5$ and $n = 0.89$); (*b*) concave profile ($k = 1.5$ and $n = 0.52$)

Figures 7.5 and 7.6 show these functions a_r, b_r, a_t and b_t versus dimensionless variable ρ, for both steel disks ($\nu = 0.3$), whose independent integrals u_1 and u_2 are shown in Fig. 7.4a, b respectively. These diagrams make it possible to calculate these functions rapidly for $0 < \rho < 1$, which is the range of variation of ρ for both converging disks of potential technical interest, which may be solid or annular, one being convex and the other concave. It should be noted that, with the exception of the convex profile, for which a_r, like u_1, has a finite value for $\rho = 1/n$, a_r and a_t tend to infinity for $\rho \to 1/n$ for all other cases of technical interest, given that $u_1 \to \infty$ at the outer edge of the pointed-tip disk ($r = R$, $\rho = 1/n$), whereas b_r and b_t, tend to infinity for $\rho \to 0$, i.e., at the disk axis, given that $u_2 \to \infty$ for $r \to 0$ ($\rho \to 0$).

The strain state related to the stress state given by relations (6.36), written in terms of ρ, is then obtained from (1.25) from which temperature terms are omitted. Radial displacement can then be determined immediately, as $u = r \cdot \varepsilon_t$.

Finally, it should be noted that all of the above relations are generalizations of those obtained for the conical disk, which is thus a specific case of the variable-thickness disk according to (7.1). These relations can be used to obtain, for $n = k = 1$, relations for the converging conical disk, and for $n = -1$ and $k = 1$, relations for the diverging conical disk.

7.2.3 General Integral and Corresponding Stress and Strain State

As was specified at the beginning of this paragraph, the general solution of the non-homogeneous differential equation (7.4) governing the displacement field of the rotating disk without thermal load is the sum of the solution of its associated homogeneous equation (7.6), given by (6.34), with u_1 and u_2 given respectively by (7.18) and (7.21), and of the particular integral u_p, previously obtained and given by (7.10). Thus, radial displacement u at the generic dimensionless radius t will be obtained from relation (6.39) written in terms of ρ.

By using the method of superposition in calculating total strains and stresses and by considering relations (7.23) as well as relations (7.11), where g_r and g_t are obtained from relations (7.12), we find that radial and hoop stresses in a rotating disk having variable profile described by means of a power of a linear function, and also subjected to surface force distribution at inner and outer radii, are still expressed by means of relations (6.40), where $\sigma_0 = \gamma \cdot \omega \cdot r_e^2$.

Here again, the strain distributions related to the stress distributions given by (6.40) are obtained from (1.25) from which temperature terms are omitted, after substituting (6.40) in them. We will then have $u = r \cdot \varepsilon_t$.

7.3 Non-Linearly Variable Thickness Disks Having Constant Density and Subjected to Thermal Load

We will now consider a variable thickness disk of constant density and subjected to thermal load only; this disk features a temperature gradient distribution along its radius expressed by relation (4.6) deriving from the function $T = T(r)$ given by (4.5). In this case, passing to variable ρ, the solving differential equation (7.5) becomes:

$$\frac{d^2u}{d\rho^2} + \left(\frac{1}{\rho} - \frac{n \cdot k}{1 - n \cdot \rho}\right) \cdot \frac{du}{d\rho} - \left[\frac{1}{\rho^2} + \frac{v \cdot n \cdot k}{\rho \cdot (1 - n \cdot \rho)}\right] \cdot u - \alpha \cdot (1 + v) \cdot r_e \cdot \left(\sum_{i=1}^{n} i \cdot k_i \cdot \rho^{i-1} - \frac{n \cdot k}{1 - n \cdot \rho} \sum_{i=1}^{n} k_i \cdot \rho^i\right) = 0. \tag{7.26}$$

The general integral of this equation is the sum of the solution of the homogeneous equation, which is already known, and of a particular integral of the non-homogeneous equation. Using the superposition method, the latter can be obtained as the linear combination of contributions deriving from component n terms $k_i \cdot \rho^i$ (with $i = 1, 2, \ldots, n$, as the constant term k_0 does not contribute to the stress state) of the function $T = T(r)$; thus, by considering the *i-th* term of this function equation (7.26) becomes:

$$\frac{d^2u}{d\rho^2} + \left(\frac{1}{\rho} - \frac{n \cdot k}{1 - n \cdot \rho}\right) \cdot \frac{du}{d\rho} - \left[\frac{1}{\rho^2} + \frac{v \cdot n \cdot k}{\rho \cdot (1 - n \cdot \rho)}\right] \cdot u + \alpha \cdot (1 + v) \cdot r_e \cdot k_i \cdot \frac{n \cdot (i + k) \cdot \rho^i - i \cdot \rho^{i-1}}{1 - n \cdot \rho} = 0. \tag{7.27}$$

To obtain a particular integral u'_p of differential eqaution (7.27), we first put

$$u'_p = a_{i+1} \cdot \rho^{i+1} + a_i \cdot \rho^i + a_{i-1} \cdot \rho^{i-1} + \ldots. , \tag{7.28}$$

where $a_{i+1}, a_i, a_{i-1}, \ldots.$ are constants; by substituting this relation, along with its first and second derivatives, in (7.27) and by equalling to zero the coefficients of various powers in the variable ρ, the following relations are obtained:

$$\begin{aligned} a_{i+1} &= \alpha \cdot r_e \cdot k_i \cdot \frac{(i + k) \cdot (1 + v)}{(i + 1) \cdot (i + 1 + k) - (1 - v \cdot k)} \\ a_i &= \alpha \cdot r_e \cdot k_i \cdot \frac{i \cdot k \cdot (1 - v^2)}{n \cdot [i \cdot (i + k) - (1 - v \cdot k)] \cdot [(i + 1) \cdot (i + 1 + k) - (1 - v \cdot k)]}, \end{aligned} \tag{7.29}$$

while for the subsequent coefficients a_{i-p}, with $p = 1, 2, \ldots,$ we can derive the general relation

$$a_{i-p} = a_{i-p+1} \cdot \frac{(i - p) \cdot (i - p + 2)}{n \cdot [(i - p) \cdot (i - p + k) - (1 - v \cdot k)]}. \tag{7.30}$$

If we take $i = p$ in relation (7.30), we obtain $a_0 = 0$. On the basis of the same relation, all coefficients after a_0 related to all negative powers of ρ are thus zero. Therefore, given i as any positive integer, the polynomial (7.28), within the first $(i + 1)$ terms starting from the linear term $a_1 \cdot \rho$, always represents a particular integral of (7.27).

By substituting the series expansion (7.28) and its first derivative in (1.27), while considering (7.29) and (7.30) as well as the *i-th* term of the function $T = T(r)$, i.e., $T = k_i \cdot \rho^i$, we obtain the following expressions of the corresponding stresses σ_r and σ_t as polynomials of grade i in ρ, with a constant, non-zero term, which are valid for any value of i:

$$\begin{cases} \sigma_r = E \cdot \alpha \cdot k_i \cdot \left(A_i \cdot \rho^i + A_{i-1} \cdot \rho^{i-1} + \ldots. + A_2 \cdot \rho^2 + A_1 \cdot \rho + A_0\right) \\ \quad = E \cdot \alpha \cdot k_i \cdot \varphi_i(\rho) \\ \sigma_t = E \cdot \alpha \cdot k_i \cdot \left(B_i \cdot \rho^i + B_{i-1} \cdot \rho^{i-1} + \ldots. + B_2 \cdot \rho^2 + B_1 \cdot \rho + B_0\right) \\ \quad = E \cdot \alpha \cdot k_i \cdot \psi_i(\rho). \end{cases} \tag{7.31}$$

The coefficients A and B appearing in expressions (7.31) are obtained from the following relations:

$$\begin{aligned} A_i &= -\frac{i}{(i+1) \cdot (i+1+k) - (1 - v \cdot k)} \\ A_{i-1} &= \frac{k \cdot i \cdot (i+v)}{n \cdot [(i+1) \cdot (i+1+k) - (1 - v \cdot k)] \cdot [i \cdot (i+k) - (1 - v \cdot k)]} \\ B_i &= -\frac{i \cdot (i+1+k)}{(i+1) \cdot (i+1+k) - (1 - v \cdot k)} \\ B_{i-1} &= \frac{k \cdot i \cdot (1 + v \cdot i)}{n \cdot [(i+1) \cdot (i+1+k) - (1 - v \cdot k)] \cdot [i \cdot (i+k) - (1 - v \cdot k)]}, \end{aligned} \tag{7.32}$$

while all the subsequent coefficients A_{i-1-p} and B_{i-1-p}, with $p = 1, 2, \ldots, (i - 1)$, are derived progressively through the relations:

$$\begin{aligned} A_{i-1-p} &= A_{i-p} \cdot \frac{(i-p+v)}{(i-p+1+v)} \cdot \frac{(i-p) \cdot (i-p+2)}{n \cdot [(i-p) \cdot (i-p+k) - (1 - v \cdot k)]} \\ B_{i-1-p} &= B_{i-p} \cdot \frac{1 + v \cdot (i-p)}{1 + v \cdot (i-p+1)} \cdot \frac{(i-p) \cdot (i-p+2)}{n \cdot [(i-p) \cdot (i-p+k) - (1 - v \cdot k)]}, \end{aligned} \tag{7.33}$$

These relations are obtained by means of simple proportions between the coefficients of the powers in ρ^{i-p-1} and ρ^{i-p}. Given that coefficients A_{i-1} and B_{i-1} are known, as directly defined from the second and the fourth expressions (7.32) respectively, relations (7.33) can be used to progressively determine all other coefficients present in (7.31). It can be readily seen that, for a specific value of i, $\sum_{i=0}^{n} A_i = 0$. These relations, in any case, clearly show that coefficients A and B appearing in polynomials $\varphi_i(\rho)$ and $\psi_i(\rho)$, are simply functions of i, v, k and n.

In general terms, for any value of i, the relations whereby functions $\varphi_i(\rho)$ and $\psi_i(\rho)$ appearing in expression (7.31) can be calculated are as follows

$$\varphi_i(\rho) = -\frac{i \cdot \rho^i}{i^2 + 2i + k \cdot (1 + i + v)} + \frac{i \cdot \prod_{j=2}^{i} (j^2 - 1)}{n^i \cdot \prod_{j=1}^{i} [j^2 + 2j + k \cdot (1 + j + v)]}$$
$$+ i \cdot \left[\prod_{j=2}^{i} (j^2 - 1)\right] \cdot \sum_{m=1}^{i-1} \frac{(m + 1 + v) \cdot k \cdot \rho^m}{n^{i-m} \cdot \left[\prod_{j=1}^{m} (j^2 + 2j)\right] \cdot \prod_{j=m}^{i} [j^2 + 2j + k \cdot (1 + j + v)]};$$

$$\psi_i(\rho) = -\frac{i \cdot (k + i + 1) \cdot \rho^i}{i^2 + 2i + k \cdot (1 + i + v)} + \frac{i \cdot \prod_{j=2}^{i} (j^2 - 1)}{n^i \cdot \prod_{j=1}^{i} [j^2 + 2j + k \cdot (1 + j + v)]}$$
$$+ i \cdot \left[\prod_{j=2}^{i} (j^2 - 1)\right] \cdot \sum_{m=1}^{i-1} \frac{[1 + (m + 1) \cdot v] \cdot k \cdot \rho^m}{n^{i-m} \cdot \left[\prod_{j=1}^{m} (j^2 + 2j)\right] \cdot \prod_{j=m}^{i} [j^2 + 2j + k \cdot (1 + j + v)]}. \tag{7.34}$$

These general relations derived by the authors [78] can be used to obtain several others of greater technical relevance for engineering calculations, viz.: $i = 1$ (linear variation of temperature with radius); $i = 2$ (temperature variation with radius according to a second-order function); $i = 3$ (temperature variation with radius according to a third-order function). Indeed, almost all functions $T = T(r)$ found in commonly used disks can be satisfactorily approximated with a third-order polynomial. The explicit relations whereby $\varphi_i(\rho)$ and $\psi_i(\rho)$ can be calculated for $i = 1, 2, 3$, are as follows

(a) For $i = 1$, $T = k_1 \cdot \rho$ and, hence:

$$\varphi_1(\rho) = \frac{1 - n \cdot \rho}{n \cdot [3 + k \cdot (2 + v)]}; \quad \psi_1(\rho) = \frac{1 - (2 + k) \cdot n \cdot \rho}{n \cdot [3 + k \cdot (2 + v)]}; \tag{7.35}$$

(b) For $i = 2$, $T = k_2 \cdot \rho^2$ and, hence:

$$\varphi_2(\rho) = \frac{2 \cdot \{-n^2\rho^2 \cdot [3 + k \cdot (2 + v)] + (2 + v) \cdot n \cdot k \cdot \rho + 3\}}{n^2 \cdot [8 + k \cdot (3 + v)] \cdot [3 + k \cdot (2 + v)]};$$
$$\psi_2(\rho) = \frac{2 \cdot \{-n^2\rho^2 \cdot (k + 3) \cdot [3 + k \cdot (2 + v)] + (1 + 2v) \cdot n \cdot k \cdot \rho + 3\}}{n^2 \cdot [8 + k \cdot (3 + v)] \cdot [3 + k \cdot (2 + v)]}; \tag{7.36}$$

(c) For $i = 3$, $T = k_3 \cdot \rho^3$ and, hence:

$$\begin{aligned}\varphi_3(\rho) &= \frac{3\cdot\{k\cdot n^2\rho^2\cdot(3+\nu)\cdot[3+k\cdot(2+\nu)]+8\cdot k\cdot n\cdot\rho\cdot(2+\nu)+24\}}{n^3\cdot[15+k\cdot(4+\nu)]\cdot[8+k\cdot(3+\nu)]\cdot[3+k\cdot(2+\nu)]} \\ &\quad -\frac{3\cdot\{n^3\rho^3\cdot[8+k\cdot(3+\nu)]\cdot[3+k\cdot(2+\nu)]\}}{n^3\cdot[15+k\cdot(4+\nu)]\cdot[8+k\cdot(3+\nu)]\cdot[3+k\cdot(2+\nu)]}; \\ \psi_3(\rho) &= \frac{3\cdot\{k\cdot n^2\rho^2\cdot(1+3\nu)\cdot[3+k\cdot(2+\nu)]+8\cdot k\cdot n\cdot\rho\cdot(1+2\nu)+24\}}{n^3\cdot[15+k\cdot(4+\nu)]\cdot[8+k\cdot(3+\nu)]\cdot[3+k\cdot(2+\nu)]} \\ &\quad -\frac{3\cdot\{n^3\rho^3\cdot(k+4)\cdot[8+k\cdot(3+\nu)]\cdot[3+k\cdot(2+\nu)]\}}{n^3\cdot[15+k\cdot(4+\nu)]\cdot[8+k\cdot(3+\nu)]\cdot[3+k\cdot(2+\nu)]}.\end{aligned} \tag{7.37}$$

Figure 7.7a, b show functions $\varphi_i(\rho)$ and $\psi_i(\rho)$ up to $i = 10$ versus dimensionless variable ρ for two steel disks ($\nu = 0.3$), one convex ($k = 0.5$ and $n = 0.89$), and the other concave ($k = 1.5$ and $n = 0.52$).

On the basis of the foregoing considerations and expressions (6.36) written in terms of ρ, the stress state can be expressed as follows where a non-zero temperature gradient is present along the radius given by relation (4.6):

$$\begin{cases}\sigma_r = A\cdot a_r + B\cdot b_r + E\cdot\alpha\cdot(k_1\cdot\varphi_1 + k_2\cdot\varphi_2 + k_3\cdot\varphi_3 + \dots.) \\ \quad = A\cdot a_r + B\cdot b_r + E\cdot\alpha\cdot\varphi(\rho) \\ \sigma_t = A\cdot a_t + B\cdot b_t + E\cdot\alpha\cdot(k_1\cdot\psi_1 + k_2\cdot\psi_2 + k_3\cdot\psi_3 + \dots.) \\ \quad = A\cdot a_t + B\cdot b_t + E\cdot\alpha\cdot\psi(\rho),\end{cases} \tag{7.38}$$

with $\varphi_i = \varphi_i(\rho)$ and $\psi_i = \psi_i(\rho)$.

Here again, the strain state associated with the stress state given by relations (7.38) is obtained from (1.25), after substituting (7.38) in them; we will then have $u = r\cdot\varepsilon_t$.

Here as elsewhere, it should be noted that all of the above relations for thermal load are generalizations of those obtained for the conical disk, which is thus a specific case of the variable-thickness disk according to (7.1). Written in terms of variable t, these relations can be used to obtain, for $n = k = 1$, relations for the converging conical disk, and for $n = -1$ and $k = 1$ relations for the diverging conical disk.

7.4 Non-Linearly Variable Thickness Disks Having Density Variation on Radius

We will now consider a disk having variable profile described by means of a power of a linear function, subjected to centrifugal load and featuring density variation along its radius expressed by the polynomial relation:

$$\gamma = \gamma_0 + \gamma_1\cdot\rho + \gamma_2\cdot\rho^2 + \dots = \sum_{i=0}^{n}\gamma_i\cdot\rho^i \tag{7.39}$$

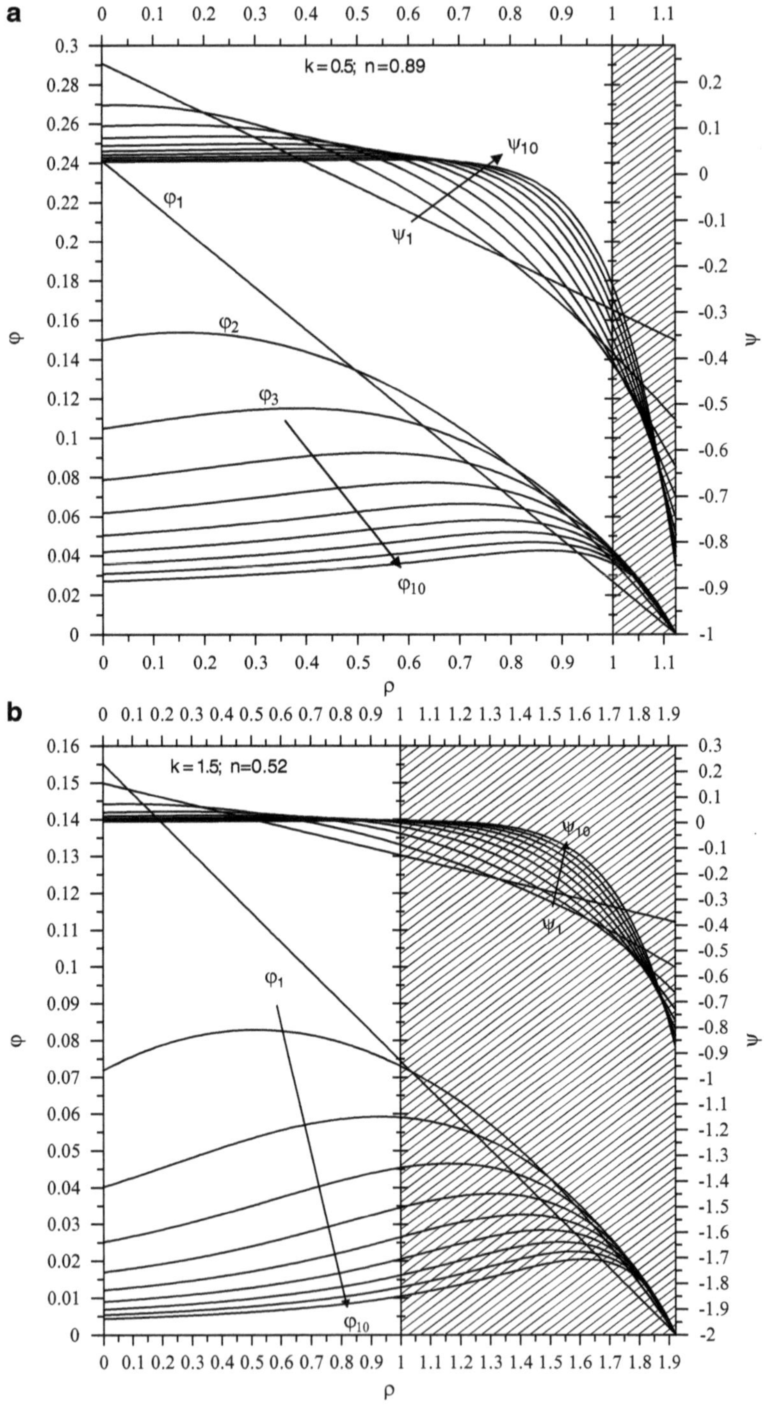

Fig. 7.7 Distribution of $\varphi_i(\rho)$ and $\psi_i(\rho)$, as functions of ρ, within the interval $0 \leq \rho \leq 1/n$, up to $i = 10$, for two steel disks ($\nu = 0.3$) having variable thickness according to (7.1): (**a**) convex profile ($k = 0.5$ and $n = 0.89$); (**b**) concave profile ($k = 1.5$ and $n = 0.52$)

in whose series γ_0 is the density of the rotor material, while $\gamma_1, \gamma_2, \ldots$ are constants, and $i = 1, 2, \ldots, n$. Function (7.39) is similar to (3.44), except that the dimensionless variable ρ takes the place of the radial coordinate r. Thus, by considering (7.39) and by applying it to solve differential equation (7.4), we obtain:

$$\frac{d^2u}{d\rho^2} + \left(\frac{1}{\rho} - \frac{n \cdot k}{1 - n \cdot \rho}\right) \cdot \frac{du}{d\rho} - \left[\frac{1}{\rho^2} + \frac{\nu \cdot n \cdot k}{\rho \cdot (1 - n \cdot \rho)}\right] \cdot u$$
$$+ \left(1 - \nu^2\right) \cdot \frac{\omega^2 \cdot r_e^3 \cdot \rho}{E} \cdot \sum_{i=0}^{n} \gamma_i \cdot \rho^i = 0. \quad (7.40)$$

In this case too, the general integral of (7.40) is the sum of the solution of the homogeneous equation, which is already known, and of a particular integral of the non-homogeneous equation. The latter can be obtained as the linear combination of contributions deriving from component n terms $\gamma_i \cdot \rho^i$ (with $i = 0, 1, 2, \ldots, n$) of the function $\gamma = \gamma\ (\rho)$; thus, by considering the *i-th* term of this function, (7.40) becomes:

$$\frac{d^2u}{d\rho^2} + \left(\frac{1}{\rho} - \frac{n \cdot k}{1 - n \cdot \rho}\right) \cdot \frac{du}{d\rho} - \left[\frac{1}{\rho^2} + \frac{\nu \cdot n \cdot k}{\rho \cdot (1 - n \cdot \rho)}\right] \cdot u$$
$$+ \left(1 - \nu^2\right) \cdot \frac{\omega^2 \cdot r_e^3}{E} \cdot \gamma_i \cdot \rho^{i+1} = 0. \quad (7.41)$$

Relation (7.41), with $i = 0$, can be used to calculate the displacement $u = u(\rho)$ related to density γ_0 of the non-linearly profile disk's basic material; the same equation, with $i = 1, 2, \ldots, n$, makes it possible to determine the contributions related to the fictitious variation of density along the radius deriving from radial blades on lateral surfaces.

The particular integrals of differential equation (7.41) can be obtained using the same procedure as that adopted for equations (7.26) and (7.27) governing thermal loading. For the sake of brevity, the analytical developments will not be specified here. Through these developments, the stress state for a non-linearly variable thickness disk with a fictitious density variation along its radius is expressed as the sum of the partial contributions of the particular integrals, and thus in the following form:

$$\begin{cases} \sigma_r = \omega^2 \cdot r_e^2 \cdot (\gamma_0 \cdot g_r + \gamma_1 \cdot \eta_1 + \gamma_2 \cdot \eta_2 + \ldots) \\ \sigma_t = \omega^2 \cdot r_e^2 \cdot (\gamma_0 \cdot g_t + \gamma_1 \cdot \zeta_1 + \gamma_2 \cdot \zeta_2 + \ldots), \end{cases} \quad (7.42)$$

where $g_r, g_t, \eta_1, \zeta_1, \eta_2, \zeta_2, \ldots$ are functions of t, ν, k and n. Functions g_r and g_t related to the particular integral in (7.41) with $i = 0$ are obtained from relations (7.12).

In general terms, for any value of i, the relations whereby functions $\eta_i(\rho)$ and $\zeta_i(\rho)$ appearing in expressions (7.42) can be calculated are as follows:

$$\eta_i(\rho) = -\frac{(i+3+v)\cdot\rho^{i+2}}{(i+4)\cdot(i+2)+k\cdot(i+3+v)} + n^{-(i+2)}\cdot\frac{(i+3+v)\cdot\prod_{j=1}^{i+1}(j^2+2j)}{\prod_{j=1}^{i+2}\left[j^2+2j+k\cdot(1+j+v)\right]}$$
$$+(i+3+v)\cdot\sum_{m=1}^{i+1}\frac{(m+1+v)\cdot k\cdot\rho^m\cdot n^{m-i-2}\cdot\prod_{j=m+1}^{i+1}(j^2+2j)}{\prod_{j=m}^{i+2}\left[j^2+2j+k\cdot(1+j+v)\right]};$$

$$\zeta_i(\rho) = -\frac{[1+(i+3)\cdot v]\cdot\rho^{i+2}}{(i+4)\cdot(i+2)+k\cdot(i+3+v)} + n^{-(i+2)}\cdot\frac{(i+3+v)\cdot\prod_{j=1}^{i+1}(j^2+2j)}{\prod_{j=1}^{i+2}\left[j^2+2j+k\cdot(1+j+v)\right]}$$
$$+(i+3+v)\cdot\sum_{m=1}^{i+1}\frac{[1+(m+1)\cdot v]\cdot k\cdot\rho^m\cdot n^{m-i-2}\cdot\prod_{j=m+1}^{i+1}(j^2+2j)}{\prod_{j=m}^{i+2}\left[j^2+2j+k\cdot(1+j+v)\right]}. \tag{7.43}$$

These general relations, which were also derived by the authors [78], can be used to obtain several others of greater technical relevance for engineering calculations, viz.: $i = 1$, $i = 2$ and $i = 3$. Indeed, almost all functions $\gamma = \gamma\ (\rho)$ found in commonly used disks can be satisfactorily approximated with a third-order polynomial.

The explicit relations whereby $\eta_i(\rho)$ and $\zeta_i(\rho)$ can be calculated for $i = 1$ are as follows (explicit expressions of $\eta_2(\rho)$, $\zeta_2(\rho)$ and $\eta_3(\rho)$, $\zeta_3(\rho)$ are omitted because they are very long):

(a) For $i = 1$, $\gamma = \gamma_I\cdot\rho$ and, hence:

$$\eta_1(\rho) = \frac{(4+v)\cdot\left\{(3+v)\cdot[3+k\cdot(2+v)]\cdot k\cdot n^2\rho^2+8\cdot(2+v)\cdot k\cdot n\cdot\rho+24\right\}}{n^3\cdot[15+k\cdot(4+v)]\cdot[8+k\cdot(3+v)]\cdot[3+k\cdot(2+v)]}$$
$$-\frac{(4+v)\cdot\left\{[8+k\cdot(3+v)]\cdot[3+k\cdot(2+v)]\cdot n^3\rho^3\right\}}{n^3\cdot[15+k\cdot(4+v)]\cdot[8+k\cdot(3+v)]\cdot[3+k\cdot(2+v)]};$$
$$\zeta_1(\rho) = \frac{-(1+4v)\cdot[8+k\cdot(3+v)]\cdot[3+k\cdot(2+v)]\cdot n^3\rho^3}{n^3\cdot[15+k\cdot(4+v)]\cdot[8+k\cdot(3+v)]\cdot[3+k\cdot(2+v)]}$$
$$+\frac{(4+v)\cdot\left\{(1+3v)\cdot[3+k\cdot(2+v)]\cdot k\cdot n^2\rho^2+8\cdot(1+2v)\cdot k\cdot n\cdot\rho+24\right\}}{n^3\cdot[15+k\cdot(4+v)]\cdot[8+k\cdot(3+v)]\cdot[3+k\cdot(2+v)]} \tag{7.44}$$

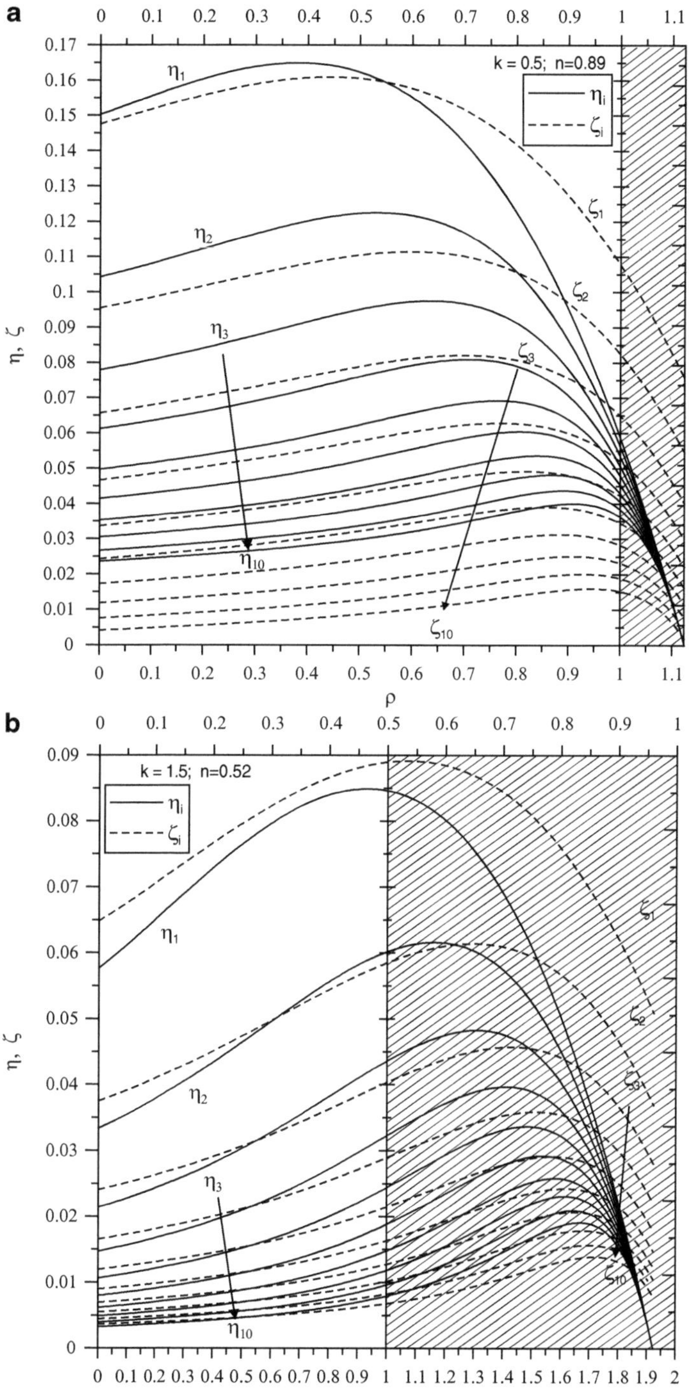

Fig. 7.8 Distribution of $\eta_i(\rho)$ and $\zeta_i(\rho)$, as functions of ρ, within the interval $0 \leq \rho \leq 1/n$, up to $i = 10$, for two steel disks ($\nu = 0.3$) having variable thickness according to (7.1): (**a**) convex profile ($k = 0.5$ and $n = 0.89$); (**b**) concave profile ($k = 1.5$ and $n = 0.52$)

Figure 7.8a, b show functions $\eta_i(\rho)$ and $\zeta_i(\rho)$ up to $i = 10$ versus dimensionless variable ρ for two steel disks ($\nu = 0.3$), one convex ($k = 0.5$ and $n = 0.89$), and the other concave ($k = 1.5$ and $n = 0.52$).

On the basis of (6.40), with $\sigma_0 = \gamma \cdot \omega^2 \cdot {r_e}^2$, and a_r, b_r, a_t and b_t given by (6.26), and (7.42), the stress state in a rotating disk having variable thickness according to a power of a linear function and featuring a fictitious density variation along the radius can be expressed as follows:

$$\begin{aligned}\sigma_r &= A \cdot a_r + B \cdot b_r + \omega^2 \cdot r_e^2 \cdot (\gamma_0 \cdot g_r + \gamma_1 \cdot \eta_1 + \gamma_2 \cdot \eta_2 + \gamma_3 \cdot \eta_3 + \ldots)\\ \sigma_t &= A \cdot a_t + B \cdot b_t + \omega^2 \cdot r_e^2 \cdot (\gamma_0 \cdot g_t + \gamma_1 \cdot \zeta_2 + \gamma_2 \cdot \zeta_2 + \gamma_3 \cdot \zeta_3 + \ldots).\end{aligned} \quad (7.45)$$

Here again, the strain state associated with the above stress state is obtained from (1.25) from which temperature terms are omitted, after substituting relations (7.45) in them; we will then have $u = r \cdot \varepsilon_t$.

Likewise, it should be noted that all of the above relations for density variation along the radius are generalizations of those obtained for the conical disk, which is thus a specific case of the variable-thickness disk according to (7.1). Written in terms of variable t, these relations can be used to obtain, for $n = k = 1$, relations for the converging conical disk, and for $n = -1$ and $k = 1$ relations for the diverging conical disk.

7.5 Non-Linearly Variable Thickness Disks with Density Variation and Subjected to Thermal and Centrifugal Loads

Lastly, we will consider a non-linearly variable thickness rotating disk subjected to thermal load and having density variation along its radius. It will be assumed that the functions $T = T(r)$ and $\gamma = \gamma(r)$ are the polynomials given by (4.5) and (7.39) respectively. Since the principle of superposition applies in the linear elastic field, it is obvious that, taking (7.31) and (7.38) into account, the stress state will be given by the relations:

$$\begin{cases}\sigma_r = A \cdot a_r + B \cdot b_r + \omega^2 \cdot r_e^2 \cdot (\gamma_0 \cdot g_r + \gamma_1 \cdot \eta_1 + \gamma_2 \cdot \eta_2 + \gamma_3 \cdot \eta_3 + \ldots)\\ \qquad + E \cdot \alpha \cdot \varphi(\rho)\\ \sigma_t = A \cdot a_t + B \cdot b_t + \omega^2 \cdot r_e^2 \cdot (\gamma_0 \cdot g_t + \gamma_1 \cdot \zeta_2 + \gamma_2 \cdot \zeta_2 + \gamma_3 \cdot \zeta_3 + \ldots)\\ \qquad + E \cdot \alpha \cdot \psi(\rho),\end{cases} \quad (7.46)$$

while the associated strain state will be given by (1.25) after substituting the above relations in them. Radial displacement, on the other hand, will be derived from the usual relation $u = r \cdot \varepsilon_t$.

7.6 Examples of Non-Linear Variable Thickness Disks

A number of numerical examples will be given below which illustrate how to proceed in calculating the stress and strain states in steel disks ($\nu = 0.3$; $E = 204$ GPa, $\gamma = 7{,}800$ kg/m^3, $\alpha = 12{\cdot}10^{-6}$ °C^{-1}) with thickness varying according to relation (7.1) and having profiles of interest for actual applications, with the exception of those discussed in Sects. 7.6.1 and 7.6.2, which are of conceptual interest: the first makes it possible to give physical meaning to the particular integral u_p.

Here again, the results for these examples are compared with those obtained using finite element models. All of the considerations presented in Sect. 6.6 apply in this connection.

7.6.1 *Rotating Solid Disk with Apex Singularity and Having Constant Density*

This type of disk, although of little technical interest because of its pointed tip at outer radius $r = r_e = R$ deserves attention inasmuch as it makes it possible to assign inherent physical significance to particular integral u_p, which directly represents radial displacement in the solid disk having non-linearly variable thickness, extended to its apex R. Indeed, as was indicated earlier, given that u_1 tends to infinity at the disk apex ($r = R, t = 1$), whereas u_2 tends to infinity at the disk axis ($r = 0, t = 0$), the constants appearing in (6.39) must be zero in order to have finite values for displacement u. In other words, it is necessary that $C_1 = C_2 = 0$.

In this connection, it should be pointed out that relation (7.8), expressing u_p as a third-order polynomial in r, and thus relations (7.12), expressing functions g_r and g_t, already fulfil the boundary conditions:

$$\begin{cases} \sigma_r = \sigma_t & \text{for } r = 0\ (\rho = 0) \\ \sigma_r = 0 & \text{for } r = R\ (\rho = 1). \end{cases} \tag{7.47}$$

Figure 7.9a, b show stress-distribution curves σ_r and σ_t and displacement-distribution curve $u = u_p$ as functions of ρ in two solid steel disks ($\nu = 0.3$; $E = 204$ GPa, $\gamma = 7{,}800$ kg/m^3), extended to the apex radius $r_e = R$ ($R = 1$ m), both rotating at angular velocity $\omega = 314$ rad/s, the first showing a convex profile ($k = 0.5$ and $n = 1$), the second a concave profile ($k = 1.5$ and $n = 1$).

The figures indicate that: (1) the curves for σ_r and σ_t follow those for functions g_r and g_t, differing for σ_0; (2) stresses σ_r exceed stresses σ_t, locally though by a small amount; (3) the maximum stress values in the concave disk, which has less mass, are significantly below those occurring, all other conditions remaining equal, in the convex disk.

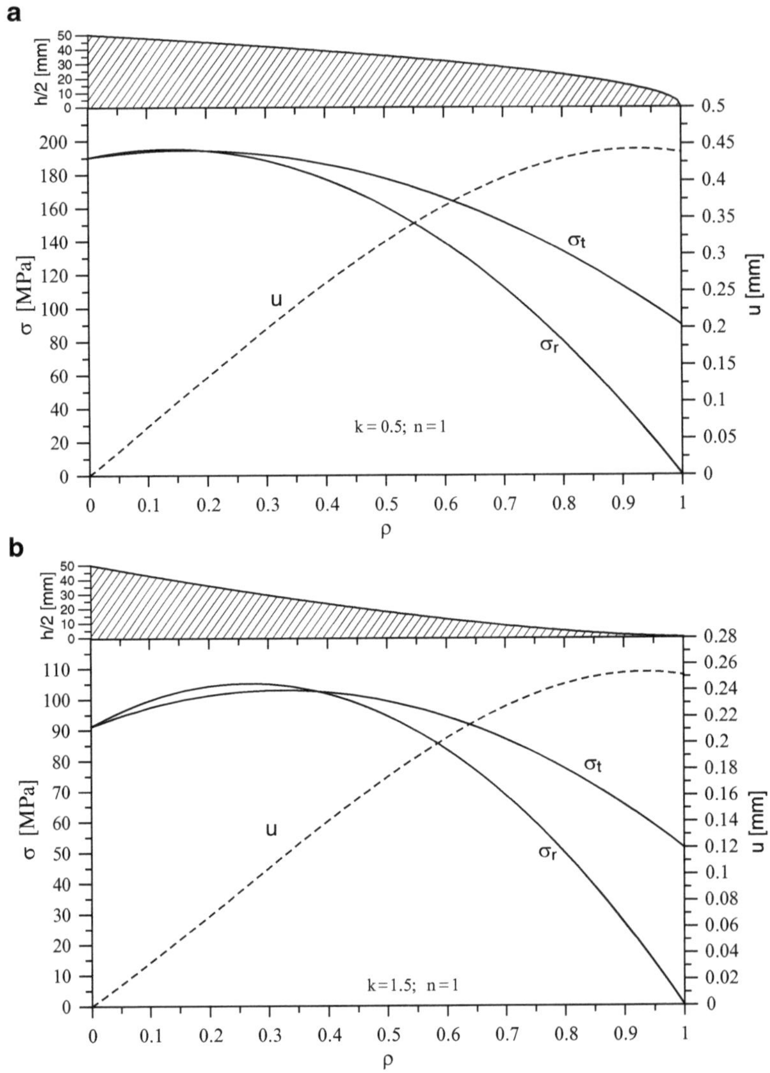

Fig. 7.9 Stress-distribution curves σ_r, σ_t and displacement-distribution curve $u = u_p$ in two solid rotating steel disks ($\nu = 0.3$) with profile variable according to (7.1) and featuring apex singularity and constant density: (**a**) convex profile ($k = 0.5$ and $n = 1$); (**b**) concave profile ($k = 1.5$ and $n = 1$)

7.6.2 Rotating Annular Disk with Apex Singularity and Having Constant Density

If the two rotating disks of the type considered in the preceding example are annular (this is another case of little design interest), the constant C_1 appearing in (6.39) must be zero in order to have a finite value for radial displacement u at the outer edge, where $r = R$ ($t = 1$); at the inner edge, given that the disks are annular, $t \neq 0$, and we will thus have $C_2 \neq 0$, as there is no singularity correlated with u_2 for $t = 0$. The relation whereby the displacement field can be calculated in this case is (6.39) from which term u_1 is omitted. The two boundary conditions are as follows:

$$\begin{cases} \sigma_r = \sigma_{ri} & \text{for } r = r_i \ (\rho = \beta) \\ \sigma_r = 0 & \text{for } r = R \ (\rho = 1). \end{cases} \tag{7.48}$$

As was shown, the second boundary condition is already fulfilled by u_p and u_2, while integration constant C_2 is obtained from the first boundary condition. Supposing that $\sigma_{ri} = 0$, i.e., that the disk inner surfaces are not loaded, the value of constant C_2 is obtained by equalling to zero (for $r = r_i$) terms in square brackets appearing in the first of (1.27), from which temperature terms are omitted, and by introducing $u = C_2 \cdot u_2 + u_p$.

As this is a case of little design interest, the developments have little to add from the conceptual standpoint and will be left to the reader.

7.6.3 Rotating Solid Disks with $r_e < R$ and Having Constant Density

In this type of disk, as u_2 tends to infinity at the disk axis ($r = 0$, $\rho = 0$), constant C_2 appearing in (6.39) must be zero in order to obtain a finite value of radial displacement u at this radius. Moreover, given that the disk is of the frustum type ($r_e < R$), $\rho = 1 < 1/n$ at the disk outer radius, and as there is no u_1-related singularity for $\rho = 1/n$, then $C_1 \neq 0$. The relation whereby the displacement field can be calculated in this case is (6.39) from which term u_2 is omitted. The boundary conditions are as follows:

$$\begin{cases} \sigma_r = \sigma_t & \text{for } r = 0 \ (\rho = 0). \\ \sigma_r = \sigma_{re} & \text{for } r = r_e \ (\rho = 1). \end{cases} \tag{7.49}$$

The first boundary condition is already fulfilled by u_p and, as was shown, by u_1, while integration constant C_1 is obtained from the second boundary condition. Supposing that $\sigma_{re} = 0$, i.e., that the disk outer surface is not loaded, the value of constant C_1 is obtained by equalling to zero (for $r = r_e$) terms in square brackets appearing in the first of (1.27), from which temperature terms are omitted, and by introducing $u = C_1 \cdot u_1 + u_p$.

Figure 7.10a, b show stress-distribution curves σ_r and σ_t and displacement-distribution curve u as functions of ρ in two solid steel disks ($\nu = 0.3$; $E = 204$ GPa, $\gamma = 7{,}800$ kg/m^3), extended to outer radius $r_e = 0.8$ m, thickness at axis $h_0 = 0.1$ m and thickness at periphery $h_e = h_0/3$, no loaded at outer radius ($\sigma_{re} = 0$), both rotating at angular velocity $\omega = 314$ rad/s, though the first shows a convex profile ($k = 0.5$ and $q = 0.89$), and the second shows a concave profile

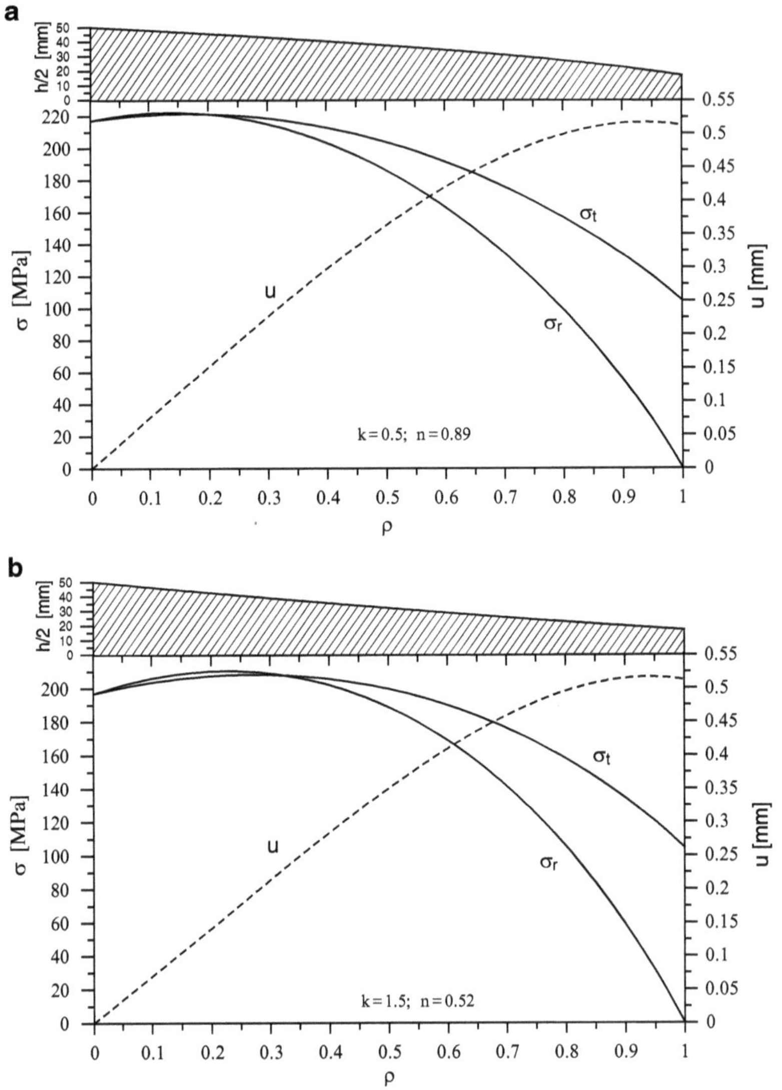

Fig. 7.10 Stress-distribution curves σ_r, σ_t and displacement-distribution curve u in two solid rotating steel disks ($\nu = 0.3$) having variable profile according to (7.1) with constant density and $r_e < R$: (**a**) convex profile ($k = 0.5$ and $n = 0.89$); (**b**) concave profile ($k = 1.5$ and n $= 0.52$)

($k = 1.5$ and $q = 0.52$). It is clear from a comparison of the two diagrams that, for actual disks featuring a finite thickness at the crown ring boundary (where $\rho = 1$), the differences between the concave and the convex profile are much less pronounced, and also that, as regards the stress state, the concave profile is slightly preferable to the convex profile.

7.6.4 Rotating Annular Disks with $r_e < R$ and Having Constant Density

As the majority of disks used in current machinery applications feature a central hole of inner radius $r_i > 0$ and an outer radius $r_e < R$, problems deriving from two singularity points $t = 0$ and $t = 1$ do not apply. In this case, radial displacement u is obtained from (6.39) complete with all terms. As there are no singularity points, it is advisable to calculate the stress field by using relations (6.40), determining integration constants A and B by establishing:

$$\begin{cases} \sigma_r = \sigma_{ri} & \text{for } \rho = \beta \\ \sigma_r = \sigma_{re} & \text{for } \rho = 1. \end{cases} \tag{7.50}$$

Constants C_1 and C_2 in (6.39) can be determined by using relations (7.25).

Figure 7.11a, b show stress-distribution curves σ_r and σ_t and displacement-distribution curve u as functions of ρ in two steel annular disks ($\nu = 0.3$) with $r_i = 0.1$ m, $r_e = 0.8$ m, $h_0 = 0.1$ m, and $h_e = h_0/3$, both rotating at angular velocity $\omega = 314$ rad/s and not loaded either at outer or inner radius ($\sigma_{ri} = 0$; $\sigma_{re} = 0$), the first showing a convex profile ($k = 0.5$ and $n = 0.89$), and the second a concave profile ($k = 1.5$ and $n = 0.52$). Here again, a comparison of the diagrams leads to the same conclusions reached in the preceding paragraph.

7.6.5 Rotating Disks with Hub and Crown Ring and Having Constant Density

If, as often occurs in machinery applications (Fig. 7.12), the disk features a crown ring and a hub, both of which are considered of constant thickness, it is first necessary to determine radial stresses $\sigma_{r,A}$ and $\sigma_{r,B}$ present in section A ($r = r_i$) and in section B ($r = r_e$) respectively, and which constitute two unknown hyperstatic values.

By imposing boundary conditions on the disk, constants A and B are first determined from the first of relations (6.40), and are expressed as functions of unknown values $\sigma_{r,A}$ and $\sigma_{r,B}$. Equations (6.40) are then used to obtain relations expressing σ_r and σ_t as functions of $\sigma_{r,A}$ and $\sigma_{r,B}$. Lastly, by imposing equality of radial displacements of the interface A, first regarded as a part of the disk and second as a part of the hub, and of the interface B, also regarded first as a part of the disk and second as a part of the crown ring, we determine unknown hyperstatic values $\sigma_{r,A}$ and $\sigma_{r,B}$ and, proceeding backwards, all other items (in this connection, see calculation example 3 of Sect. 2.5.3).

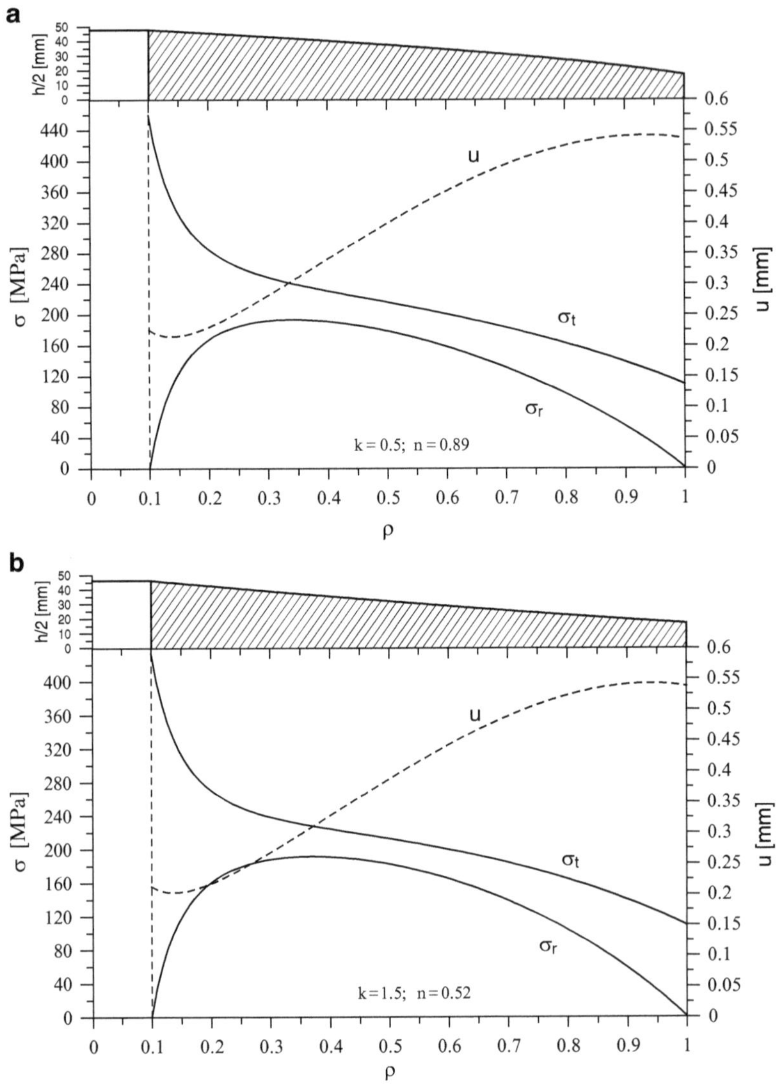

Fig. 7.11 Stress-distribution curves σ_r, σ_t and displacement-distribution curve u in two annular rotating steel disks ($\nu = 0.3$) having variable profile according to (7.1) with constant density: (**a**) convex profile ($k = 0.5$ and $n = 0.89$); (**b**) concave profile ($k = 1.5$ and $n = 0.52$)

It should be noted that, by appropriately selecting the disk profile, along with crown ring and hub, stresses σ_r and σ_t can be made almost constant; in other words, a disk of uniform strength can be approximated within certain limits.

In the practical examples in Fig. 7.12a, b, the crown ring is simulated by a radial stress at interface $B(\sigma_{rB} = 120$ MPa), while the hub may assume three configurations which differ in mass. For each of these configurations, the

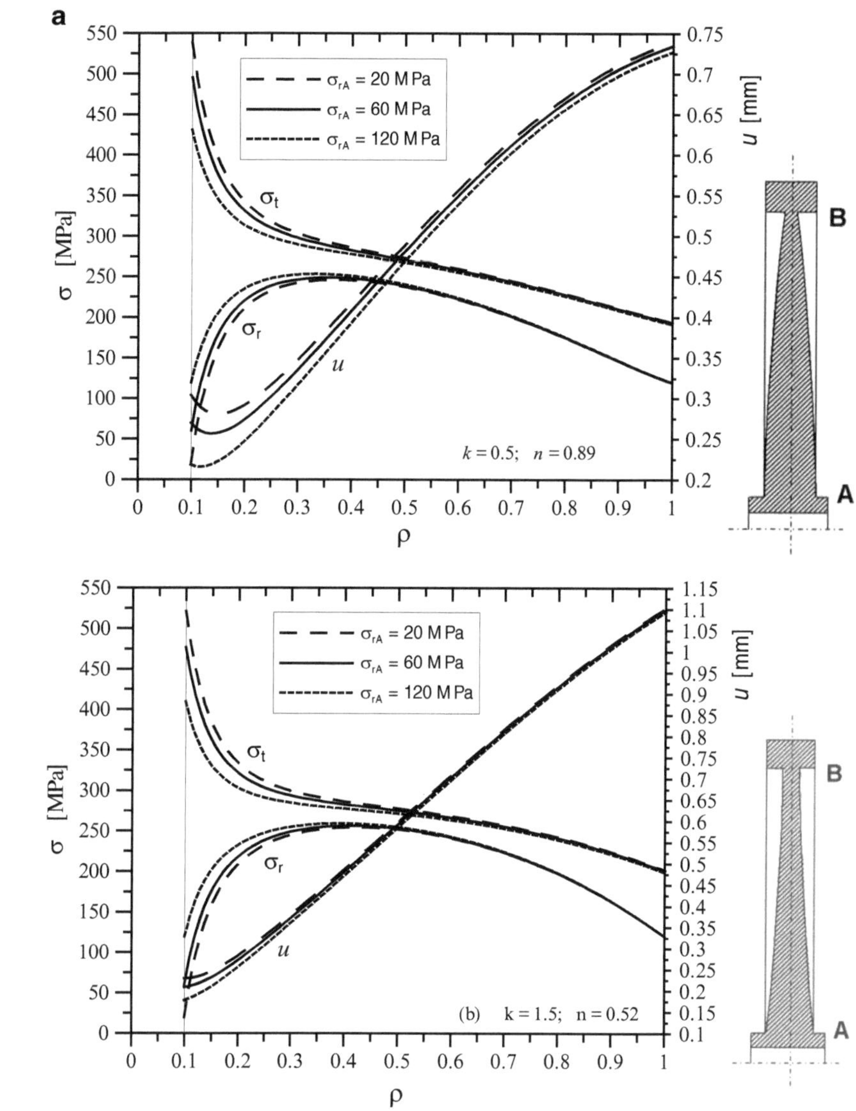

Fig. 7.12 Stress-distribution curves σ_r, σ_t and displacement-distribution curve u in two steel disks ($\nu = 0.3$) having variable thickness according to (7.1), with hub and crown ring, for three different hub configurations, each showing a specific interface stress distribution $\sigma_{r,A}$: (**a**) convex profile ($k = 0.5$ and $n = 0.89$); (**b**) concave profile ($k = 1.5$ and $n = 0.52$)

corresponding interface stresses $\sigma_{r,A}$ are indicated, but not the geometry. Figure 7.12a, b show stress-distribution curves σ_r and σ_t and displacement-distribution curve u as functions of ρ for two disks, the first showing a convex profile ($k = 0.5$ and $n = 0.89$), and the second a concave profile ($k = 1.5$ and $n = 0.52$), clarifying that the most advantageous stress distribution is obtained with the hub having greater mass.

7.6.6 Annular Disks Having Constant Density and Subjected to Temperature Gradient

We will now examine two non-linearly variable thickness disks which, like those analysed in Sect. 7.6.4, are of constant density and have no singularity points. Here, however, the disks are not rotating and are subjected to thermal load only, with temperature varying according to relation (4.5), with $i = 3$. In this case, the stress state is obtained from relations (7.38), while integration constants A and B are determined by imposing boundary conditions (7.50) of Sect. 7.6.4.

Figure 7.13a, b show stress-distribution curves σ_r and σ_t and displacement-distribution curve u as functions of ρ in two annular disks, featuring the same geometry and made of the same material ($\nu = 0.3$ and $\alpha = 12{\cdot}10^{-6}\ {}^{\circ}\mathrm{C}^{-1}$) as the ones examined in Sect. 7.6.4, not loaded at either inner or outer edge and subjected to a temperature gradient along the radius according to function $T = T_0 + k_1 r + k_2 r^2 + k_3 r^3$. The same figure shows curves of the temperature distribution, T being dimensionless relative to reference temperature T_0, as well as coefficients of the third-order polynomial function. A comparison of the diagrams shows no substantial differences between the concave and convex profile disks.

7.6.7 Rotating Annular Disks Having Density Variation on Radius

We will now examine two rotating disks which, like those analysed in Sect. 7.6.4, have no singularity points, are not subjected to thermal load, and have variable density along the radius according to relation (7.39), with $i = 3$. Here, however, the stress state is obtained from relations (7.42), while integration constants A and B are determined by imposing boundary conditions (7.50).

Figure 7.14a, b show stress-distribution curves σ_r and σ_t and displacement-distribution curve u as functions of ρ in the two annular steel disks ($\nu = 0.3$) featuring the same geometry and made of the same material as the ones examined in Sect. 7.6.4, rotating at angular velocity $\omega = 314$ rad/s, not loaded at either inner or outer edge, but featuring variable density along the radius according to function $\gamma = \gamma_0 + \gamma_1 \cdot \rho + \gamma_2 \cdot \rho^2 + \gamma_3 \cdot \rho^3$. The same figure shows the curve of the density variation, γ being dimensionless relative to the basic material density γ_0, as well as coefficients of the third-order polynomial function.

A comparison of the diagrams shows minor differences between the concave and the convex profile disk.

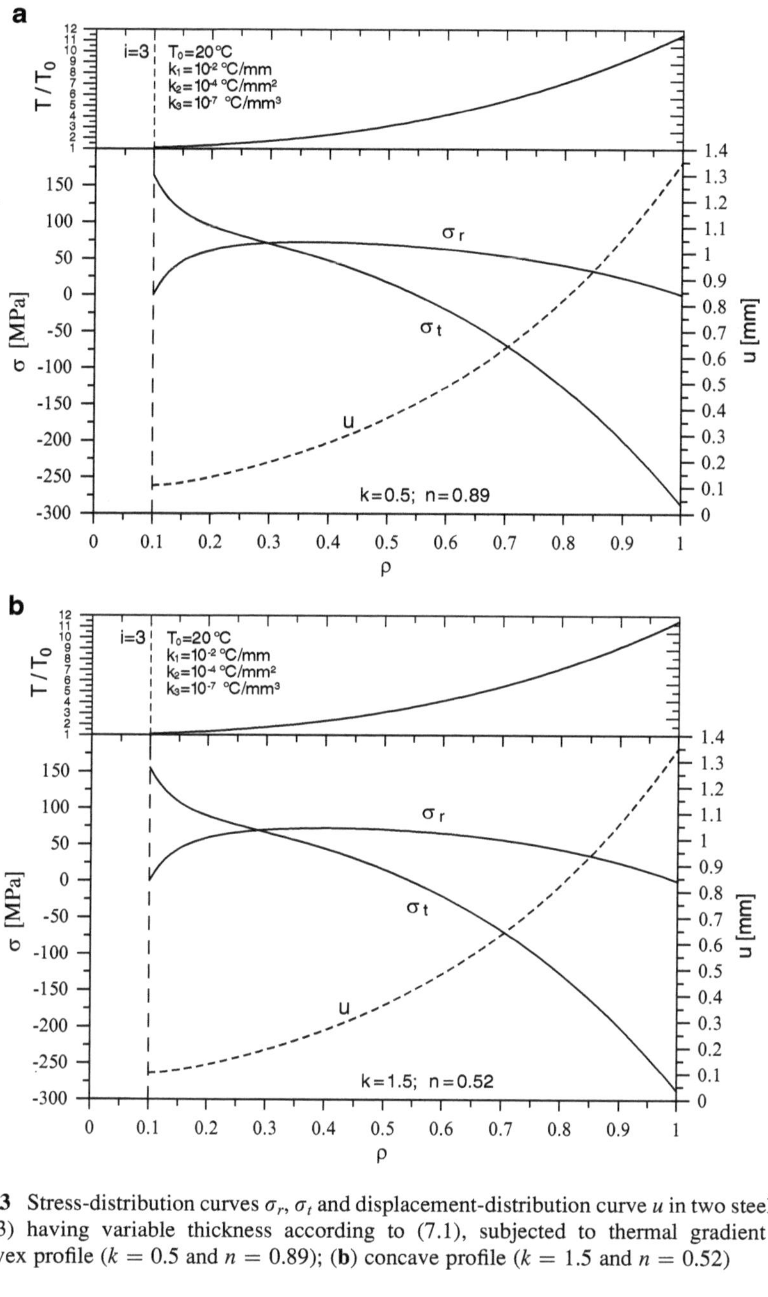

Fig. 7.13 Stress-distribution curves σ_r, σ_t and displacement-distribution curve u in two steel disks ($\nu = 0.3$) having variable thickness according to (7.1), subjected to thermal gradient only: (**a**) convex profile ($k = 0.5$ and $n = 0.89$); (**b**) concave profile ($k = 1.5$ and $n = 0.52$)

a

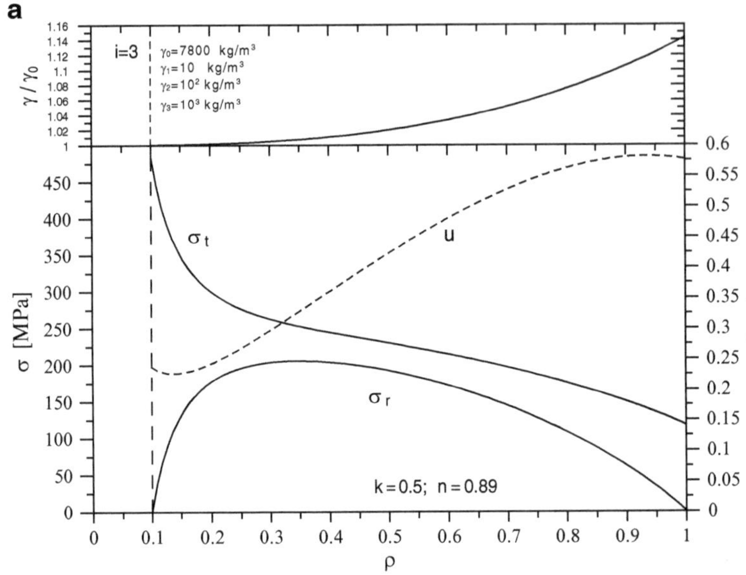

b

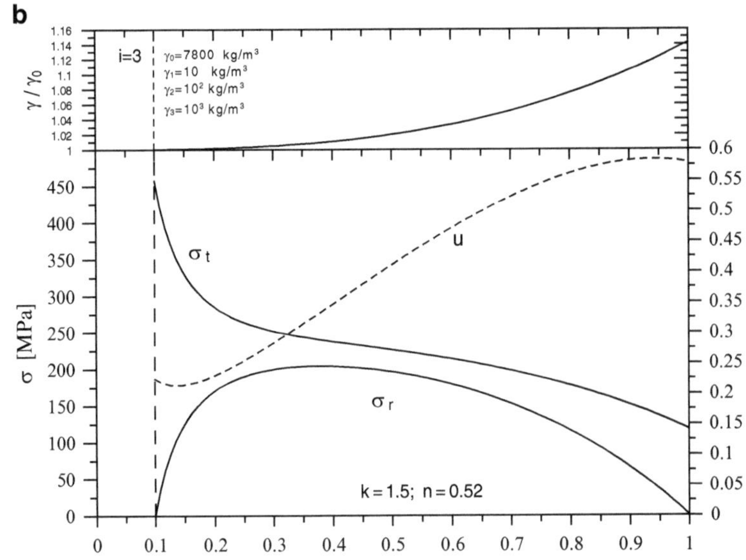

Fig. 7.14 Stress-distribution curves σ_r, σ_t and displacement-distribution curve u in two annular steel disks ($\nu = 0.3$) having variable thickness according to (7.1), showing variable density along the radius: (**a**) convex profile ($k = 0.5$ and $n = 0.89$); (**b**) concave profile ($k = 1.5$ and $n = 0.52$)

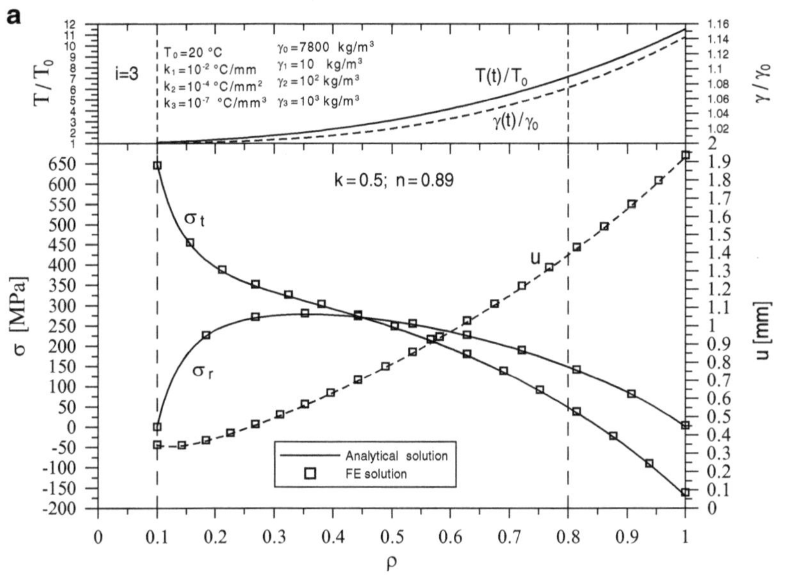

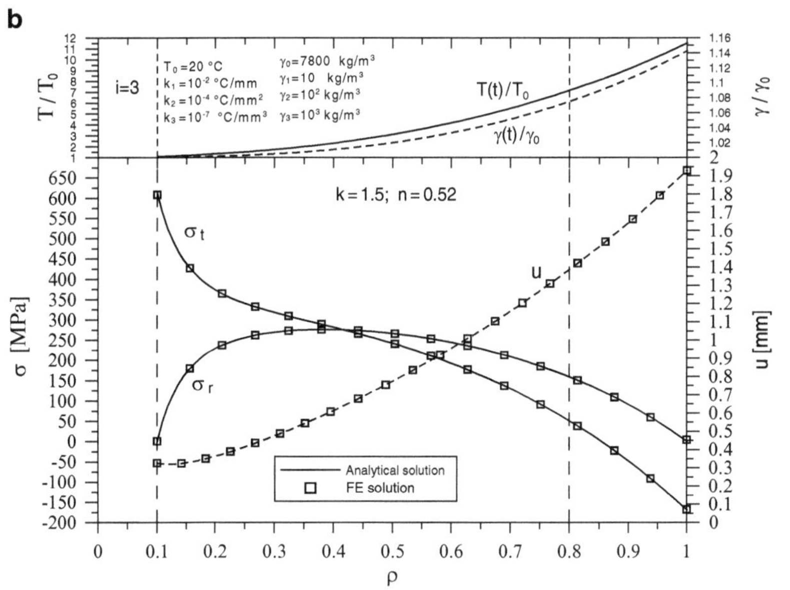

Fig. 7.15 Stress-distribution curves σ_r, σ_t and displacement-distribution curve u in two annular steel disks ($\nu = 0.3$) having variable thickness according to (7.1), subjected to thermal gradient and showing variable density along the radius, and comparison with FE-obtained results: (**a**) convex profile ($k = 0.5$ and $n = 0.89$); (**b**) concave profile ($k = 1.5$ and $n = 0.52$)

7.6.8 Rotating Annular Disks Having Density Variation on Radius and Subjected to Thermal Load

Here, the same disks featured in the previous two examples will be examined, but rotating and subjected to thermal load, with temperature varying according to relation (4.5), with $i = 3$, and having variable density along the radius according to expression (7.39), still with $i = 3$. In this case, the stress state is obtained from relations (7.46), while integration constants A and B are determined by imposing boundary conditions (7.50).

Figure 7.15a, b show stress-distribution curves σ_r and σ_t and displacement-distribution curve u as functions of ρ in the two annular steel disks ($\nu = 0.3$) featuring the same geometry and made of the same material as the ones examined in the previous two examples, not loaded at either inner or outer edge, subject to a temperature gradient along the radius and featuring variable density along the radius. The functions of variation for T and γ are the cubical ones referred to in the two preceding examples, and shown in the diagram in dimensionless form. A comparison of the diagrams shows minor differences between the concave and the convex profile disks.

The figure also compares the results obtained by applying the analytical method proposed here and the numerical results obtained, all other conditions remaining unchanged, through FEM. The diagrams clearly show that the results of the numerical model and those of the analytical model match perfectly.

7.6.9 Comparing Various Disk Types

Four solid disk configurations will now be compared *ceteris paribus*, respectively with conical ($k = 1$ and $n = 0.67$), convex ($k = 0.5$ and $n = 0.89$) and concave ($k = 1.5$ and $n = 0.52$) profiles and the fourth with uniform strength profile. All four rotating disks are subjected to a stress distribution σ_{re} at the outer radius r_e.

All four disks taken into consideration fall into the category discussed in Sect. 7.6.3. Accordingly, the conditions indicated in said section also apply here, the only variation being $\sigma_r = \sigma_{re} \neq 0$ at the outer radius r_e. By imposing this boundary condition and bearing in mind that $C_2 = B = 0$ and $u = C_1u_1 + u_p$, the value of constant A will be obtained from the first of relations (6.40) and, consequently, the value of the constant C_1 will be obtained from the first of relations (7.25).

Figure 7.16 shows the profile geometries, as well as curves of stress-distribution curves σ_r/σ_{re}, σ_t/σ_{re} and displacement-distribution curve u/u_0 as functions of ρ in the four solid disks. Stress values are dimensionless relative to stress value σ_{re}, whereas displacement values are dimensionless relative to $u_0 = r_e\, \sigma_{re}\, (1 - \nu)/E$.

Comparison of these diagrams shows that the convex profile most closely matches the stress state of the uniform-strength disk, whereas the concave profile features the

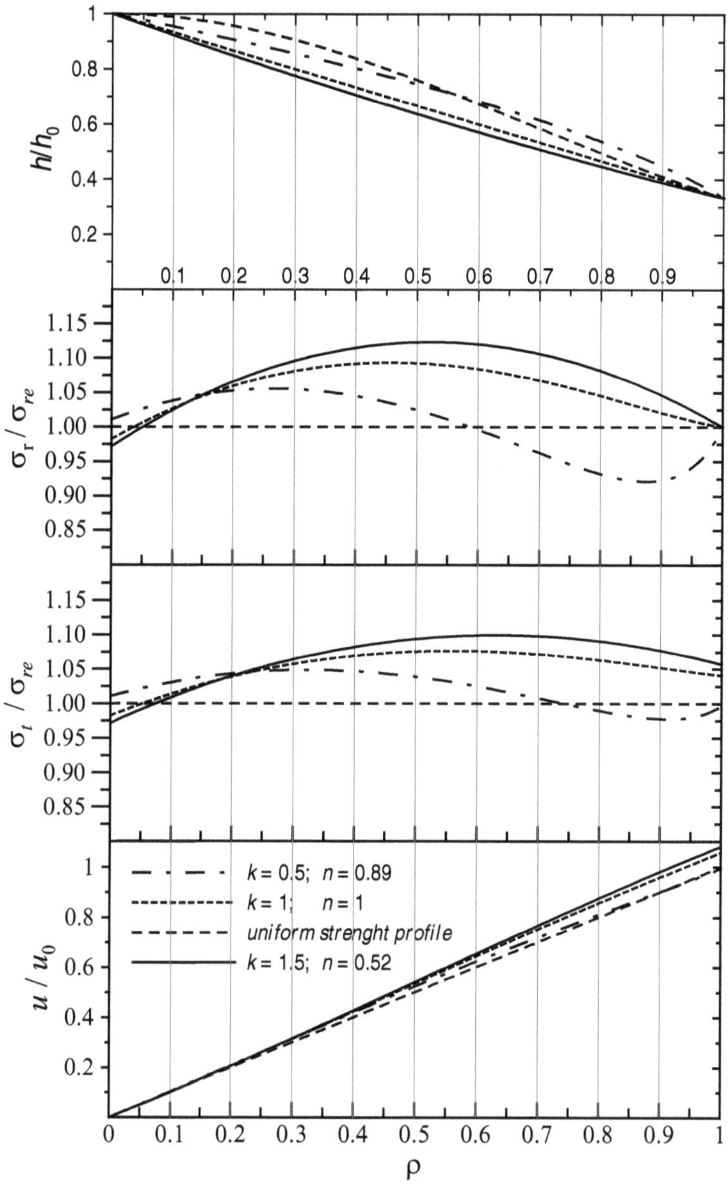

Fig. 7.16 Comparison of four solid, variable-profile disks. From the top: profile geometries; σ_r/σ_{re} versus ρ; σ_t/σ_{re} versus ρ; u/u_0 versus ρ

most substantial discrepancies. Behaviour of the conical profile is midway between that of the concave and the convex profiles. As regards radial displacement, there are no differences of design significance between the four profiles.

7.7 Non-Linearly Variable Thickness Disks Subjected to Angular Acceleration

If the disk has a thickness that varies according to a power of a linear function expressed by relation (7.1) and is subjected to angular acceleration, the solution does not involve difficulties comparable to those encountered above in analysing equilibrium and compatibility in the radial direction. In this case, the solution of (2.74) is as follows:

$$\tau_{rt} = \dot{\omega} r_e^2 \gamma \frac{(n\rho - 1)}{\rho^2 n^4} \times \frac{\left[\rho^3 n^3 (k+3)(k+2)(k+1) + 3\rho^2 n^2 (k+2)(k+1) + 6\rho n(k+1) + 6\right]}{(k+4)(k+3)(k+2)(k+1)} + \frac{C}{\rho^2 (n\rho - 1)^k}, \tag{7.51}$$

where C is an integration constant that can be determined by establishing that the shear stress at the outer radius assumes the value $(\tau_{rt})_e$, deriving from the application of a driving or braking torque at the rim of the disk. By setting this boundary condition, we obtain the relation:

$$\begin{aligned}\tau_{rt} = {} & \frac{\gamma \dot{\omega} r_e^2}{\rho^2 n^4} \cdot \frac{(n\rho - 1)}{(k+4)\cdot(k+3)\cdot(k+2)\cdot(k+1)} \\ & \times \left\{ \left[\rho^3 n^3 (k+3)\cdot(k+2)\cdot(k+1) + 3\rho^2 n^2 (k+2)\cdot(k+1) + 6\rho n\cdot(k+1) + 6\right] + \right. \\ & + \frac{(n-1)^{k+1}}{(n\rho - 1)^{k+1}} \left[n^3 (k+3)\cdot(k+2)\cdot(k+1) + 3n^2 (k+2)\cdot(k+1) + 6n\cdot(k+1) + 6\right] + \\ & \left. + \frac{(n-1)^k}{(n\rho - 1)^{k+1}} \cdot \frac{(\tau_{rt})_e}{\gamma \cdot \dot{\omega} \cdot r_e^2} \cdot n^4 (k+4)\cdot(k+3)\cdot(k+2)\cdot(k+1) \right\}.\end{aligned} \tag{7.52}$$

If no driving or braking torque is applied at the outer radius of a constant density disk whose thickness varies according to (7.1), relation (7.51) gives $C = 0$. Consequently, as $(\tau_{rt})_e = 0$, we will have:

$$\begin{aligned}\tau_{rt} = {} & \frac{\gamma \dot{\omega} r_e^2}{\rho^2 n^4} \cdot \frac{(n\rho - 1)}{(k+4)\cdot(k+3)\cdot(k+2)\cdot(k+1)} \\ & \times \left\{ \left[\rho^3 n^3 (k+3)\cdot(k+2)\cdot(k+1) + 3\rho^2 n^2 (k+2)\cdot(k+1) + 6\rho n\cdot(k+1) + 6\right] + \right. \\ & \left. + \frac{(n-1)^{k+1}}{(n\rho - 1)^{k+1}} \left[n^3 (k+3)\cdot(k+2)\cdot(k+1) + 3n^2 (k+2)\cdot(k+1) + 6n\cdot(k+1) + 6\right] \right\}.\end{aligned} \tag{7.53}$$

It is obvious that if we set $k = 1$, the above relations give the shear stress resulting from angular acceleration in a conical disk whose profile varies along with parameter n, as shown in Fig. 7.1a.

Chapter 8
Disk Having Arbitrary Profile

In cases where the disk has a generic profile which cannot be likened to those considered earlier, which are defined by analytical functions and cannot be dealt with using theoretical methods, it is necessary to use numerical solutions, which will be straightforward because of the many simplifications introduced. The problem of a rotating disk of arbitrary profile, but still symmetrical with respect to a plane perpendicular to the axis, has been studied by a number of authors (see, for example, [11, 73, 34, 16]). From the application standpoint, the methods that are most frequently used employ approximate procedures and generally follow the approach formulated by Timoshenko [73], which is largely based on the analyses carried out and the results obtained by Donath [12], Martin [49], Stodola [70] and Biezeno and Grammel [5]. With this approximate method, analysis is performed by considering the actual disk as consisting of adjacent partial annular disks having constant thickness.

A detailed analysis of the stress and strain state in rotating disks of arbitrary profile or in complex structures such as a turbine rotor, which consists of a hub, variable-thickness disk and crown ring (where the latter is not axisymmetric, as it features slots serving as seats for the blade attachments), is clearly not possible with the theoretical/analytical methods available today. Such analyses are thus carried out by means of numerical methods including, in order of introduction, the finite difference method, the finite element method, and the boundary element method.

Noteworthy finite difference methods include those proposed by Stanley Thompson and Lester [67] and by Manson [48]. Stanley Thompson and Lester developed a calculation procedure for turbine disks which takes into account the point-to-point variation in disk thickness, temperature and all physical properties except Poisson's ratio. Manson's more general method, which will be described in detail below, is a finite difference method that takes into account point-to-point variations in disk thickness, in elastic modulus, in temperature, in coefficient of thermal expansion, in material density and in Poisson's ratio. More recently, Sterner et al. [69] presented another finite difference method for analysing rotating disks of arbitrary configuration, including linearly tapered disks, in which the governing equations are written in terms of the radial stress and simulation is

V. Vullo and F. Vivio, *Rotors: Stress Analysis and Design*,
Mechanical Engineering Series, DOI 10.1007/978-88-470-2562-2_8,

© Springer-Verlag Italia 2013

performed based on repeated applications of a truncated Taylor's expansion with a number of terms such as to guarantee compatibility within a predetermined tolerance range.

From the 1960s onwards, new numerical techniques were developed and applied, including specifically the finite element method (see, for example, Zienckiewicz 1977, [19] and Cook 1981) and the boundary element method (in this connection, see Brebbia 1978 and [4]). The finite element method is a numerical technique that is as general as it is powerful, and which can be considered as an approximate Ritz method combined with a variational principle applied to continuum mechanics. It makes it possible to calculate the stress and strain state of engineering structures of any complexity with unprecedented ease and precision. Consequently, it has gained preference over other numerical approaches. The literature provides many examples of the use of this method in analysing stress distributions in actual turbine disks consisting of an integral variable thickness disk with hub and crown ring (the latter featuring slots spaced at equal angles for the blade attachments), both arranged to the disk with predetermined fillet radii to reduce any stress concentrations that occurring in these joining regions (see, for example, [17]). The boundary elements method or the boundary integral equation method, on the other hand, is advantageous because only the boundary of the axisymmetric model need be discretized (see [1]).

Of the many procedures used in this field, we will concentrate on those proposed by Timoshenko on the basis of Grammel's formulation, and by Manson, as they can be applied without losing sight of the steps involved in the general disk theory presented in the preceding sections.

8.1 Timoshenko-Grammel's Method

Timoshenko-Grammel's method will be discussed here more for its methodological value than for any interest it might have in the actual design of a disk of arbitrary thickness: from the operational standpoint, it is clear that the simple finite element models in use today are faster and more versatile. The method is based on dividing the disk into a certain number of partial annular disks whose thickness varies with a function for which the solution of differential equation (1.28) is known. For the sake of simplicity, the disk is almost always divided into partial disks having constant thickness, but partial disks with a hyperbolic profile can also be used. In dividing the disk into partial disks, the actual profile is replaced with a discretized profile having constant thickness parts, taking care to use a more closely spaced division where disk thickness varies more rapidly.

For a disk of constant thickness, the relations (2.28) apply, and are rewritten here in terms of r^2 rather than ρ^2

$$\begin{cases} \sigma_r = A - B/r^2 - C \cdot r^2 \\ \sigma_t = A + B/r^2 - D \cdot r^2. \end{cases} \tag{8.1}$$

It should also be noted that the constants B, C and D appearing in relations (8.1) differ from the corresponding constants appearing in relations (2.28) by a factor r_e^2.

It is obvious that if the values of radial and hoop stresses are known in a point of any of the partial annular disks into which the disk is discretized, it will be possible to determine integration constants A and B for the partial disk to which the point with known stress belongs. Thus, if in accordance with the procedure proposed by Grammel [31], we set:

$$\begin{cases} s = \sigma_r + C \cdot r^2 \\ t = \sigma_t + D \cdot r^2, \end{cases} \tag{8.2}$$

and assuming that the values of σ_r and σ_t at any radius r of the generic partial disk are known, we have:

$$A = \frac{1}{2}(t + s) \; ; \quad B = \frac{1}{2}(t - s) \cdot r^2. \tag{8.3}$$

The calculation procedure is the same as that described for the uniform strength disk with crown ring, except that there the unknown was the geometry, i.e., the radius at the disk/crown ring/interface, whereas here the unknowns are the values of the radial and hoop stresses. If the radial stresses at the inner edge of the first partial annular disk (the partial disks are generally counted starting from the rotational axis) and at the outer edge of the last of the n partial disks into which the disk is divided are known, there are a total of $(2n)$ unknowns, or in other words the $(n - 1)$ values of radial stresses at the interfaces connecting two adjoining partial disks of different thickness and the $(n + 1)$ values of the hoop stresses at the same interfaces, as well as at the inner edge of the first partial disk and the outer edge of the last partial disk. Overall, however, we can write $(2n)$ equations, of which $(n - 1)$ are compatibility equations which express the equality of radial displacements at the two sides of each interface, and $(n + 1)$ are equilibrium equations which establish that the product $(\sigma_r \cdot h)$ is constant on each interface, first and last included (in this connection, it should be borne in mind that the actual profile has been substituted with an average broken line); the problem is thus univocally solved.

However, rather than writing the system of $(2n)$ equations in $(2n)$ unknowns, we have preferred to follow the Timoshenko-Grammel original method and so to proceed in steps, starting the calculation from the inner edge of the first partial disk or the outer edge of the last partial disk, where the value of σ_r is known and where a tentative value σ_{t0} is assumed for σ_t. Supposing, for example, that we start the calculation from the inner edge of the first partial disk (interface 0), with σ_{r0} at this radius being known and establishing the tentative value σ_{t0}, the values of s and

t at this radius are determined by means of relations (8.2), and (8.3) are then used to determine constants A and B for the first partial disk. Once constants A and B are known, the values of stresses σ_r and σ_t at the outer radius of the first partial disk can be calculated, either directly using relations (8.1), or by first calculating values s and t for outer radius r_e by means of relations (8.3), and then calculating stresses σ_r and σ_t by means of (8.2). It should be noted, however, that since the thickness at the inner radius of the first partial disk will differ from that at the inner radius of the real profile, it is more exact to say that value σ_{r0} corresponding to the condition at the inner boundary must be corrected by means of the first of relations (8.6) in order to take thickness variation Δh into account.

To pass from one partial disk to the next, whose inner radius is the outer radius of the previous partial disk, it is necessary to consider the variation in stresses σ_r and σ_t resulting from the abrupt change in thickness at the interface. Let Δh, $\Delta\sigma_r$ and $\Delta\sigma_t$ be the abrupt changes in h, σ_r and σ_t in the transition from one partial disk to the next, and let

$$\overline{h} = h + \Delta h; \quad \overline{\sigma}_r = \sigma_r + \Delta\sigma_r; \quad \overline{\sigma}_t = \sigma_t + \Delta\sigma_t \tag{8.4}$$

be the thickness and the radial and hoop stresses in the next partial disk.

At the interface between the two adjoining partial disks, the radial forces per unit of length must be equal for there to be equilibrium, while the radial displacements of the two adjoining faces must be equal for there to be compatibility, then:

$$\sigma_r \cdot h = \overline{\sigma}_r \cdot \overline{h}; \quad u = r \cdot \varepsilon_t = \frac{r}{E} \cdot (\sigma_t - \nu\sigma_r) = \frac{r}{E} \cdot (\overline{\sigma}_t - \nu\overline{\sigma}_r). \tag{8.5}$$

Substituting relations (8.4) in relations (8.5) yields:

$$\Delta\sigma_r = -\sigma_r \cdot \frac{\Delta h}{h + \Delta h}; \quad \Delta\sigma_t = \nu \cdot \Delta\sigma_r. \tag{8.6}$$

But since the products $C \cdot r^2$ and $D \cdot r^2$ do not change in passing from one partial disk to the next at the same radius, (8.2) give:

$$\Delta s = \Delta\sigma_r; \quad \Delta t = \Delta\sigma_t. \tag{8.7}$$

Adding the stresses σ_r and σ_t calculated at the outer radius of the first partial disk to the increments from (8.6) gives stresses σ_r and σ_t at the inner radius of the second partial disk. This procedure is then followed, partial disk by partial disk, as far as the outer radius of the last partial disk. At this point, we can check whether or not the tentative value σ_{t0} assumed initially was correct. If we are fortunate, and the initial tentative value σ_{t0} was right, the stress σ_r obtained at the outer radius of the last partial disk will be the known value deriving from the boundary condition at that radius; more precisely, because the thickness at the outer radius of the last partial disk will generally differ from the outer thickness of the real profile, the calculated stress σ_r corrected using the first of relations (8.6) to allow for thickness variation Δh must be equal to stress σ_r corresponding to the outer boundary condition, i.e., to σ_{re}. In reality, however, as it is extremely unlikely that the tentative value σ_{t0} assigned at the outset will be correct, the radial stress at the outer edge will differ from that given by the boundary condition.

The calculation described above will thus be repeated for $C = D = 0$, or in other words assuming that disk is not rotating. It will also be assumed that radial stress at the inner edge is zero and that the hoop stress, again at the inner edge, is σ_{t0}, i.e., equal to that assumed with the first iteration, whereby we arrived at a system of stresses σ'_r and σ'_t, differing from that which was sought. Starting from this new condition, and repeating the preceding calculation, we arrive at a new system of stresses σ''_r and σ''_t which satisfy equilibrium and compatibility conditions, with stresses proportional to the initial value σ_{t0} at all points.

However, the desired stress state can be found as the linear combination of two stress states satisfying equilibrium and compatibility conditions. Considering that the two stress states calculated earlier meet these requirements, the desired stress state will be expressed by:

$$\begin{aligned} \sigma_r &= \sigma'_r + K \cdot \sigma''_r \\ \sigma_t &= \sigma'_t + K \cdot \sigma''_t. \end{aligned} \tag{8.8}$$

The first of these relations makes it possible to calculate constant K, as the following boundary condition must be respected at the outer edge

$$\sigma_{re} = \sigma'_{re} + K \cdot \sigma''_{re}\ . \tag{8.9}$$

Once the constant K which univocally solves the problem has been determined, the stress state in the entire disk can be calculated using relations (8.8).

As we saw earlier, if the disk is solid, we will have $B = 0$ and $\sigma_r = \sigma_t$ at the axis. In this case, the tentative calculation will start having the value $\sigma_{t0} = \sigma_{r0}$ at the centre of the disk and the calculation will be repeated starting with the same value for the stationary disk.

Grammel's method can be extended to disks having arbitrary profile subjected to a non-zero temperature gradient along the radius. In this case, relations (3.8), (3.15) and (3.20) will apply respectively for the three temperature variation functions given by (3.1), (3.2) and (3.3). Here, these relations are rewritten in terms of r rather than ρ:

$$\begin{cases} \sigma_r = A - B/r^2 - C' \cdot r^n \\ \sigma_t = A + B/r^2 - D' \cdot r^n. \end{cases} \tag{8.10}$$

$$\begin{cases} \sigma_r = A - \dfrac{B}{r^2} - \alpha \cdot E \cdot \displaystyle\sum_{i=1}^{n} k_i \cdot \dfrac{r^i}{i+2} \\ \sigma_t = A + \dfrac{B}{r^2} - \alpha \cdot E \cdot \displaystyle\sum_{i=1}^{n} \dfrac{i+1}{i+2} \cdot k_i \cdot r^i \end{cases} \tag{8.11}$$

$$\begin{cases} \sigma_r = A - \dfrac{B}{r^2} - \alpha \cdot E \cdot \dfrac{1}{r^2} \cdot \displaystyle\int_{r_i}^{r} T \cdot r \cdot dr \\ \sigma_t = A + \dfrac{B}{r^2} + \alpha \cdot E \cdot \left(\dfrac{1}{r^2} \cdot \displaystyle\int_{r_i}^{r} T \cdot r \cdot dr - T \right). \end{cases} \tag{8.12}$$

Once again, it will be necessary to introduce constants B, C' and D' which differ from those appearing in relations (3.8) by a factor r_e^2 and r_e^nrespectively. We again use relations (8.2), obviously substituting the second terms in the second member with $C' \cdot r^n$ and $D' \cdot r^n$ for function (3.1), with $\alpha \cdot E \cdot \sum_{i=1}^{n} k_i \cdot \frac{r^i}{i+2}$ and $\alpha \cdot E \cdot \sum_{i=1}^{n} \frac{i+1}{i+2} \cdot k_i \cdot r^i$ for function (3.2), and with $\alpha \cdot E \cdot \frac{1}{r^2} \cdot \int_{r_i}^{r} T \cdot r \cdot dr$ and $-\alpha \cdot E \cdot \left(\frac{1}{r^2} \cdot \int_{r_i}^{r} T \cdot r \cdot dr - T \right)$ for function (3.3). On the basis of (8.3), we proceed from partial disk to partial disk as described above.

When using this procedure, a simplifying assumption that is frequently introduced is to consider that the temperature variation is linear in the individual constant-thickness partial disks into which the actual disk of arbitrary profile is divided. This is equivalent to substituting the curve for temperature variation along to the radius with a discretized curve with straight section. In this case, the exponents n and i appearing in relations (8.10) and (8.11) respectively are unity, k is the ratio of the difference in temperatures to the difference in radii at the ends, while the last terms appearing in relations (8.10) and (8.11) assume the following values:

$$\begin{aligned} C' \cdot r^n &= \alpha \cdot E \cdot \sum_{i=1}^{n} k_i \cdot \frac{r^i}{i+2} = \frac{1}{3} \cdot E \cdot k \cdot \alpha \cdot r; \\ D' \cdot r^n &= \alpha \cdot E \cdot \sum_{i=1}^{n} \frac{i+1}{i+2} \cdot k_i \cdot r^i = \frac{2}{3} \cdot E \cdot k \cdot \alpha \cdot r. \end{aligned} \tag{8.13}$$

The same result is reached if we start from relations (8.12). In this case, as $T = T_i + k \cdot r$, with $k = (T_{i+1} - T_i)/(r_{i+1} - r_i)$, elaborating the last terms in relations (8.12) and neglecting the constant terms which, as was demonstrated, do not contribute to the stress state, leads to the conclusion that they can still be expressed as the terms corresponding to them in relations (8.10) and (8.11), i.e.:

$$\begin{aligned} \alpha \cdot E \cdot \frac{1}{r^2} \cdot \int_{r_i}^{r} T \cdot r \cdot dr &= \frac{1}{3} \cdot E \cdot k \cdot \alpha \cdot r; \\ \alpha \cdot E \cdot \left(\frac{1}{r^2} \cdot \int_{r_i}^{r} T \cdot r \cdot dr - T \right) &= -\frac{2}{3} \cdot E \cdot k \cdot \alpha \cdot r. \end{aligned} \tag{8.14}$$

If the variability of E and α along the radius is also to be taken into account, values of E and α must be introduced for each partial disk which correspond to the mean temperature in that partial disk. Here again, at the interface between the two adjoining partial disks, the radial forces per unit of length must be equal for there to be equilibrium, while the radial displacements of the two adjoining faces must be equal for there to be compatibility. Consequently, we must have:

$$\begin{aligned} \sigma_r \cdot h = \bar{\sigma}_r \cdot \bar{h}; \quad u = r \cdot \varepsilon_t &= \frac{r}{E} \cdot [(\sigma_t - \nu\sigma_r) + E \cdot \alpha \cdot T] \\ &= \frac{r}{E'} \cdot \{[(\sigma_t + \Delta\sigma_t) - \nu(\sigma_r + \Delta\sigma_r)] + E'\alpha' \cdot T'\}. \end{aligned} \tag{8.15}$$

These relations yield:

$$\begin{cases} \Delta\sigma_r = -\sigma_r \cdot \dfrac{\Delta h}{h + \Delta h} \\ \Delta\sigma_t = \nu\Delta\sigma_r - (\sigma_t - \nu\sigma_r) \cdot \left(1 - \dfrac{E'}{E}\right) + E' \cdot (\alpha T - \alpha' T'), \end{cases} \tag{8.16}$$

where E, α and T and E', α' and T' are respectively the Young's moduli, the coefficient of thermal expansion and the mean temperature in the two adjoining partial disks.

As regards the variation of E and α with temperature, with reference to the temperature differentials typical of gas turbine disks, the following relations are usually assumed:

$$\begin{aligned} dE/dT &= -64{,}7\ \mathrm{N/mm^2/^\circ C} \\ d\alpha/dT &= 1{,}1 \cdot 10^{-8}/^\circ\mathrm{C}. \end{aligned} \tag{8.17}$$

By contrast, the variation of ν with temperature is generally neglected, as it does not have an appreciable influence on thermal stresses: even if ν is varied over a very

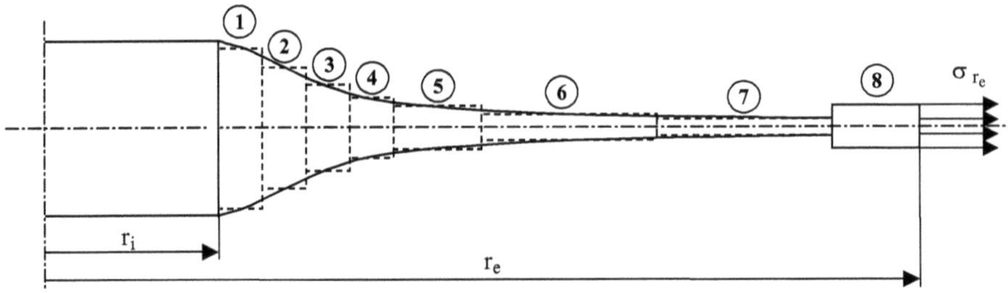

Fig. 8.1 Variable profile disk and breakdown into eight partial disks having constant thickness (projection lines are omitted)

Table 8.1 Geometry of constant-thickness partial disks, operating conditions and material properties

$E = 2.1{\cdot}10^2$ GPa		$\gamma = 7{,}800$ kg/m^3	
$\nu = 0.30$		$\omega = 314$ rad/s	
$r_e = 0.5$ m		$\sigma_{re} = 10$ MPa	
$r_i = 0.1$ m		$\sigma_{ri} = 0$	
Partial disk	Inner radius (m)	Outer radius (m)	Thickness (m)
1	0.100	0.125	0.100
2	0.125	0.150	0.070
3	0.150	0.175	0.050
4	0.175	0.200	0.035
5	0.200	0.250	0.025
6	0.250	0.350	0.015
7	0.350	0.450	0.010
8	0.450	0.500	0.025

wide range (from $\nu = 0.2$ to $\nu = 0.5$ corresponding to plastic flow), the variations in stress will be insignificant.

8.1.1 Example

We will consider a steel disk having variable profile as represented with a solid line in Fig. 8.1. This disk is divided into eight partial disks of constant thickness having the geometry (inner and outer radii, and axial thickness) indicated in Table 8.1, which also shows material properties and operating conditions (angular velocity ω and stress σ_{re} applied at the outer radius). We will evaluate the stress state using Timoshenko-Grammel's method.

Figure 8.2, which summarizes the results achieved with Timoshenko-Grammel's method, shows radial stress σ_r and hoop stress σ_t versus radius. In the intermediate zone of the disk, where the highest stress field values occur, radial stresses can be

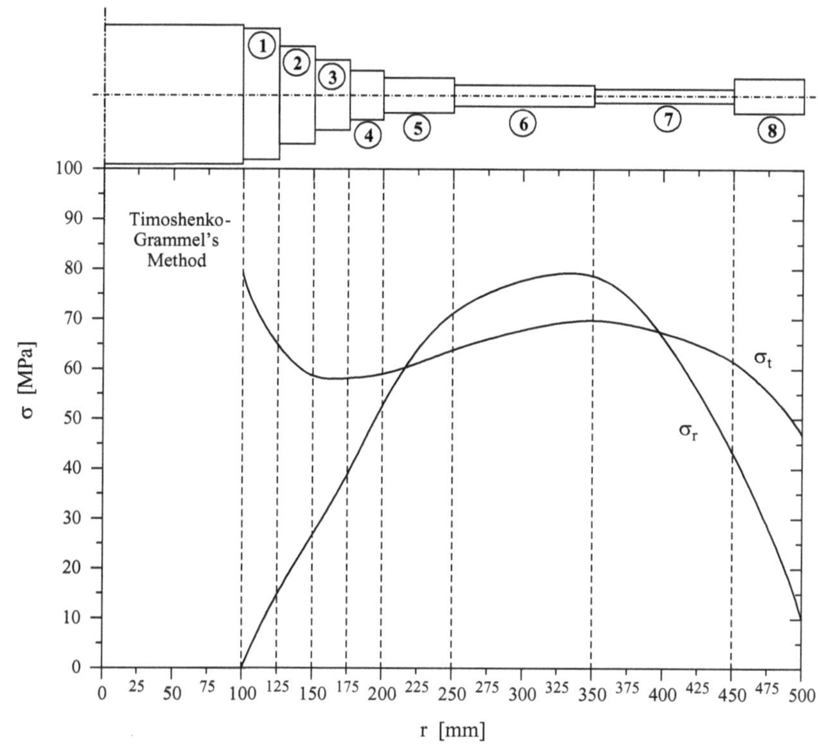

Fig. 8.2 Distribution of radial and hoop stresses versus radius

seen to exceed hoop stresses. However, the hoop stress at the inner radius assumes a value comparable to (and in this specific case, slightly higher than) that of the radial stress in the intermediate zone.

8.2 Manson's Method

Finite difference Manson's method makes it possible to determine the stress state in the elastic field of axisymmetric rotating disks that are also subjected to a temperature gradient. It is a numerical method, based on the replacement of equilibrium and compatibility differential equations by the corresponding finite-difference equations. The method is sufficiently versatile that account can be not only be taken of variations along the radius in disk thickness, but also in temperature, elastic modulus, Poisson's ratio, the material's density, and its coefficient of thermal expansion. In addition, as will be shown in Chap. 12, the method can be modified to calculate the stresses in disks subjected to plastic flow and creep.

Given the variability of E, ν and α with the radius, the compatibility equation to be used is (1.24), which Manson writes in the following form:

$$\frac{d}{dr}\left(\frac{\sigma_t}{E}\right)-\frac{d}{dr}\left(\frac{\nu\cdot\sigma_r}{E}\right)+\frac{d}{dr}(\alpha\cdot T)-\frac{(1+\nu)\cdot(\sigma_r-\sigma_t)}{E\cdot r}=0. \tag{8.18}$$

Were Manson's method to be extended to axisymmetric orthotropic materials, it would be necessary to use a somewhat more complex form of the compatibility equation, which can be derived directly from relation (1.22) and relation (1.17) without introducing further simplifying assumptions. In this connection, since the matrix of material compliances is symmetrical with respect to the principal diagonal, and the first term ν_{rt}/E_r appearing in the second line of relation (1.22) must thus be substituted by ν_{tr}/E_t, giving $\nu_{rt} = \nu_{tr}(E_r/E_t)$, the compatibility equation can be written in the following form:

$$\frac{d}{dr}\left(\frac{\sigma_t-\nu_{tr}\cdot\sigma_r}{E_t}+\alpha_t\cdot T\right)=\frac{1}{r}\cdot\left[\frac{\sigma_r}{E_r}\left(1+\frac{E_r}{E_t}\cdot\nu_{tr}\right)-\frac{\sigma_t}{E_t}\cdot(1+\nu_{tr})+(\alpha_r-\alpha_t)\cdot T\right], \tag{8.19}$$

in which only ν_{tr} appears.

To solve the problem, we must simultaneously consider both the equilibrium equation given by the first expression (1.10), and the compatibility equation written either in the form (8.18) or in the form (8.19), depending on the case involved. Writing these equations in terms of dimensionless variable ρ, for a homogeneous and isotropic material, with E, ν and α varying along the radius, we obtain:

$$\begin{cases}\dfrac{d}{d\rho}(\sigma_r\cdot\rho\cdot h)-\sigma_t\cdot h+\gamma\cdot\omega^2\cdot r_e^2\cdot\rho^2\cdot h=0\\ \dfrac{d}{d\rho}\left(\dfrac{\sigma_t}{E}\right)-\dfrac{d}{d\rho}\left(\dfrac{\nu\cdot\sigma_r}{E}\right)+\dfrac{d}{d\rho}(\alpha\cdot T)-\dfrac{(1+\nu)\cdot(\sigma_r-\sigma_t)}{E\cdot\rho}=0.\end{cases} \tag{8.20}$$

For an axisymmetric orthotropic material whose elastic and thermophysical properties also vary with the radius, we have:

$$\begin{cases}\dfrac{d}{d\rho}(\sigma_r\cdot\rho\cdot h)-\sigma_t\cdot h+\gamma\cdot\omega^2\cdot r_e^2\cdot\rho^2\cdot h=0\\ \dfrac{d}{d\rho}\left(\dfrac{\sigma_t-\nu_{tr}\cdot\sigma_r}{E_t}+\alpha_t\cdot T\right)+\dfrac{1}{\rho}\cdot\left[\dfrac{\sigma_t}{E_t}\cdot(1+\nu_{tr})-\sigma_r\left(\dfrac{1}{E_r}+\dfrac{\nu_{tr}}{E_t}\right)+(\alpha_t-\alpha_r)\cdot T\right]=0.\end{cases} \tag{8.21}$$

We then establish $(n + 1)$ values of the radius starting from inner radius r_0 up to outer radius r_n. With $y(\rho)$ designating a generic variable in the section from radius r_{i-1} to radius r_i (i being an integer variable from 0 to n), the incremental ratio $(y_i - y_{i-1})/(\rho_i - \rho_{i-1})$ is substituted for the derivative $dy/d\rho$ in equations (8.20) or (8.21) and the mean value of the quantity $(y_i + y_{i-1})/2$ is substituted for the local value y of the same quantity. Employing, for example, equations (8.20), we obtain n groups of two equations of the following type:

$$\begin{cases} \dfrac{\sigma_{ri}\cdot\rho_i\cdot h_i-\sigma_{ri-1}\cdot\rho_{i-1}\cdot h_{i-1}}{\rho_i-\rho_{i-1}}-\dfrac{\sigma_{ti}\cdot h_i+\sigma_{ti-1}\cdot h_{i-1}}{2} \\ \quad +\dfrac{\omega^2\cdot r_e^2}{2}\cdot\left(\gamma_i\cdot\rho_i^2\cdot h_i+\gamma_{i-1}\cdot\rho_{i-1}^2\cdot h_{i-1}\right)=0 \\ \dfrac{\sigma_{ti}/E_i-\sigma_{ti-1}/E_{i-1}}{\rho_i-\rho_{i-1}}-\dfrac{\nu_i\cdot\sigma_{ri}/E_i-\nu_{i-1}\cdot\sigma_{ri-1}/E_{i-1}}{\rho_i-\rho_{i-1}}+\dfrac{\alpha_i\cdot T_i-\alpha_{i-1}\cdot T_{i-1}}{\rho_i-\rho_{i-1}}+ \\ \quad -\dfrac{1}{2}\left[\dfrac{(1+\nu_i)\cdot(\sigma_{ri}-\sigma_{ti})}{E_i\cdot\rho_i}+\dfrac{(1+\nu_{i-1})\cdot(\sigma_{ri-1}-\sigma_{ti-1})}{E_{i-1}\cdot\rho_{i-1}}\right]=0 \end{cases} \tag{8.22}$$

which reduce to n groups of two equations of the following type:

$$\begin{cases} C_i\cdot\sigma_{ri}+D_i\cdot\sigma_{ti}+F_i\cdot\sigma_{ri-1}+G_i\cdot\sigma_{ti-1}=H_i \\ C'_i\cdot\sigma_{ri}+D'_i\cdot\sigma_{ti}+F'_i\cdot\sigma_{ri-1}+G'_i\cdot\sigma_{ti-1}=H'_i. \end{cases} \tag{8.23}$$

With the more general equation (8.21), the coefficients appearing in equations (8.23) are:

$$\begin{cases} C_i=2\cdot\rho_i\cdot h_i \\ D_i=-h_i\cdot(\rho_i-\rho_{i-1}) \\ F_i=-2\cdot\rho_{i-1}\cdot h_{i-1} \\ G_i=-h_{i-1}\cdot(\rho_i-\rho_{i-1}) \\ H_i=-\omega^2\cdot r_e^2\cdot(\rho_i-\rho_{i-1})\cdot\left(\gamma_i\cdot\rho_i^2\cdot h_i+\gamma_{i-1}\cdot\rho_{i-1}^2\cdot h_{i-1}\right) \\ C'_i=-\dfrac{1}{E_{ti}}\cdot\left[2\cdot\nu_i+(\rho_i-\rho_{i-1})\cdot\dfrac{\mu_i+\nu_i}{\rho_i}\right] \\ D'_i=\dfrac{1}{E_{ti}}\cdot\left[2+(\rho_i-\rho_{i-1})\cdot\dfrac{1+\nu_i}{\rho_i}\right] \\ F'_i=\dfrac{1}{E_{ti-1}}\cdot\left[2\cdot\nu_{i-1}-(\rho_i-\rho_{i-1})\cdot\dfrac{\mu_{i-1}+\nu_{i-1}}{\rho_{i-1}}\right] \\ G'_i=-\dfrac{1}{E_{ti-1}}\cdot\left[2-(\rho_i-\rho_{i-1})\cdot\dfrac{1+\nu_{i-1}}{\rho_i}\right] \\ H'_i=-2\cdot(\alpha_{ti}\cdot T_i-\alpha_{ti-1}\cdot T_{i-1})+(\rho_i-\rho_{i-1}) \\ \quad\cdot\left[(\alpha_{ti}-\alpha_{ri})\cdot\dfrac{T_i}{\rho_i}+(\alpha_{ti-1}-\alpha_{ri-1})\cdot\dfrac{T_{i-1}}{\rho_i}\right], \end{cases} \tag{8.24}$$

with:

$$\mu_i=E_{ti}/E_{ri};\ \ \nu_i=\nu_{tri}.$$

These are ten constants depending on known quantities (disk geometry, angular velocity, the material's mechanical and thermophysical properties, and temperature). For a homogeneous and isotropic material, whose equilibrium and compatibility equations are given by equations (8.20), the ten constants appearing in (8.23) are obtained from relations (8.24), where we set

$$\mu_i = 1,\ E_{ti} = E_{ri} = E_i,\ E_{ti-1} = E_{ri-1} = E_{i-1},\ \alpha_{ti} = \alpha_{ri} \text{ and } \alpha_{ti-1} = \alpha_{ri-1}.$$

The calculation presents no difficulties. We first select the $(n + 1)$ radii r_i (with $i = 0, 1, \ldots n$), noting that in order to ensure a good approximation with a limited number of radii, it is necessary to use a closer spacing wherever there are sharp changes in any of the quantities involved. We then calculate the $(10n)$ constants given by relations (8.24) and we write the system of $(2n)$ equations obtained by ordering the n groups of relations (8.23). There are only $(2n)$ unknowns, even though $2(n + 1)$ stress values appear in the system of $(2n)$ equations represented by relations (8.23), viz., $(n + 1)$ radial stresses σ_{ri} and $(n + 1)$ hoop stresses σ_{ti}; two stress values, usually σ_{r0} and σ_{rn} are known, and constitute the boundary conditions. In the case examined here for a disk with central hole, the system of $(2n)$ equations can be solved by inverting the matrix of coefficients:

$$\begin{Bmatrix} \sigma_{t0} \\ \sigma_{r1} \\ \sigma_{t1} \\ \sigma_{r2} \\ \sigma_{t2} \\ -- \\ -- \\ \sigma_{rn-1} \\ \sigma_{tn-1} \\ \sigma_{tn} \end{Bmatrix} = \begin{bmatrix} G_1 & C_1 & D_1 & 0 & 0 & \ldots & 0 & 0 & 0 \\ G'_1 & C'_1 & D'_1 & 0 & 0 & \ldots & 0 & 0 & 0 \\ 0 & F_2 & G_2 & C_2 & D_2 & \ldots & 0 & 0 & 0 \\ 0 & F'_2 & G'_2 & C'_2 & D'_2 & \ldots & 0 & 0 & 0 \\ 0 & 0 & 0 & F_3 & G_3 & \ldots & 0 & 0 & 0 \\ -- & -- & -- & -- & -- & -- & -- & -- & -- \\ -- & -- & -- & -- & -- & -- & -- & -- & -- \\ 0 & 0 & 0 & 0 & 0 & \ldots & C'_{n-1} & D'_{n-1} & 0 \\ 0 & 0 & 0 & 0 & 0 & \ldots & F_n & G_n & D_n \\ 0 & 0 & 0 & 0 & 0 & \ldots & F'_n & G'_n & D'_n \end{bmatrix}^{-1}$$

$$\times \begin{Bmatrix} -F_1 \cdot \sigma_{r0} + H_1 \\ -F'_1 \cdot \sigma_{r0} + H'_1 \\ H_2 \\ H'_2 \\ H_3 \\ -- \\ -- \\ H'_{n-1} \\ H_n - C_n \cdot \sigma_{rn} \\ H'_n - C'_n \cdot \sigma_{rn} \end{Bmatrix}. \tag{8.25}$$

If the disk does not have a central hole, stress σ_{r0} at radius $r_0 = 0$ is one of the unknowns. The boundary conditions for the centre of the disk can be expressed by imposing that the radial and hoop stresses coincide at that point, i.e., setting $\sigma_{r0} = \sigma_{t0}$. In this case, it is sufficient to substitute $(F_1 + G_1)$ for G_1 and $(F'_1 + G')$ for G'_1 in the matrix of coefficients and eliminate the σ_{r0} terms from the vector in the second member of (8.25). The matrix solution obviously calls for inverting a matrix of order $(2n)$ through the use of appropriate computational tools.

In Manson's original formulation, however, the method was developed for tabular computation and proceeds along different lines as presented below. The stress state at the generic radius r_i is written as a function of the unknown σ_{t0} in the following form:

$$\begin{cases} \sigma_{ri} = A_{ri} \cdot \sigma_{t0} + B_{ri} \\ \sigma_{ti} = A_{ti} \cdot \sigma_{t0} + B_{ti}, \end{cases} \tag{8.26}$$

where constants A_{ri}, A_{ti}, B_{ri} and B_{ti}, for radius r_i can be calculated on the basis of constants A_{ri-1}, A_{ti-1}, B_{ri-1} and B_{ti-1} for radius r_{i-1} by means of the relations:

$$\begin{cases} A_{ri} = k_i \cdot A_{ri-1} + L_i \cdot A_{ti-1} \\ A_{ti} = k'_i \cdot A_{ri-1} + L'_i \cdot A_{ti-1} \\ B_{ri} = k_i \cdot B_{ri-1} + L_i \cdot B_{ti-1} + M_i \\ B_{ti} = k'_i \cdot B_{ri-1} + L'_i \cdot B_{ti-1} + M'_i, \end{cases} \tag{8.27}$$

with

$$\begin{cases} k_i = (D_i \cdot F'_i - D'_i \cdot F_i)/Z_i \\ k'_i = (C'_i \cdot F_i - C_i \cdot F'_i)/Z_i \\ L_i = -(D'_i \cdot G_i - D_i \cdot G'_i)/Z_i \\ L'_i = -(C_i \cdot G'_i - C'_i \cdot G_i)/Z_i \\ M_i = (D'_i \cdot H_i - D_i \cdot H'_i)/Z_i \\ M'_i = (C_i \cdot H'_i - C'_i \cdot H_i)/Z_i \\ Z_i = C_i \cdot D'_i - C'_i \cdot D_i. \end{cases} \tag{8.28}$$

The values of A_{r0}, A_{t0}, B_{r0} and B_{t0} at radius r_0 needed to start calculating A_{ri}, A_{ti}, B_{ri} and B_{ti} are determined by setting boundary conditions, and we find (for a disk only rotating $\sigma_{r0} = 0$):

$$A_{r0} = B_{t0} = 0; \quad A_{t0} = 1; \quad B_{r0} = \sigma_{r0}, \tag{8.29}$$

for the disk with a central hole, and:

$$A_{r0} = A_{t0} = 1; \quad B_{r0} = B_{t0} = 0, \tag{8.30}$$

for the solid disk.

We then proceed with successive applications of (8.27) to determine values A_{ri}, A_{ti}, B_{ri} and B_{ti}, until we obtain A_{rn} and B_{rn}. As the radial stress at the outer edge is known, the first of relations (8.26) can be used to find σ_{t0}:

$$\sigma_{t0} = (\sigma_{rn} - B_{rn})/A_{rn} \ . \tag{8.31}$$

Equations (8.26) provide the entire stress state in the disk. This type of procedure has the advantage of not requiring matrix inversion, as well as of occupying less memory space in automatic computation, since only four vectors of order $n + 1$ need be stored, rather than a matrix of order $(2n \times 2n)$.

In general, Manson's method entails a more closely spaced division of the disk than that required by Timoshenko-Grammel's method, but leads to much simpler equations in each partial disk. It is better suited to automatic computation, and is more versatile since, as we will see in Chap. 12, it can be adapted with very few modifications to take the effects of plastic flow and creep into account.

It should also be pointed out that Manson's method is also applicable to composite structure turbine disks i.e., disks consisting of a heat-resisting outer portion and an inner portion for which high temperature resistance is not required. These two portions are generally shrink-fit and then welded together. If it is assumed that shrink fitting is carried out by maintaining the inner portion 1 and the outer portion 2 at constant temperatures of T_{1c} and T_{2c} respectively for a sufficient period of time to ensure that the stress state due to thermal coaction in these portions is zero, and that the shrink-fit temperatures are maintained constant throughout the welding process, and if it is further assumed that the residual stresses correlated with welding are negligible and that both portions have the same width at the mating face, the procedure described above can be applied with the following changes.

The disk is first divided into a larger number of radii on both sides of the mating interface with the welded joint. As the densities of the two materials and functions $\nu = \nu\,(T)$ and $E = E(T)$ are known, the values of temperatures T_i at the *i-th* radius are determined by subtracting the shrink-fit temperatures, at which there is zero thermal stress just before cooling produces shrink interference, from the actual local temperatures. T_{1c} and T_{2c} are thus the reference temperatures for the two portions of the disk. Again at the *i-th* radius, the values of the coefficients of thermal expansion to be considered in the calculations are the averages between actual local temperatures and the reference temperature. In other words, for example, if room temperature T_a is assumed as the reference temperature, the average value α_{1-2} of the coefficient of thermal expansion at the two generic temperatures T_1 and T_2 is given by the relation:

$$\alpha_{1-2} = \frac{\alpha'_2 \cdot T'_2 - \alpha'_1 \cdot T'_1}{(T_2 - T_1) \cdot (1 + \alpha'_1 \cdot T'_1)}, \tag{8.32}$$

where α'_1 and α'_2 are the average coefficients of thermal expansion between room temperature T_a and temperatures T_1 and T_2, respectively, while $T'_2 = T_2 - T_a$ and $T'_1 = T_1 - T_a$.

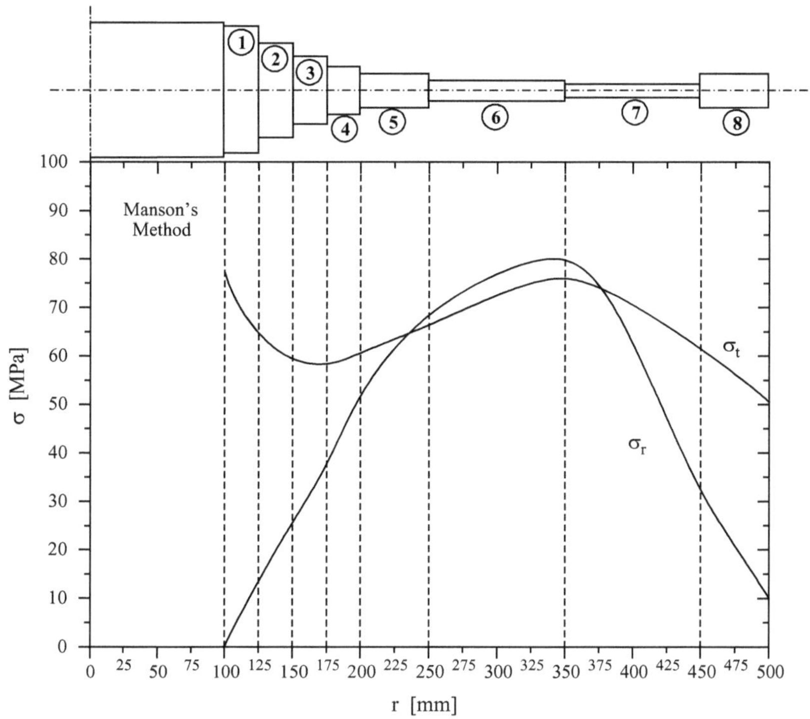

Fig. 8.3 Distribution of radial and hoop stresses versus radius

Without going into the details of the calculations, which present no difficulties and are left to the reader, it should be noted that the shrink-fit between the two portions of a composite structure turbine disk gives rise to a pre-existing stress state which will be superimposed on those due to centrifugal load and thermal load in the actual operating conditions in which the disk fulfils its design functions. The shrink-fit can be optimized to reduce stress peaks in actual service conditions to the greatest possible extent. However, the optimization procedure becomes more complex if the shrink interference is so high that it generates a stress field with stresses beyond yielding. This topic will be covered in Chaps. 11 and 12.

8.2.1 Example

We will now consider the same variable-profile disk shown in Fig. 8.1, with the same division into partial disks having constant thickness indicated therein. The nine radius values delimiting the eight partial disks into which the disk is divided are thus those given in Table 8.1. We will evaluate the stress state using Manson's method.

Figure 8.3, which summarizes the results achieved with Manson's method, shows radial stress σ_r and hoop stress σ_t versus radius.

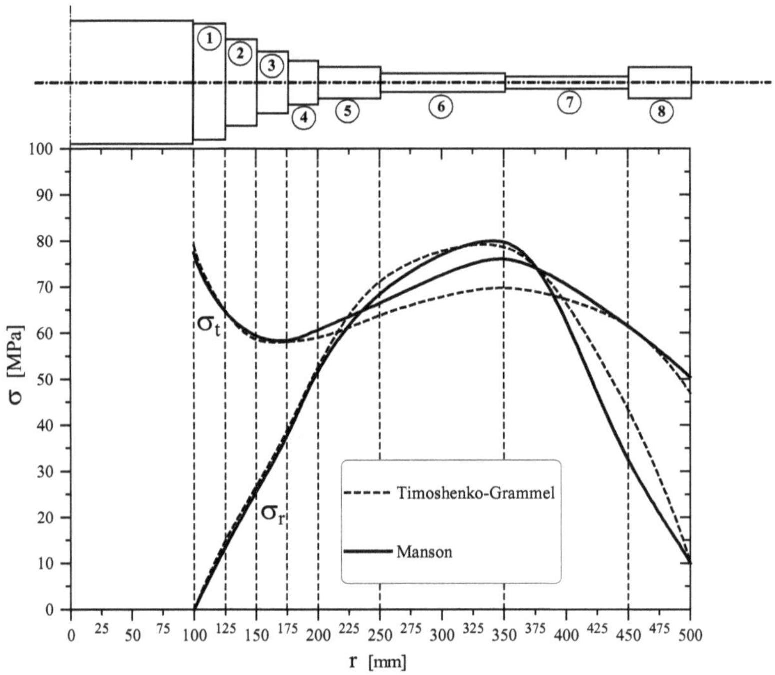

Fig. 8.4 Comparison between the results obtained with the Timoshenko-Grammel and Manson methods

These results confirm that the highest stress field values determined using Manson's method are once again those in the intermediate zone of the disk, where the radial stress is higher than the hoop stress. For the calculation example examined here, the only appreciable difference between this and Timoshenko-Grammel's method is that the latter gives a hoop stress at the inner radius which is slightly higher than the maximum radial stress in the central zone, whereas Manson's method gives a hoop stress at the inner radius which is slightly below the maximum radial stress.

These conclusions, however, cannot be generalized, as calculation results are influenced by the substantial difference in the approach taken by the two methods (the first considers the various quantities involved at the interfaces, while the second considers their averages between two adjoining radii), as well as the effect of the geometry of the individual partial disks into which the disk is divided. It must also be borne in mind that to obtain more accurate results, Manson's methods requires a finer discretization of the disk than that needed with Timoshenko-Grammel's method. In this connection, Manson himself suggests using a more closely spaced division at the inner radius for disks with a central hole.

The results obtained with Timoshenko-Grammel's method and Manson's method for the same disk and the same discretization are compared in Fig. 8.4.

Chapter 9
Design of Rotating Disks and Stress Concentrations

9.1 Failure Criteria

With the usual assumptions of a plane stress state ($\sigma_z = 0$) made for rotating disks, the stress field is generally biaxial. Uniaxial stress conditions can only occur locally: at the inner edge of non-rotating disks loaded at the outer edge (Figs. 2.1 and 2.2) and of only rotating disks (Fig. 2.6), and at the outer edge of non-rotating disks loaded at the inner edge (Figs. 2.3 and 2.4), of rotating annular disks (Fig. 2.6) and rotating solid disks (Fig. 2.15). Biaxial tensile, tensile-compressive and tensile or compressive stress occur respectively at the centre of only rotating disks (Fig. 2.15), at the inner edge of non-rotating annular disks loaded at the inner edge (Figs. 2.3 and 2.4) and at the outer edge of non-rotating annular disks loaded at the outer edge (Figs. 2.1 and 2.2). Biaxial tensile, compressive or tensile-compressive stress conditions occur in the intermediate sections of the disk.

From this overview of stress conditions, we can conclude that strength evaluation of rotating disks calls for the use of a failure criterion. In this area, the criteria in current use are the maximum normal stress criterion (σ_{max}), the maximum strain criterion (ε_{max}), the maximum shear stress criterion (τ_{max}) and the maximum distortion energy criterion.

With these criteria, bearing in mind that radial stress σ_r and hoop stress σ_t are principal stresses, the equivalent or ideal stress σ_e is given by:

- σ_{max} criterion (Coulomb or Rankine criterion):

$$\sigma_e = max(|\sigma_t|;|\sigma_r|;|\sigma_z|)\ . \tag{9.1}$$

- ε_{max} criterion (Saint Venant criterion):

$$\sigma_e = max[|\sigma_t - \nu \cdot (\sigma_r + \sigma_z)|\,;\,|\sigma_r - \nu \cdot (\sigma_t + \sigma_z)|\,;\,|\sigma_z - \nu \cdot (\sigma_r + \sigma_z)|]\ . \tag{9.2}$$

V. Vullo and F. Vivio, *Rotors: Stress Analysis and Design*,
Mechanical Engineering Series, DOI 10.1007/978-88-470-2562-2_9,
© Springer-Verlag Italia 2013

- τ_{max} criterion (Tresca or Guest criterion):

$$\sigma_e = max(|\sigma_t - \sigma_r|;|\sigma_t - \sigma_z|;|\sigma_r - \sigma_z|). \tag{9.3}$$

- Maximum distortion energy criterion (Maxwell or Von Mises criterion):

$$\sigma_e^2 = \sigma_t^2 + \sigma_r^2 + \sigma_z^2 - \sigma_t \cdot \sigma_r - \sigma_r \cdot \sigma_z - \sigma_z \cdot \sigma_t. \tag{9.4}$$

To understand which failure criterion best interprets the experimental results, a few words are in order concerning the latter. Most available experimental results involve biaxial stress states. To compare data calculated with theoretical models and experimental data, it is necessary to consider plane (σ_r, σ_t). For the sake of convenience, it is also advisable to make stresses dimensionless by relating them to the equivalent stress σ_e, so that plane (σ_r, σ_t) becomes $(\sigma_r/\sigma_e, \sigma_t/\sigma_e)$.

The formulation of the maximum distortion energy criterion (the Von Mises criterion) in plane (σ_r, σ_t) translates into the equation of an ellipse, with semi-major and semi-minor diameters equal to $\sigma_e/\sqrt{2}$ and $\sigma_e\sqrt{2/3}$ respectively, positioned so that they bisect dihedra (σ_r, σ_t) and $(\sigma_r, -\sigma_t)$. This ellipse is represented in Fig. 9.1 on coordinates $(\sigma_r/\sigma_e, \sigma_t/\sigma_e)$.

As can be seen from the curve in Fig. 2.4, stress σ_t in an annular disk subjected only to negative internal pressure $\sigma_{ri} = -p_i$ is always tensile, while stress σ_r is always compressive, except at the outer radius where it is zero. At any radius, the values assumed by hoop stress are always greater than the absolute value of radial stress, or in other words the ratio $\sigma_t/|\sigma_r|$ is above unity at all times. Consequently, the arc of ellipse of interest to us here is that between the line bisecting dihedron $(-\sigma_r/\sigma_e, \sigma_t/\sigma_e)$ and the axis of stresses σ_t/σ_e i.e., the arc $\widehat{BC}$ whose correspondent in the fourth quadrant is arc $\widehat{B'C'}$.

As the curve in Fig. 2.2 shows, both stresses σ_r and σ_t in a radial thick-walled disk subjected only to external pressure $\sigma_{re} = -p_e$ are compressive (an exception is radial stress at the inner radius, which is zero). Here again, at any radius, the absolute value of hoop stress is always greater than the absolute value of radial stress, or in other words the ratio $|\sigma_t|/|\sigma_r|$ is above unity at all times. In this case, the arc of ellipse of interest to us here is that between the line bisecting dihedron $(-\sigma_r/\sigma_e, -\sigma_t/\sigma_e)$ and the negative part of the σ_t/σ_e axis, i.e., the arc $\widehat{A'B'}$, whose correspondent in the first quadrant is arc $\widehat{AB}$.

If the failure criterion used is that of maximum shear stress (the Tresca or Guest criterion), instead of having the Von Mises ellipse on coordinates (σ_r, σ_t), we will have the failure polygon (hexagon) as shown on coordinates $(\sigma_r/\sigma_e, \sigma_t/\sigma_e)$ in Fig. 9.1. With the same approach as used for the maximum-distortion-energy failure criterion, we can conclude that the parts of this failure polygon of interest in the rotors and disks design are respectively the straight-line $\overline{BF}$, which corresponds to straight-line $\overline{DF'}$, and straight-line $\overline{A'B'}$ corresponding to straight-line $\overline{AB}$.

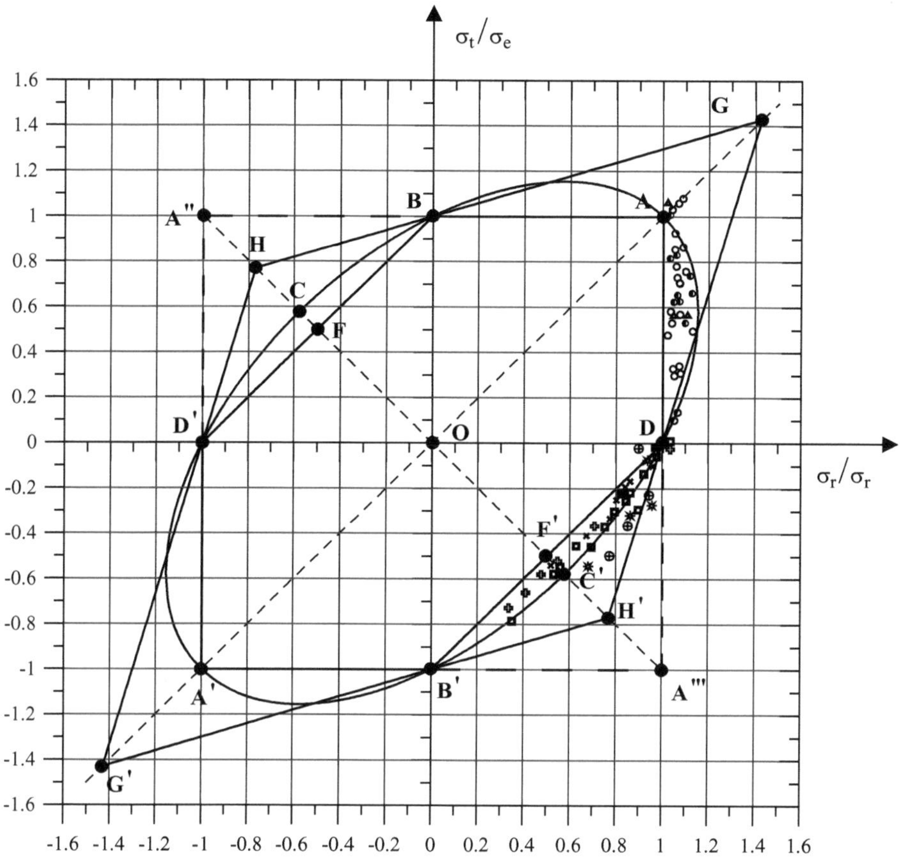

Fig. 9.1 Plane stress failure curves of maximum-distortion energy ellipse, maximum-shear-stress hexagon, maximum-normal-stress square and maximum-normal-strain rhombus on coordinates $(\sigma_r/\sigma_e, \sigma_t/\sigma_e)$

It should be noted that the maximum distortion energy failure criterion can be regarded as the generalization of the τ_{max} criterion, in that it also considers the influence exerted by the third principal stress on yielding. To be handled as a generalization of the τ_{max} criterion, it must be seen as a maximum octahedral shear stress criterion $\tau_{oct,max}$. As we know from continuum mechanics, the stresses acting in the eight octahedral planes, each of which is obtained by cutting across one of the corners of a principal element (a cubical element stressed by the three principal stresses $\sigma_1, \sigma_2, \sigma_3$), have interesting and significant properties. In fact, the same normal stress acts on each of the eight octahedral planes delimiting the octahedron thus obtained. This stress state, a hydrostatic stress state, tends to compress or enlarge the octahedron homothetically, without distorting it, with a consequent variation in volume but not in shape. The shear stresses acting on each of the eight octahedral planes are also equal, but they tend to distort the octahedron without changing its volume. Although the octahedral shear stress is less than the maximum principal shear stress (in this connection, it should be borne in mind that in Mohr's plane $\tau = \tau\ (\sigma)$, the octahedral stresses are represented by the points outside the two smaller Mohr's circles and inside the principal Mohr's circle, or in other words lie in the area between these circles), it is a single value which depends on all three principal stresses $(\sigma_1, \sigma_2, \sigma_3)$.

The following relations express the octahedral normal stress and the octahedral shear stress as a function of the three principal stresses.

$$\sigma_{oct} = \frac{1}{3}(\sigma_1 + \sigma_2 + \sigma_3) \tag{9.5}$$

$$\tau_{oct} = \frac{1}{3}\left[(\sigma_1 - \sigma_2)^2 + (\sigma_2 - \sigma_3)^2 + (\sigma_3 - \sigma_1)^2\right]^{1/2}. \tag{9.6}$$

The octahedral-shear-stress can also be expressed as a function of the three principal-shear-stresses $\tau_{1,2}$, $\tau_{1,3}$, $\tau_{2,3}$ (i.e., the maximum values of the shear stresses correlated with the three Mohr's circles in Mohr's plane) as follows:

$$\tau_{oct} = \frac{2}{3}\left[\tau_{1,2}^2 + \tau_{1,3}^2 + \tau_{2,3}^2\right]^{1/2}. \tag{9.7}$$

According to the maximum octahedral shear stress criterion, the strength limit condition will be reached when

$$(\sigma_1 - \sigma_2)^2 + (\sigma_2 - \sigma_3)^2 + (\sigma_3 - \sigma_1)^2 = 2\sigma_e^2. \tag{9.8}$$

This relation, where $\sigma_1 = \sigma_t, \sigma_2 = \sigma_z$ and $\sigma_3 = \sigma_r$, is linked to relation (9.4), or in other words to the maximum distortion energy failure criterion.

Figure 9.1 also shows the geometric loci which represent the other two failure criteria considered, viz., the σ_{max} and the ε_{max} criteria, on coordinates $(\sigma_r/\sigma_e, \sigma_t/\sigma_e)$. In this plane, the maximum-normal-stress yield criterion translates into the failure square obtained by extending, pairwise, sides $\overline{AB}$ and $\overline{A'B'}$ and sides $\overline{AD}$ and $\overline{A'D'}$ until they meet. Lastly, the maximum normal strain failure criterion translates into the failure rhombus whose major and minor diagonals are superimposed on the semi-major and semi-minor diameters of the failure ellipse, and whose four sides pass through points B, D, B' and D' and converge pairwise on the points lying on the major diagonal having coordinates $[1/(1-\nu), 1/(1-\nu)]$ and $[-1/(1-\nu), -1/(1-\nu)]$ respectively.

Mohr's failure criterion assumes that the strength limit conditions occurs when the stress state is represented by the points in Mohr's plane $\tau = \tau\ (\sigma)$ that lie on the envelope of the infinite Mohr's circles for stress states at the limit of failure onset. In Mohr's plane, this envelope is represented by two nonlinear curves that are symmetrical with respect to the axis of the abscissas (σ axis) and converge on a point of the latter with a positive abscissa, where the material breaks as a result of cohesive failure (Fig. 9.2). These curves become two straight lines (simplified Mohr's criterion), again symmetrical with respect to the σ axis and converging on a point

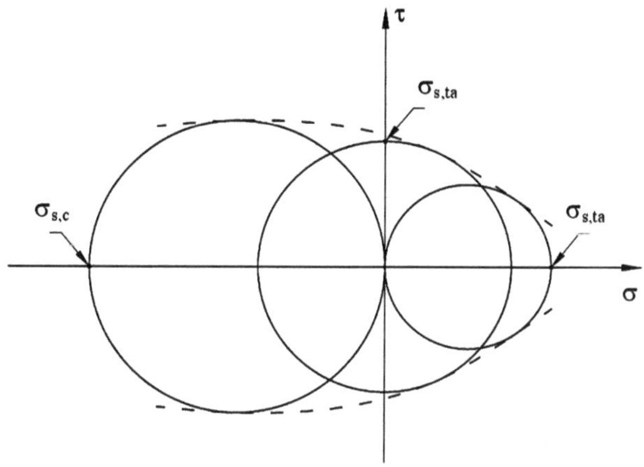

Fig. 9.2 Mohr's criterion for biaxial stress represented on Mohr plane $\tau = \tau(\sigma)$

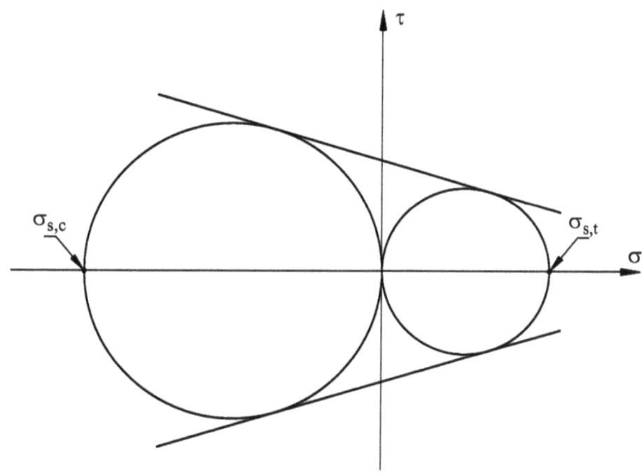

Fig. 9.3 Simplified Mohr's criterion for biaxial stress represented on Mohr plane $\tau = \tau(\sigma)$

of the latter having coordinates ($\sigma > 0$, $\tau = 0$), in the case where the experimental data used are provided by two uniaxial tests, the first uniaxial tension test and the second uniaxial compression test (Fig. 9.3). From this standpoint, Mohr's criterion can be considered as a modification of the τ_{max} criterion.

In plane (σ_1, σ_2), Mohr's criterion for the biaxial stress state results in two nonlinear curves that are symmetrical with respect to the diagonal bisecting the dihedra of the first and third quadrants. The points where these two curves (Fig. 9.4) intersect the coordinate axes and the shear diagonal (the straight-line bisecting the dihedra of the second and fourth quadrants) can be readily identified by means of uniaxial tension and compression data and pure shear data. The upper curve will thus intersect the axis of the abscissa at the point whose coordinates are ($-\sigma_{f,c}$, 0), the axis of the ordinate at the point whose coordinates are (0, $\sigma_{f,t}$) and the shear diagonal at the point whose coordinates are ($\sigma_{f,ta}$, $\sigma_{f,ta}$), where $\sigma_{f,t}$, $\sigma_{f,c}$ and $\sigma_{f,ta}$ are the failure limit stresses in uniaxial tension test, compression test and pure shear test, respectively. For the lower curve, the points of intersection with the three straight lines can be identified through the symmetry indicated above.

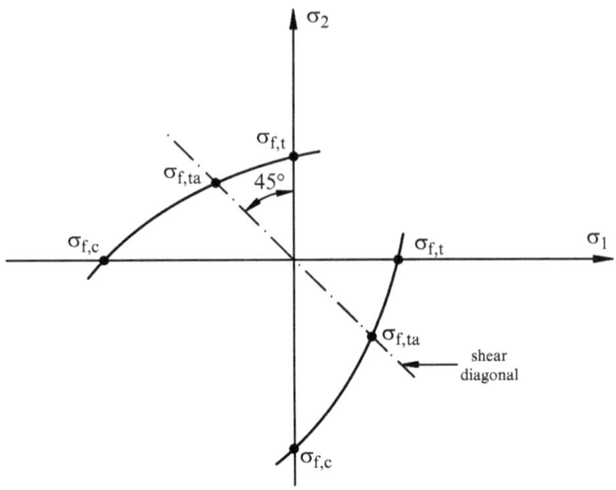

Fig. 9.4 Mohr's criterion for biaxial stress represented on coordinates (σ_1, σ_2)

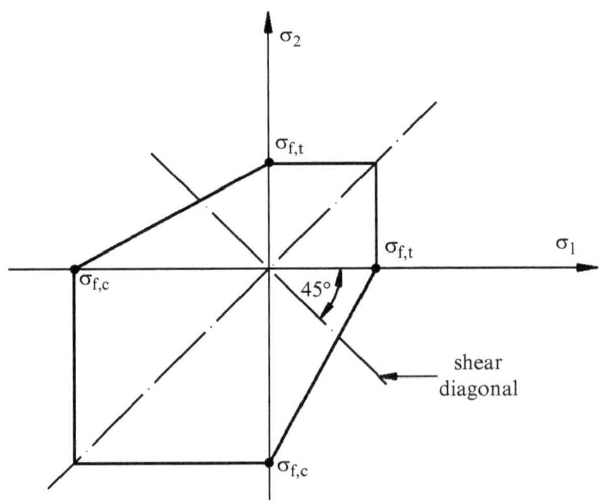

Fig. 9.5 Failure strength closed irregular polygon of the simplified Mohr's criterion for biaxial stress represented on coordinates (σ_1, σ_2)

Just as the two straight lines in Fig. 9.3, which are symmetrical with respect to the axis of the abscissa, are an approximate representation of the two curves in Fig. 9.2 for Mohr's criterion in Mohr's plane, in plane (σ_1, σ_2), the two curves in Fig. 9.4 can be represented by a closed irregular hexagon, whose sides differ in the first and third quadrants (Fig. 9.5). This is the strength limit hexagon for Mohr's simplified criterion. From this standpoint as well, Mohr's criterion can be considered as a modification of the τ_{max} criterion.

In other words, the $\tau = \tau(\sigma)$ and $\sigma_2 = \sigma_2(\sigma_1)$ curves for Mohr's criterion can be obtained starting from the corresponding curves for the τ_{max} criterion, considering and including the effect of internal friction. For this reason Mohr's criterion is also called the internal-friction-criterion. Proposed in 1897 by Reyto, this criterion assumes that sliding along the slip planes is inhibited by the friction forces associated with the compressive forces acting orthogonally to said planes.

In plane (σ_1, σ_2), Mohr's criterion in its generalized formulation, i.e., that using comprehensive experimental data, leads to a closed curve like the failure strength ellipse, but asymmetrical with respect to the shear diagonal. It is thus an ovoid, larger in the third quadrant and smaller in the first quadrant, but still symmetrical with respect to the straight-line bisecting the first and third quadrants. Mohr's criterion is thus more general than the maximum distortion energy criterion, in that it can account for and simulate the material's differing behaviour in compression and tension.

9.2 General Considerations on the Use of Failure Criteria

The main failure criteria applied in the design sector of interest to us here were discussed in the preceding section. To choose the best criterion for use, however, they must be re-examined in the light of the experimental results obtained by the various researchers who have addressed this topic. To achieve the desired biaxial stress state, many of these researchers have used thin-walled steel, copper and aluminium tubes subjected simultaneously to internal pressure and to axial and torsional loading, or tubes under combined tensile and torsional loading. Most of the results obtained, all from tests at ambient temperature, consist of points representing stress states in the first and fourth quadrants, and which are denser around the maximum-distortion-energy failure ellipse or, at most, fall in the zones between the straight lines of the strength limit hexagon and the corresponding curve segments of the strength limit ellipse. The location of these experimental points, which are shown in Fig. 9.1, demonstrates unequivocally that the failure criterion that best interprets experimental findings is that of maximum-distortion-energy.

For certain load combinations, several of the four main failure criteria (σ_{max}, τ_{max}, ε_{max} and maximum-distortion-energy) provide the same results, and are equally advantageous. In other cases, the results of the theoretical models differ, and the problem thus arises of selecting the criterion that best interprets the actual risk of failure in the material. Generally speaking, it must be recognized that there is no single criterion: we can only identify the criterion that is best for a particular purpose. The same criterion applied in different situations may result in errors. For example, repeated experimental tests have demonstrated that yield strength in torsional is about 12–15 % higher than the shear strength determined through tensile testing. This discrepancy, though it can be explained theoretically, would not arise if the τ_{max} criterion were correct. The maximum-distortion-energy criterion does not suffer from this limitation, and its accuracy is borne out both by theoretical considerations and experimental findings, making it the optimal choice among the different failure criteria at least for ductile materials.

A number of tests carried out on specimens consisting of cast iron or other brittle materials have led to the conclusion that the σ_{max} failure criterion is better for brittle materials than the other criteria. However, data from more accurate tests conducted in recent years have clearly demonstrated that the behaviour of brittle materials can be effectively interpreted by the maximum distortion energy criterion, providing that account is made for the effects of stress concentrations resulting from the inclusions that often cause a structural component to lose ductility.

Lastly, the ε_{max} failure criterion, widely used in the nineteenth century and the first half of the twentieth, particularly in the design of artillery and of porcelain and reinforced concrete components, has been largely abandoned today. Here, it is not used.

From the general design standpoint, in any case, the σ_{max} failure criterion can also be used with success, providing that the stress state is represented by points falling in the first and third quadrants of Fig. 9.1. In the first quadrant, both stresses are positive and the σ_{max} and τ_{max} criteria lead to the same results. When the principal stresses are all equal, the three σ_{max}, τ_{max} and maximum distortion energy criteria lead to the same results. If the stress state is represented by points falling in the second and third quadrants, it is obvious that only the τ_{max} and maximum distortion energy criteria can be used.

In mechanical design, the material's failure criterion is selected and used on the basis of more general considerations. In these pages, we are dealing with operating conditions that do not involve excessive elastic displacements, instability, creep, impact and fatigue, all of which call for further and more specific considerations. Notch effects and stress concentrations are considered as localised effects in the areas affected by the geometrical irregularities. The fatigue aspects are not considered. Attention focuses here on static and quasi-static.

Where a preliminary review of actual design conditions indicates that component failure may take place during service with modes that differ from those found for standard specimens, it is obvious that the failure criteria discussed above cannot be applied. To give an example, if it is suspected that brittle fracture could occur in a component consisting of normally ductile steel, none of these criteria is capable of linking this condition to the data obtained from standard tensile tests. In such a situation, it will be necessary to use more advanced theoretical considerations and experimental data from specific tests (impact tests, for instance) carried out under conditions that approximate those encountered during actual service to the greatest possible extent.

We will now focus attention on cases where the material failure criteria can be correctly applied, starting from their use in predicting ductile yielding. Here, the criterion that interprets experimental results most accurately is that of the maximum distortion energy. It is thus recommended that it be used, though with some reservations. For the biaxial stress state, when good experimental data are available that go beyond those from simple tensile tests, Mohr's criterion is to be preferred. In this connection, however, a possible weakness of Mohr's criterion should be noted: as it is difficult to obtain highly accurate compression and tension data (i.e., giving as general an idea as possible regarding the stress state in experimental conditions), the errors may greater than those associated with the maximum distortion energy criterion.

For triaxial stress states, the amount of data available is relatively modest. The maximum distortion energy criterion can nevertheless be applied with good results, but there is also no lack of grounds for recommending that the more conservative internal-friction-criterion be used. For both biaxial and triaxial stress states, it is still fairly common to use the simple and slightly conservative τ_{max} criterion together with data from standard tensile tests. Many regulations and calculation standards are based on this criterion.

The predictive criteria for ductile fracture coincide in general with the predictive criteria for yielding. However, the experimental data available in this connection are not only more limited numerically, but also show greater scatter. For these reasons, it is sometimes recommended that the internal-friction-criterion be used instead of the maximum-distortion-energy criterion.

To predict the failure strength of materials that can be affected by brittle fractures in simple tension and compression, it is once again necessary to have reliable experimental data. For biaxial stress states, Mohr's criterion can be used. The internal-friction-criterion can be employed if uniaxial tension and compression data are available, and is reasonably accurate. It is often recommended for this reason, but the results obtained are in general conservative. The σ_{max} criterion is occasionally used, but it often leads to results that, for biaxial tension-compression stress states, can be affected by significant errors, as they are heavily skewed towards the non-conservative side. Experimental results obtained by a number of researchers on the basis of tests conducted on grey cast iron are clearly included between the theoretical curves for the σ_{max} criterion and the internal-friction-criterion, which are thus in agreement when a biaxial stress state is considered.

9.3 Effects of Stress Concentrations

It was mentioned earlier that rotors generally have irregularities in form, most frequently because of holes provided for purposes such as retaining the rotor to the machine, attaching parts to the disk or channeling cooling fluids. If the hole is small by comparison with the size of the disk, or if the distance between two successive holes is much larger than their radius, the effect of the holes can be considered as a local disturbance of the stress field calculated assuming axisymmetry.

For an isolated hole of small diameter $2{\cdot}r_i$ centred along the axis of a thin-walled rectangular plate, having short sides of dimension $2{\cdot}r_e >> 2{\cdot}r_i$ and stressed on the mid-plane of the short sides by a uniform surface force distribution σ (Fig. 9.6), an elegant theoretical solution presented by Kirsch [38], which has been confirmed experimentally through strain gauge and photoelastic analyses, can be used to determine the stress state on the hole's periphery through the superposition of two stress states (Fig. 9.7) for the plate with central hole whose outer radius $2{\cdot}r_e$ is indicated by the dashed line in Fig. 9.6.

The first stress state sees the plate as subjected in its mid-plane and at its outer radius to a uniform radial stress distribution $\sigma/2$, while the second sees the same plate as simultaneously subjected, again in its mid-plane and at its outer radius, to a radial stress distribution $(\sigma/2){\cdot}cos\ 2\vartheta$ and a shear stress distribution $-(\sigma/2){\cdot}sin\ 2\vartheta$

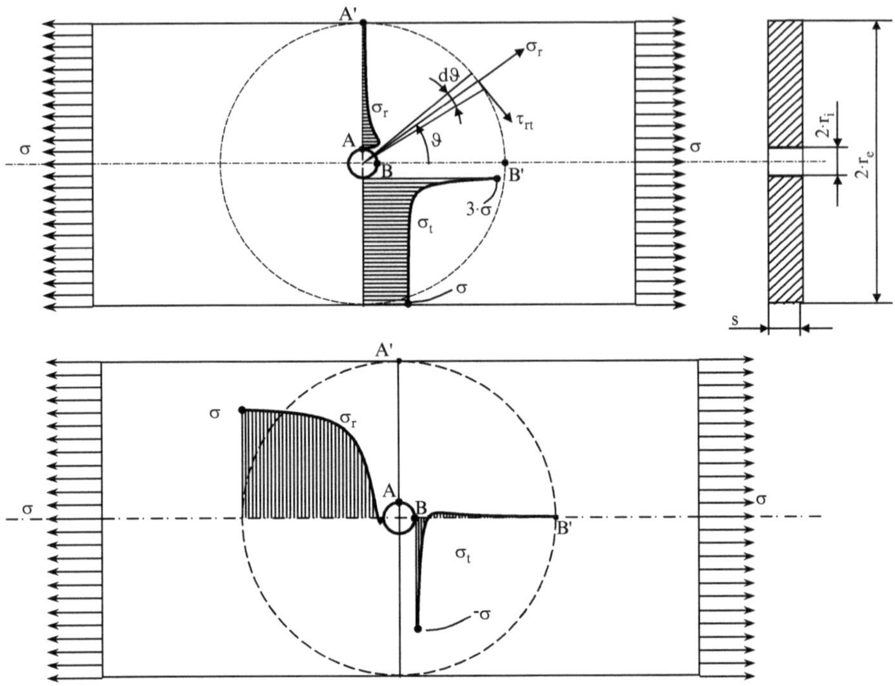

Fig. 9.6 Thin rectangular sheet with central hole whose radius r_i is small compared to r_e, stressed on the short sides and in the mid-plane by a surface force distribution σ

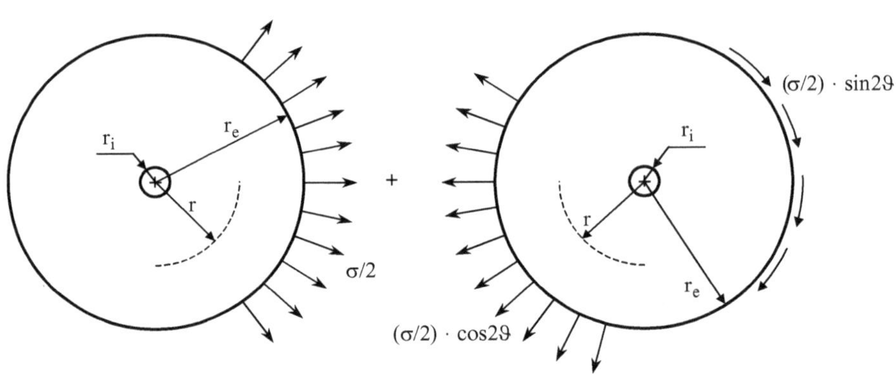

Fig. 9.7 Superposition of two stress states for analysing the stress state of the circular plate take from the rectangular plate shown in Fig. 9.6

(the minus sign indicates that, as illustrated in Fig. 9.6, the shear stress acts in the opposite direction to the increasing angular coordinates ϑ).

The results obtained by Kirsch through the modelling procedure described above enable us to express the stress state at the generic radius r and at the generic angular coordinate ϑ with the following relations:

$$\begin{cases} \sigma_r = \frac{\sigma}{2} \cdot \left(1 - \frac{r_i^2}{r^2}\right) + \frac{\sigma}{2} \cdot \left(1 + \frac{3 \cdot r_i^4}{r^4} - \frac{4 \cdot r_i^2}{r^2}\right) \cdot cos\, 2\vartheta \\ \sigma_t = \frac{\sigma}{2} \cdot \left(1 + \frac{r_i^2}{r^2}\right) - \frac{\sigma}{2} \cdot \left(1 + \frac{3 \cdot r_i^4}{r^4}\right) \cdot cos\, 2\vartheta \\ \tau_{rt} = -\frac{\sigma}{2} \cdot \left(1 - \frac{3 \cdot r_i^4}{r^4} + \frac{2 \cdot r_i^2}{r^2}\right) \cdot sin\, 2\vartheta. \end{cases} \tag{9.9}$$

These relations derive from the more general relations, neglecting the ratio r_i^2/r_e^2 and, even more importantly, the ratio r_i^4/r_e^4 in the latter. It can thus be concluded that the stress state in points A, A', B and B' in Fig. 9.6 is characterized by the following values:

$$\begin{array}{llll} A(r = r_i; \vartheta = \pi/2): & \sigma_r = 0; & \sigma_t = 3 \cdot \sigma; & \tau_{rt} = 0 \\ A'(r = r_e; \vartheta = \pi/2): & \sigma_r = 0; & \sigma_t = \sigma; & \tau_{rt} = 0 \\ B(r = r_i; \vartheta = 0): & \sigma_r = 0; & \sigma_t = -\sigma; & \tau_{rt} = 0 \\ B'(r = r_e; \vartheta = 0): & \sigma_r = \sigma; & \sigma_t = 0; & \tau_{rt} = 0. \end{array}$$

These results show that along the periphery of the hole, at the ends of the diameter perpendicular to the direction of loading, the only non-zero stress is the hoop stress, which is $3{\cdot}\sigma$ at A and σ at A'. At the ends of the diameter parallel to the direction of loading, stresses at the hole periphery, i.e., at B, are $\sigma_t = -\sigma$ and $\sigma_r = \sigma$ at B', while all other stresses are zero. Figure 9.6 shows hoop stress σ_t and radial stress σ_r versus radius, both in direction A-A' perpendicular to the loading axis, and in direction B-B' parallel to the direction of stresses σ. The curve for σ_t shows that we have $\sigma_t = 3{\cdot}\sigma$ at A and $\sigma_t = -\sigma$ in B. The stress concentration resulting from the hole thus leads to a threefold increase in the value of applied stress σ localized at A.

For a plate subjected to uniaxial stress along the direction indicated in Fig. 9.6, we have the result summarized in Fig. 9.8a. If the same plate is stressed at its mid-plane, along two mutually orthogonal directions (A-A' and B-B' in Fig. 9.6) by uniformly distributed surface forces σ_1 and σ_2, we can apply the principle of superposition to obtain the result shown in Fig. 9.8b.

However, the problem of interest to us here is that of a rotating disk featuring a small-diameter eccentric hole with axis parallel to that of the rotor and located at a distance r from it. Once the stresses σ_r and σ_t that we would have at the position occupied by the centre of the hole if the latter did not exist have been calculated, it is obvious that a rectangular plate element such as that represented in Fig. 9.6 can be identified, with direction A-A' along the radial passing through the centre of the hole. In this case, bearing Fig. 9.8b in mind, we have $\sigma_1 = \sigma_t$ and $\sigma_2 = \sigma_r$.

Applying Kirsch's theory to this specific case thus enables us to quantify the maximum stresses at the periphery of the hole as follows:

$$\begin{aligned} \sigma_{t,max} &= 3 \cdot \sigma_t - \sigma_r \\ \sigma_{r,max} &= 3 \cdot \sigma_r - \sigma_t, \end{aligned} \tag{9.10}$$

where, as we have said, σ_t and σ_r are the stresses that would occur at the position occupied by the centre of the hole if the hole did not exist. Stresses (9.10) act

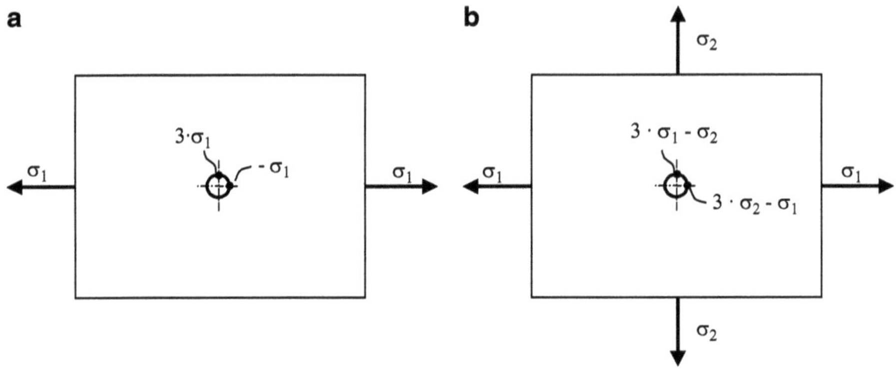

Fig. 9.8 Values of hoop stress σ_t at points A and B indicated in Fig. 9.6: (**a**) uniaxial stress; (**b**) biaxial stress

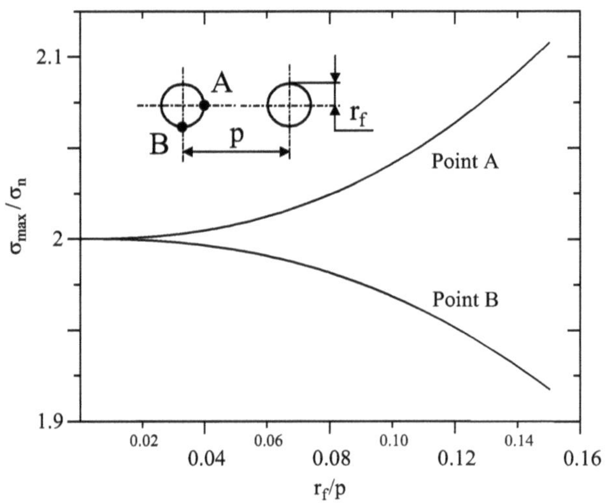

Fig. 9.9 Ratio σ_{max}/σ_n at edge of holes versus ratio r_f/p

tangentially to the edge of the hole and are found at the ends of the diameter whose extension passes through the centre of the disk and at the ends of the diameter perpendicular to the first.

The small-diameter central hole can be regarded as the limiting case of an eccentric hole as the distance of its axis from the disk axis tends to zero. In this case, σ_r and σ_t tend to be equal and the value at the edge of the hole tends to a value which is twice that which it would assume if there were no hole.

It also follows from relations (9.10) that making a hole in a uniform strength disk ($\sigma_t = \sigma_r = \sigma$) doubles the value of the stress, regardless of point where the hole is located.

If there is an array of holes, it should be borne in mind that the increase in stress will be greater than that which would occur with a single hole. By way of example, Fig. 9.9 shows the ratio (σ_{max}/σ_n) of maximum stress at the edge of the holes to nominal stress σ, versus ratio r_f/p of hole radius to spacing for a uniform strength disk. Consequently, it is good design practice to avoid holes in favour of other methods of securing highly stressed disks to the shaft.

To evaluate stress concentrations in fillets, notches and any other irregularities in form, it is necessary to use numerical discretization techniques such as the finite element method.

Chapter 10
Stress Analysis of Rotating Cylinders in the Linear Elastic Field

10.1 Principles and General Equations

We will consider a circular cylinder of any radial thickness (Fig. 10.1a) having theoretically infinite length and geometry of revolution, consisting of homogeneous and isotropic material subjected in the linear elastic field to surface forces, body forces and thermal loads which are all axisymmetric and invariant in the direction of rotational axis z. In these conditions, taking for example a rotor having heavy-walled thickness, each elementary portion of the cylinder located between two cross sections having longitudinal coordinates z and $z + dz$ (Fig. 10.1a) would, if considered in isolation, behave as a disk, with a resulting strain state that would show a longitudinal unit elongation ε_z which is a function of the radius (Fig. 10.1b).

However, in long (or infinitely long) circular cylinder subjected to centrifugal load and to thermal load, it is not conceivable that longitudinal unit elongations and displacements can vary with the radius, except at the ends, which are assumed to be free. The fact that it is impossible to have a longitudinal unit elongations ε_z that differs with radius r (expansion cannot proceed freely in a so loaded continuous body) gives rise to a state of longitudinal coaction characterised by a longitudinal normal stress σ_z which is a function of the radius. By comparison with a disk subjected to the same load conditions, where a plane stress state is found, the problem examined here entails either a plane strain state ($\varepsilon_z = 0$), if the rotor is constrained at its ends or indefinitely extended in the direction of its axis, or a strain state characterised by a uniform axial translation of a generic cross section ($\varepsilon_z = const$) if the rotor has finite axial length and is free at the ends. Both cases of strain characterised by $\varepsilon_z = 0$ and $\varepsilon_z = const$ will be examined here in turn.

The general treatment of this case assumes that the end sections of the solid are not loaded. The stress state resulting from any longitudinal external loads acting on the ends can be evaluated separately and then added to those analysed below, as the principle of superposition applies in the linear elastic field assumed here. As for the disk, the radial, tangential and axial directions are the principal directions of both

V. Vullo and F. Vivio, *Rotors: Stress Analysis and Design*,
Mechanical Engineering Series, DOI 10.1007/978-88-470-2562-2_10,
© Springer-Verlag Italia 2013

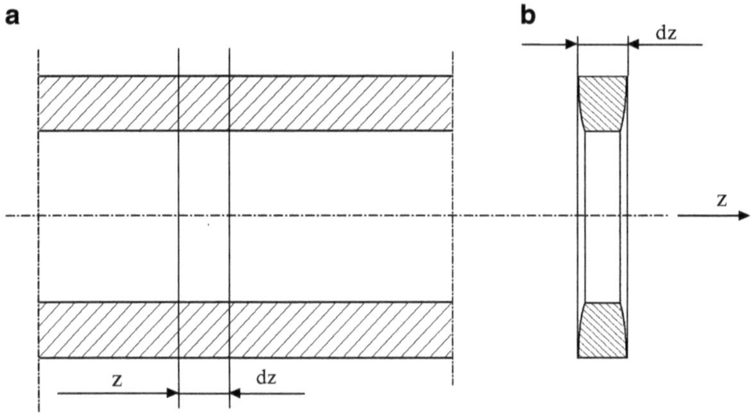

Fig. 10.1 (**a**) Heavy-walled cylinder extended indefinitely in the direction of the z axis; (**b**) isolated elementary portion of dimension dz, its longitudinal deformation and distortion of its end sections

stresses and strains, as the rotor is loaded axisymmetrically in the radial direction and uniformly in the axial direction.

As a general premise, it should first be borne in mind that the equilibrium equation in the radial direction (the first of (1.10)) continues to apply, and can be simplified here, given that when we are dealing with a part of a cylinder having unitary axial length located at a sufficient distance from the ends to be unaffected by perturbations at the edges, the thickness h disappears from the equation, which thus becomes:

$$\frac{d}{dr}(\sigma_r \cdot r) - \sigma_t + \gamma \cdot \omega^2 \cdot r^2 = 0. \tag{10.1}$$

The first two geometric relations (1.14) which give normal strain components ε_r and ε_t as function of radial displacement $u = u(r)$ also continue to hold, and are given here:

$$\varepsilon_r = \frac{du}{dr}; \qquad \varepsilon_t = \frac{u}{r}, \tag{10.2}$$

as is the compatibility equations (1.17):

$$\varepsilon_r = \frac{d}{dr}(r \cdot \varepsilon_t). \tag{10.3}$$

To solve the problem, two further relations must be added to those indicated above, viz.:

$$\varepsilon_z = \frac{dw}{dz} = \varepsilon_0 = const, \tag{10.4}$$

and

$$F_z = \int_A \sigma_z dA = 0. \tag{10.5}$$

Relation (10.4) expresses the plane strain state condition, in cases where the constant is zero, or the strain state characterised by a uniform axial translation of a generic cross section of the cylinder in cases where the constant is not zero; w denotes the displacement of the generic cross section of abscissa z, in the direction of the rotational axis z. Relation (10.5), which applies to the cylinder with free unloaded ends, expresses the global equilibrium condition of a generic cross section A in the direction of the rotational axis; dA is the area element of this cross section, while F_z is the axial load.

The thermoelasticity relations which link strain components (normal strain components and shear strain components) to stress components (normal stress components and shear stress components) and express Hooke's constitutive laws for the triaxial stress state in the linear elastic field with thermal loading are as follows:

$$\begin{cases} \varepsilon_r = \dfrac{1}{E} \cdot [\sigma_r - \nu(\sigma_t + \sigma_z)] + \alpha \cdot T \\ \varepsilon_t = \dfrac{1}{E} \cdot [\sigma_t - \nu(\sigma_r + \sigma_z)] + \alpha \cdot T \\ \varepsilon_z = \dfrac{1}{E} \cdot [\sigma_z - \nu(\sigma_t + \sigma_r)] + \alpha \cdot T \\ \gamma_{rt} = \dfrac{\tau_{rt}}{G} \\ \gamma_{tz} = \dfrac{\tau_{tz}}{G} \\ \gamma_{zr} = \dfrac{\tau_{zr}}{G}. \end{cases} \tag{10.6}$$

The last three of these relations, i.e., those linking shear strain components to shear stress components, are clearly not influenced by temperature, as unconstrained thermal expansion does not case angular distortions in an isotropic material.

From the first three thermoelasticity relations (10.6), introducing the first invariant of the strain tensor, i.e., volume expansion

$$e = \varepsilon_r + \varepsilon_t + \varepsilon_z = \frac{1 - 2\nu}{E} J_1 + 3\alpha T, \tag{10.7}$$

with $J_1 = (\sigma_r + \sigma_t + \sigma_z)$ being the first invariant of the stress tensor, we obtain the following general relations, which are also called thermoelasticity relations and express the three principal radial, hoop and axial stresses as a function of the principal strains:

$$\begin{cases} \sigma_r = \frac{E}{(1+v)} \cdot \left[\frac{v \cdot e}{1-2v} + \varepsilon_r\right] - \frac{E}{1-2v}\alpha \cdot T \\ \sigma_t = \frac{E}{(1+v)} \cdot \left[\frac{v \cdot e}{1-2v} + \varepsilon_t\right] - \frac{E}{1-2v}\alpha \cdot T \\ \sigma_z = \frac{E}{(1+v)} \cdot \left[\frac{v \cdot e}{1-2v} + \varepsilon_z\right] - \frac{E}{1-2v}\alpha \cdot T \end{cases} \tag{10.8}$$

These relations can also be written in the following more compact form:

$$\begin{cases} \sigma_r = \lambda \cdot e + 2 \cdot G \cdot \varepsilon_r - \frac{E}{1-2v}\alpha \cdot T \\ \sigma_t = \lambda \cdot e + 2 \cdot G \cdot \varepsilon_t - \frac{E}{1-2v}\alpha \cdot T \\ \sigma_z = \lambda \cdot e + 2 \cdot G \cdot \varepsilon_z - \frac{E}{1-2v}\alpha \cdot T \end{cases} \tag{10.9}$$

where $\lambda = v\,E/[(1+v)(1-2v)]$ is Lamé's constant and $G = E/[2(1+v)]$ is the modulus of elasticity in shear or modulus of rigidity.

Relation (10.4) must be respected for any constraint condition at the ends of the circular cylindrical body considered in the direction of the axis of revolution, along which the temperature gradient is assumed to be zero, and axisymmetry is also assumed. Consequently, deriving the first of relations (10.8) or of relations (10.9) by respect to the radius gives:

$$\frac{d\sigma_r}{dr} = \frac{E}{1+v} \cdot \left[\frac{v}{1-2v} \cdot \left(\frac{d^2u}{dr^2} + \frac{1}{r} \cdot \frac{du}{dr} - \frac{u^2}{r^2}\right) + \frac{d^2u}{dr^2}\right] - \frac{E}{1-2v} \cdot \alpha \cdot \frac{dT}{dr}. \tag{10.10}$$

Expanding the equilibrium equation (10.1) which holds for axisymmetric circular cylindrical bodies and introducing the first and second of relations (10.8) or of relations (10.9) as well as relation (10.10) yields the following second order differential equation in terms of displacement $u = u\,(r)$:

$$\frac{d^2u}{dr^2} + \frac{1}{r} \cdot \frac{du}{dr} - \frac{u^2}{r^2} - \alpha \cdot \frac{1+v}{1-v} \cdot \frac{dT}{dr} + \frac{(1+v) \cdot (1-2v)}{(1-v)} \cdot \frac{\gamma \cdot \omega^2 \cdot r}{E} = 0. \tag{10.11}$$

This can also be obtained from (2.1) after substituting (as well be demonstrated in the following section) α with $\alpha(1+v)$, v with $v/(1-v)$ and E with $E/(1-v^2)$ in the latter. Without assuming any function for temperature distribution along the radius in order to ensure the most general possible treatment for the problem, two successive integrations of relation (10.11) lead to the relation:

$$u = \alpha \cdot \frac{1+\nu}{1-\nu} \cdot \frac{1}{r} \cdot \int_{r_i}^{r} T \cdot r \cdot dr - \frac{(1+\nu)\cdot(1-2\nu)}{(1-\nu)} \cdot \frac{\gamma \cdot \omega^2 \cdot r^3}{8E} + C_1 \cdot \frac{r}{2} + \frac{C_2}{r} \tag{10.12}$$

and then, through derivation, to:

$$\frac{du}{dr} = \alpha \cdot \frac{1+\nu}{1-\nu} \cdot T - \alpha \cdot \frac{1+\nu}{1-\nu} \cdot \frac{1}{r^2} \cdot \int_{r_i}^{r} T \cdot r \cdot dr - \frac{(1+\nu)\cdot(1-2\nu)}{(1-\nu)} \cdot \frac{3}{8} \cdot \frac{\gamma \cdot \omega^2 \cdot r^2}{E} + \frac{C_1}{2} - \frac{C_2}{r^2} \tag{10.13}$$

10.2 Circular Cylindrical Body with Clamped Ends or Indefinitely Extended in the Direction of its Axis Subjected to Centrifugal and Thermal Loads

For a circular cylindrical body of finite length clamped at the ends and for a circular cylindrical body indefinitely extended in the direction of its axis, we necessarily have $\varepsilon_z = 0$. Accordingly, the third relation (10.6) gives:

$$\sigma_z = \nu \cdot (\sigma_r + \sigma_t) - E \cdot \alpha \cdot T. \tag{10.14}$$

Substituting the σ_z thus determined in the first two thermoelasticity relations given by equations (10.6) yields the following expressions for normal strain components ε_r and ε_t:

$$\begin{cases} \varepsilon_r = \dfrac{1-\nu^2}{E} \cdot \left(\sigma_r - \dfrac{\nu}{1-\nu}\sigma_t\right) + (1+\nu) \cdot \alpha \cdot T \\ \varepsilon_t = \dfrac{1-\nu^2}{E} \cdot \left(\sigma_t - \dfrac{\nu}{1-\nu}\sigma_r\right) + (1+\nu) \cdot \alpha \cdot T. \end{cases} \tag{10.15}$$

Returning to the corresponding relations for ε_r and ε_t obtained earlier for the disk, for which a plane stress state ($\sigma_z = 0$) is assumed, viz.:

$$\begin{cases} \varepsilon_r = \dfrac{1}{E} \cdot [\sigma_r - \nu \cdot \sigma_t] + \alpha \cdot T \\ \varepsilon_t = \dfrac{1}{E} \cdot [\sigma_t - \nu \cdot \sigma_r] + \alpha \cdot T \end{cases} \tag{10.16}$$

and comparing them with expressions (10.15), we can see that passing from the disk to the long circular cylindrical body subjected to the same loads (centrifugal load and thermal load) involves the following variations, which concern only the material's thermomechanical properties:

$$\begin{aligned} \sigma_z = 0 \quad &\rightarrow \quad \varepsilon_z = 0 \\ \alpha \quad &\rightarrow \quad \alpha \cdot (1+\nu) \\ 1/E \quad &\rightarrow \quad (1-\nu^2)/E \\ \nu \quad &\rightarrow \quad \nu/(1-\nu). \end{aligned} \tag{10.17}$$

To take advantage of expressions (10.17), we will now return to differential equation (2.1) for the constant thickness disk ($\sigma_z = 0$) subjected simultaneously to centrifugal load and to a temperature gradient along the radius, which we will rewrite in the following form:

$$\frac{d}{dr}\left[\frac{1}{r} \cdot \frac{d}{dr}(u \cdot r)\right] = \alpha \cdot (1+\nu) \cdot \frac{dT}{dr} - (1-\nu^2) \cdot \frac{\gamma \cdot \omega^2 \cdot r}{E}, \tag{10.18}$$

and, without assuming any function for temperature distribution along the radius in order to ensure the most general possible treatment for the problem, we proceed to integrate the equation. In successive steps, we obtain:

$$\begin{aligned} u &= \frac{\alpha \cdot (1+\nu)}{r} \cdot \int_{r_i}^{r} T \cdot r \cdot dr - \frac{(1-\nu^2) \cdot \gamma \cdot \omega^2}{E} \cdot \frac{r^3}{8} + C_1 \cdot \frac{r}{2} + \frac{C_2}{r} \\ \frac{du}{dr} &= \alpha \cdot (1+\nu) \cdot T - \frac{\alpha \cdot (1+\nu)}{r^2} \cdot \int_{r_i}^{r} T \cdot r \cdot dr \\ &\quad - \frac{(1-\nu^2) \cdot \gamma \cdot \omega^2}{E} \cdot \frac{3 \cdot r^2}{8} + \cdot \frac{C_1}{2} - \frac{C_2}{r^2}. \end{aligned} \tag{10.19}$$

Solving (10.16) for stresses σ_r and σ_t gives:

$$\begin{cases} \sigma_r = \dfrac{E}{1-\nu^2} \cdot [\varepsilon_r + \nu \cdot \varepsilon_t - \alpha \cdot (1+\nu) \cdot T] \\ \sigma_t = \dfrac{E}{1-\nu^2} \cdot [\varepsilon_t + \nu \cdot \varepsilon_r - \alpha \cdot (1+\nu) \cdot T]. \end{cases} \tag{10.20}$$

Bearing (10.2) in mind and substituting relations (10.19) in (10.20) gives the following relations for σ_r and σ_t as a function of radius r:

$$\begin{cases} \sigma_r = -\dfrac{\alpha \cdot E}{r^2} \cdot \displaystyle\int_{r_i}^{r} T \cdot r \cdot dr - \dfrac{(3+\nu)}{8} \cdot \gamma \cdot \omega^2 \cdot r^2 + \dfrac{E \cdot C_1}{2 \cdot (1-\nu)} - \dfrac{E \cdot C_2}{(1+\nu) \cdot r^2} \\ \sigma_t = \dfrac{\alpha \cdot E}{r^2} \cdot \displaystyle\int_{r_i}^{r} T \cdot r \cdot dr - \alpha \cdot E \cdot T - \dfrac{(1+3\nu)}{8} \cdot \gamma \cdot \omega^2 \cdot r^2 + \dfrac{E \cdot C_1}{2 \cdot (1-\nu)} + \dfrac{E \cdot C_2}{(1+\nu) \cdot r^2}. \end{cases} \tag{10.21}$$

Finally, introducing the dimensionless variable ρ and taking relations (2.9) as well as relations (2.27) into account, relations (10.21) become:

$$\begin{cases} \sigma_r = A - B/\rho^2 - C \cdot \rho^2 - \left(\alpha \cdot E/\rho^2\right) \cdot \int_\beta^\rho T \cdot \rho \cdot d\rho \\ \sigma_t = A + B/\rho^2 - D \cdot \rho^2 + \alpha \cdot E \cdot \left(\frac{1}{\rho^2} \int_\beta^\rho T \cdot \rho \cdot d\rho - T \right). \end{cases} \tag{10.22}$$

As was mentioned earlier, passing from the plane stress state ($\sigma_z = 0$) to the plane strain state ($\varepsilon_z = 0$) entails the variations (10.17). Consequently, relations (10.21) applying to the disk change as follows for the long circular cylindrical body:

$$\begin{cases} \sigma_r = -\frac{\alpha \cdot E}{(1-\nu) \cdot r^2} \cdot \int_{r_i}^{r} T \cdot r \cdot dr - \frac{(3-2\nu)}{8 \cdot (1-\nu)} \cdot \gamma \cdot \omega^2 \cdot r^2 + \frac{E \cdot C_1}{2 \cdot (1+\nu) \cdot (1-2\nu)} - \frac{E \cdot C_2}{(1+\nu) \cdot r^2} \\ \sigma_t = \frac{\alpha \cdot E}{(1-\nu) \cdot r^2} \cdot \int_{r_i}^{r} T \cdot r \cdot dr - \frac{\alpha \cdot E \cdot T}{(1-\nu)} - \frac{(1+2\nu)}{8 \cdot (1-\nu)} \cdot \gamma \cdot \omega^2 \cdot r^2 + \frac{E \cdot C_1}{2 \cdot (1+\nu) \cdot (1-2\nu)} + \frac{E \cdot C_2}{(1+\nu) \cdot r^2}. \end{cases} \tag{10.23}$$

Obviously, relations (10.23) can be obtained directly from the first two thermoelasticity relations (10.8) or (10.9) by substituting relations (10.12) and (10.13) in them and merging the constant $\varepsilon_z = \varepsilon_0$ in constant C_1.

Subsequently, by introducing the dimensionless variable ρ, with the meaning and the expression of σ_0 given by the first relation (2.27) remaining unchanged, and by setting:

$$\bar{A} = \frac{E \cdot C_1}{2 \cdot (1 + \nu) \cdot (1 - 2\nu)} \qquad \bar{B} = B = \frac{E \cdot C_2}{(1 + \nu) \cdot r_e^2}$$

$$\bar{C} = \frac{3 - 2\nu}{8 \cdot (1 - \nu)} \cdot \sigma_0 \qquad \bar{D} = \frac{1 + 2\nu}{8 \cdot (1 - \nu)} \cdot \sigma_0 \, ,$$

we obtain the following relations from (10.23):

$$\begin{cases} \sigma_r = \bar{A} - \frac{\bar{B}}{\rho^2} - \bar{C} \cdot \rho^2 - \frac{\alpha \cdot E}{(1 - \nu) \cdot \rho^2} \cdot \int_\beta^\rho T \cdot \rho \cdot d\rho \\ \sigma_t = \bar{A} + \frac{\bar{B}}{\rho^2} - \bar{D} \cdot \rho^2 + \frac{\alpha \cdot E}{(1 - \nu)} \cdot \left(\frac{1}{\rho^2} \int_\beta^\rho T \cdot \rho \cdot d\rho - T \right). \end{cases} \tag{10.24}$$

For a hollow circular cylindrical body subjected simultaneously to centrifugal load and a temperature gradient along the radius, we can impose the following boundary conditions:

$$\begin{cases} \sigma_r = 0 & \text{for} \quad \rho = 1 \\ \sigma_r = 0 & \text{for} \quad \rho = \beta, \end{cases} \tag{10.25}$$

to obtain values for constants $\bar{A}$ and $\bar{B}$:

$$\begin{aligned} \bar{A} &= \frac{3-2v}{8\cdot(1-v)}\cdot\sigma_0\cdot\left(1+\beta^2\right)+\frac{\alpha\cdot E}{(1-v)\cdot\left(1-\beta^2\right)}\cdot\int_\beta^1 T\cdot\rho\cdot d\rho \\ \bar{B} &= \beta^2\cdot\left[\frac{3-2v}{8\cdot(1-v)}\cdot\sigma_0+\frac{\alpha\cdot E}{(1-v)\cdot\left(1-\beta^2\right)}\cdot\int_\beta^1 T\cdot\rho\cdot d\rho\right]. \end{aligned} \tag{10.26}$$

Substituting the constants thus determined in relations (10.24) yields the following expressions for stresses σ_r and σ_t:

$$\begin{cases} \sigma_r = \dfrac{3-2v}{8\cdot(1-v)}\cdot\sigma_0\cdot\left(1+\beta^2-\dfrac{\beta^2}{\rho^2}-\rho^2\right)+\dfrac{\alpha\cdot E}{\rho^2\cdot(1-v)}\cdot\left[\dfrac{\rho^2-\beta^2}{1-\beta^2}\cdot\displaystyle\int_\beta^1 T\cdot\rho\cdot d\rho-\int_\beta^\rho T\cdot\rho\cdot d\rho\right] \\ \sigma_t = \dfrac{3-2v}{8\cdot(1-v)}\cdot\sigma_0\cdot\left(1+\beta^2+\dfrac{\beta^2}{\rho^2}-\dfrac{1+2v}{3-2v}\rho^2\right)+\dfrac{\alpha\cdot E}{\rho^2\cdot(1-v)}\cdot\left[\dfrac{\rho^2+\beta^2}{1-\beta^2}\cdot\displaystyle\int_\beta^1 T\cdot\rho\cdot d\rho+\right. \\ \left.+\displaystyle\int_\beta^\rho T\cdot\rho\cdot d\rho\right]-\dfrac{\alpha\cdot E\cdot T}{(1-v)}. \end{cases} \tag{10.27}$$

Subsequently, substituting relations (10.27) in relation (10.14) gives the following general expression of axial stress σ_z:

$$\begin{aligned} \sigma_z = {} & \frac{3-2v}{4\cdot(1-v)}\cdot v\cdot\sigma_0\cdot\left(1+\beta^2-\frac{2\cdot\rho^2}{3-2v}\right)+\frac{\alpha\cdot E}{(1-v)} \\ & \times\left[\frac{2\cdot v}{1-\beta^2}\cdot\int_\beta^1 T\cdot\rho\cdot d\rho-T\right]. \end{aligned} \tag{10.28}$$

In conclusion, the three principal stresses that define the stress state in a circular cylindrical body with clamped ends or indefinitely extended in the direction of its axis, with no load at either the inner or outer radius and subjected to both centrifugal load and a temperature gradient along the radius according to any function $T = T(\rho)$ are as follows:

$$\begin{cases} \sigma_r = \dfrac{3-2\nu}{8\cdot(1-\nu)}\cdot\sigma_0\cdot\left(1+\beta^2-\dfrac{\beta^2}{\rho^2}-\rho^2\right)+\dfrac{\alpha\cdot E}{\rho^2\cdot(1-\nu)}\cdot\left[\dfrac{\rho^2-\beta^2}{1-\beta^2}\cdot\displaystyle\int_\beta^1 T\cdot\rho\cdot d\rho-\int_\beta^\rho T\cdot\rho\cdot d\rho\right] \\ \sigma_t = \dfrac{3-2\nu}{8\cdot(1-\nu)}\cdot\sigma_0\cdot\left(1+\beta^2+\dfrac{\beta^2}{\rho^2}-\dfrac{1+2\nu}{3-2\nu}\rho^2\right)+\dfrac{\alpha\cdot E}{\rho^2\cdot(1-\nu)}\cdot\left[\dfrac{\rho^2+\beta^2}{1-\beta^2}\cdot\displaystyle\int_\beta^1 T\cdot\rho\cdot d\rho+\right. \\ \left.+\displaystyle\int_\beta^\rho T\cdot\rho\cdot d\rho\right]-\dfrac{\alpha\cdot E\cdot T}{(1-\nu)}. \\ \sigma_z = \dfrac{3-2\nu}{4\cdot(1-\nu)}\cdot\nu\cdot\sigma_0\cdot\left(1+\beta^2-\dfrac{2\cdot\rho^2}{3-2\nu}\right)+\dfrac{\alpha\cdot E}{(1-\nu)}\cdot\left[\dfrac{2\cdot\nu}{1-\beta^2}\cdot\displaystyle\int_\beta^1 T\cdot\rho\cdot d\rho-T\right]. \end{cases} \tag{10.29}$$

On the other hand, the radial displacement $u = u(r)$ is linked to tangential strain component ε_t by relation $u = r\cdot\varepsilon_t$; given the second relation (10.6), we thus have:

$$u = r\cdot\varepsilon_t = \frac{r}{E}\cdot[\sigma_t - \nu\cdot(\sigma_r+\sigma_z)+\alpha\cdot E\cdot T]. \tag{10.30}$$

Introducing the dimensionless variable ρ and substituting relations (10.29) in relation (10.30) yields:

$$u = \frac{r_e\cdot\rho}{E}\cdot\frac{(1+\nu)}{(1-\nu)}\left\{\frac{(3-2\nu)}{8}\cdot\sigma_0\cdot\left[(1+\beta^2)\cdot(1-2\nu)+\frac{\beta^2}{\rho^2}-\frac{(1-2\nu)}{(3-2\nu)}\cdot\rho^2\right]+\frac{\alpha\cdot E}{(1-\beta^2)}\right.$$
$$\left.\times\left[(1-2\nu)+\frac{\beta^2}{\rho^2}\right]\cdot\int_\beta^1 T\cdot\rho\cdot d\rho+\frac{\alpha\cdot E}{\rho^2}\cdot\int_\beta^\rho T\cdot\rho\cdot d\rho\right\}. \tag{10.31}$$

Obviously, in the case considered here of a circular cylindrical body constrained at the ends or indefinitely extended in the direction of its axis, the axial displacement w of any cross section of abscissa z will be zero.

If the same body is also subjected to surface force distributions acting at the inner and outer radii, the stress state and the resulting radial displacement are given by the corresponding relations found for the disk having the same cross sectional geometry. This, as has already been demonstrated, assuming a plane stress state, a plane strain state, or a strain state characterised by uniform axial translation of a generic cross section will all lead to the same result in this case. It goes without saying that the overall stress state and the total radial displacement can be determined by means of the method of superposition, that is as the sum of the contributions made by the various loads, separately acting on the body, as it as been assumed that Hooke's law holds (linear elastic field).

Relations (10.29) and (10.31) also apply if the cylinder is not hollow. In this case, as we are dealing with a solid cylindrical rod, it is sufficient to set $\beta = 0$ in these relations. For the hollow cylindrical body, for $\beta \to 0$, the triaxial problem also gives rise to the singularity encountered in the biaxial problem; here again, in calculating the values at the inner radius for radial and hoop stresses and radial displacement, the singularity requires us to consider the ratio β^2/ρ^2 as unity.

10.2.1 Circular Cylindrical Body with Clamped Ends or Indefinitely Extended in the Direction of its Axis Subjected to Centrifugal Load

If the temperature terms are omitted from relations (10.29) and (10.31), we obtain the following relations expressing the stress field and radial displacement in the hollow cylinder with clamped ends or indefinitely extended in the direction of its axis, subjected only to centrifugal load:

$$\begin{cases} \sigma_r = \dfrac{(3-2\nu)}{8\cdot(1-\nu)}\cdot\sigma_0\cdot\left(1+\beta^2-\dfrac{\beta^2}{\rho^2}-\rho^2\right) \\ \sigma_t = \dfrac{(3-2\nu)}{8\cdot(1-\nu)}\cdot\sigma_0\cdot\left(1+\beta^2+\dfrac{\beta^2}{\rho^2}-\dfrac{1+2\nu}{3-2\nu}\cdot\rho^2\right) \\ \sigma_z = \dfrac{(3-2\nu)}{4\cdot(1-\nu)}\cdot\nu\cdot\sigma_0\cdot\left(1+\beta^2-\dfrac{2}{3-2\nu}\cdot\rho^2\right) \\ u = \dfrac{r_e\cdot\rho}{8E}\cdot\dfrac{(1+\nu)}{(1-\nu)}\cdot(3-2\nu)\cdot\sigma_0\cdot\left[(1+\beta^2)\cdot(1-2\nu)+\dfrac{\beta^2}{\rho^2}-\dfrac{(1-2\nu)}{(3-2\nu)}\cdot\rho^2\right]. \end{cases} \tag{10.32}$$

The first two of these relations can also be obtained from relations (2.34) by substituting ν with $\nu/(1 - \nu)$ in the latter. As relations (10.32) are specializations of relations (10.29) and (10.31), they also apply if the cylinder body only rotating is not hollow; in this case, it is sufficient to set $\beta = 0$. The problem regarding singularity obviously remains unchanged.

Figure 10.2 shows the distribution curves of the three principal stresses σ_r, σ_t and σ_z along the dimensionless radius of the annular cylindrical body considered here. As can be seen, all three principal stresses remain positive throughout the variation range of ρ, obviously with zero radial stresses at the inner and outer radii. Stress at the outer radius reaches its maximum value:

$$\sigma_{r,max} = \frac{(3-2\nu)}{8\cdot(1-\nu)}\cdot\sigma_0\cdot(1-\beta)^2 \tag{10.33}$$

for $\rho = \sqrt{\beta}$.

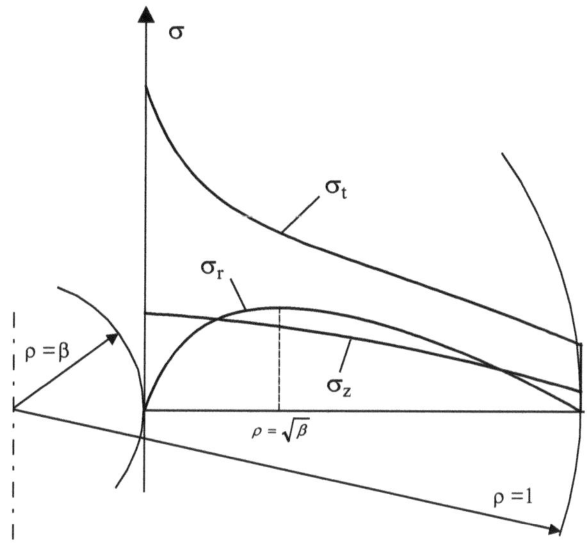

Fig. 10.2 Distribution curves of the three principal stresses σ_r, σ_t and σ_z, along the dimensionless radius of a rotating hollow cylinder with clamped ends

The stress states at the inner radius ($\rho = \beta$) and at the outer radius ($\rho = 1$) are given by the following two sets of three relations respectively:

$$\begin{cases} \sigma_r = 0 \\ \sigma_t = \dfrac{(3-2\nu)}{4\cdot(1-\nu)}\cdot\sigma_0\cdot\left(1+\dfrac{1-2\nu}{3-2\nu}\cdot\beta^2\right) & \text{for } \rho=\beta \\ \sigma_z = \nu\cdot\sigma_t \end{cases} \tag{10.34}$$

and

$$\begin{cases} \sigma_r = 0 \\ \sigma_t = \dfrac{(3-2\nu)}{4\cdot(1-\nu)}\cdot\sigma_0\cdot\left(\beta^2+\dfrac{1-2\nu}{3-2\nu}\right) & \text{for } \rho=1 \\ \sigma_z = \nu\cdot\sigma_t. \end{cases} \tag{10.35}$$

For a hollow cylindrical body with clamped ends or indefinitely extended in the direction of its axis and subjected only to surface force distributions at the inner and/or outer radius, we can start from relations (2.16) and (2.19), using the similarities indicated above. But (2.16) and (2.19) do not contain quantities E, ν and α, which thus do not change. Consequently, we are dealing in this case with the generalised plane strain state, for which the results do not differ from those for the plane stress state.

10.2.1.1 Example

A hollow cylindrical rotor made of unquenched AISI 1060 steel, with $\sigma_y =$ 480 MPa, rotates at an angular velocity $\omega = 750$ rad/s and has clamped ends and inner and outer radii unloaded. The rotor features outer and inner radii of 600 mm and 300 mm respectively and is 3 m long.

We will calculate the principal stresses at the inner and outer radii and the radial displacements, also at the inner and outer radii, in the cross section located at an equal distance from the two end constraints. In addition, we will use the maximum distortion energy failure criterion to determine the safety factor for the material at the inner radius under stationary conditions.

Bearing in mind that $v = 0.3$, $\beta = r_i/r_e = 0.5$ and $E = 210$ GPa, relations (10.34) give the following values for the three principal stresses at the inner radius:

$$\sigma_r = 0$$
$$\sigma_t = 353 \text{ MPa}$$
$$\sigma_z = 106 \text{ MPa}.$$

Relations (10.35) give the following values for the three principal stresses at the outer radius:

$$\sigma_r = 0$$
$$\sigma_t = 141 \text{ MPa}$$
$$\sigma_z = 42 \text{ MPa}.$$

The last of relations (10.32) gives the following values for radial displacements in the cross section located at an equal distance from the end constraints, for $\rho = \beta = 0.5$ and for $\rho = 1$ respectively:

$$(u)_{\rho=\beta} = 2.3 \times 10^{-4} \text{ m}; \ (u)_{\rho=1} = 1.8 \times 10^{-4} \text{ m}.$$

From relation $\sigma_e = (\sigma_t^2 + \sigma_z^2 - \sigma_t\sigma_z)^{1/2}$ calculated for $\rho = \beta$, we find that the equivalent stress at the inner radius is equal to 313 MPa. Consequently, the safety factor for the material is 1.53.

10.2.2 *Circular Cylindrical Body with Clamped Ends or Indefinitely Extended in the Direction of its Axis Subjected to Thermal Load*

If the terms for centrifugal loading are omitted from relations (10.29) and (10.31), we obtain the following relations expressing the stress state and radial displacement in the hollow cylinder with clamped ends or indefinitely extended in the direction of its axis, subjected only to thermal load:

$$\begin{cases} \sigma_r = \dfrac{\alpha \cdot E}{\rho^2 \cdot (1-\nu)} \cdot \left[\dfrac{\rho^2 - \beta^2}{1-\beta^2} \cdot \displaystyle\int_\beta^1 T \cdot \rho \cdot d\rho - \int_\beta^\rho T \cdot \rho \cdot d\rho \right] \\ \sigma_t = \dfrac{\alpha \cdot E}{\rho^2 \cdot (1-\nu)} \cdot \left[\dfrac{\rho^2 + \beta^2}{1-\beta^2} \cdot \displaystyle\int_\beta^1 T \cdot \rho \cdot d\rho + \int_\beta^\rho T \cdot \rho \cdot d\rho \right] - \dfrac{\alpha \cdot E \cdot T}{(1-\nu)} \\ \sigma_z = \dfrac{\alpha \cdot E}{(1-\nu)} \cdot \left[\dfrac{2 \cdot \nu}{1-\beta^2} \cdot \displaystyle\int_\beta^1 T \cdot \rho \cdot d\rho - T \right] \\ u = r_e \cdot \rho \cdot \alpha \cdot \dfrac{(1+\nu)}{(1-\nu)} \cdot \left[\dfrac{1}{1-\beta^2} \cdot \left(1 - 2\nu + \dfrac{\beta^2}{\rho^2} \right) \cdot \displaystyle\int_\beta^1 T \cdot \rho \cdot d\rho + \dfrac{1}{\rho^2} \cdot \int_\beta^\rho T \cdot \rho \cdot d\rho \right]. \end{cases} \tag{10.36}$$

Here again, the first two of these relations can be obtained from relations (3.24) by substituting α with $\alpha \cdot (1 + \nu)$, ν with $\nu/(1 - \nu)$ and E with $E/(1 - \nu^2)$. As relations (10.36) are specializations of relations (10.29) and (10.31), they also apply if the non-rotating cylindrical body subject to a temperature gradient along its radius is not hollow; the solid cylindrical body equations can thus be obtained from relations (10.36), setting $\beta = 0$ in it. The problem regarding singularity obviously remains unchanged.

By contrast with the procedure employed for disks, we will now assume that the function $T = T(\rho)$ of temperature distribution with the radius is Fourier's law of heat-flow by conduction in cylindrical bodies of circular or annular cross section having inside radius r_i ($\rho = \beta$) and outside radius r_e ($\rho = 1$). For long circular bodies such as heat exchanger tubes, smokestacks and long rotors, heat flow is essentially radial, except at the ends. In the steady-state conditions considered here, the heat flow passes through a portion of the generic cylindrical element surface having unitary axial length, inner radius ρ and outer radius $\rho + d\rho$, and is expressed by Fourier's equation:

$$Q = -2\pi k \rho \frac{dT}{d\rho}, \tag{10.37}$$

where Q is the heat flow, that is the instantaneous rate at which heat is transferred in direction normal to flow surface (Q is considered positive when in the centrifugal direction), k is the material's coefficient of thermal conductivity, $2\pi\rho$ is the flow surface having unitary axial length, and the $dT/d\rho$ is the instantaneous temperature gradient in the direction normal to flow surface (the minus sign indicates that heat flow is in the direction of decreasing temperatures, from the inner to the outer radius).

Thermal conductivity is a thermophysical property of the material which, in general, depends on temperature. For anisotropic materials, k also depends on the direction of heat flow. In the case of isotropic materials, the coefficient of thermal conductivity k can be expressed by means of a power series in temperature T and, in many cases of design interest, can be approximated by either a constant or a linear function of T.

For one-dimensional heat flow under steady-state conditions in a homogeneous material where there are no energy sources and no energy sinks and for which k is a known function of temperature and direction, Q is constant and (10.37) can be integrated provided that wall temperatures T_i and T_e at the body's inner and outer radii are known. Barring exceptional cases, which will be briefly discussed in a later section, steady-state heat flow conditions are almost always respected in design. Indeed, in the problems of greatest interest to the structural designer, variations of temperature over time can almost always be ignored, and the temperature profile can be considered as a succession of states, each under steady-state conditions, for which the relations given below apply.

By separating the variables and integrating (10.37) between the inner and the outer radii, with the general assumption that k is a function of temperature, we obtain:

$$\frac{Q}{2\pi}\ln\frac{r_e}{r_i} = -\int_{T_i}^{T_e} k \cdot dT. \tag{10.38}$$

Multiplying and dividing the right-hand side member of relation (10.38) by the temperature difference $(T_e - T_i)$, this relation becomes:

$$\frac{Q}{2\pi}\ln\frac{r_e}{r_i} = -\left[\frac{\int_{T_i}^{T_e} k \cdot dT}{T_e - T_i}\right](T_e - T_i). \tag{10.39}$$

The term in square brackets, indicated with

$$\bar{k} = \frac{\int_{T_i}^{T_e} k \cdot dT}{T_e - T_i}, \tag{10.40}$$

represents the average value of k in the temperature range $(T_e - T_i)$. When this range is relatively small, a constant coefficient of thermal conductivity k, equal to $\bar{k}$ and calculated as the arithmetic mean of the values of k at temperatures T_i and T_e, is usually introduced in Fourier's equation (10.37). This is the assumption that will be made henceforth.

With this assumption, integrating Fourier's equation between the generic radius ρ and the outer radius (the same result can be reached by integrating (10.37) and calculating the integration constant with the condition that $T = T_e$ at the outer radius, i.e., at $\rho = 1$) gives the following function $T = T(\rho)$:

$$T = T_e - \frac{Q}{2\pi k} \cdot \ln \rho, \tag{10.41}$$

where $T_e = T_{\rho=1}$ is the temperature at the outer radius.

For $\rho = \beta$, i.e., for $T = T_i$, where $T_i = T_{\rho=\beta}$ is the temperature at the inner radius, relation (10.41) gives:

$$-\frac{Q}{2\pi k} = \frac{T_i - T_e}{\ln \beta}, \tag{10.42}$$

Consequently, relation (10.41) can be rewritten in the following form:

$$T = T_e + \frac{T_i - T_e}{\ln \beta} \ln \rho. \tag{10.43}$$

By introducing relation (10.43) in relations (10.36) and performing integrations, we then obtain the following relations:

$$\begin{cases} \sigma_r = \dfrac{\alpha E}{2(1-\nu)} \dfrac{T_i - T_e}{\ln \beta} \left(\dfrac{\beta^2}{\rho^2} \dfrac{1-\rho^2}{1-\beta^2} \ln \beta - \ln \rho \right) \\ \sigma_t = \dfrac{\alpha E}{2(1-\nu)} \dfrac{T_i - T_e}{\ln \beta} \left(-1 - \dfrac{\beta^2}{\rho^2} \dfrac{1+\rho^2}{1-\beta^2} \ln \beta - \ln \rho \right) \\ \sigma_z = -\dfrac{\alpha E}{2(1-\nu)} \left\{ \dfrac{T_i - T_e}{\ln \beta} \left[\nu + 2\nu \dfrac{\beta^2}{1-\beta^2} \ln \beta + 2 \ln \rho \right] + 2T_e(1-\nu) \right\} \\ u = r_e \rho \alpha \dfrac{(1+\nu)}{(1-\nu)} \left\{ \dfrac{T_i - T_e}{2 \ln \beta} \left[-(1-\nu) + \ln \rho - \dfrac{\beta^2}{1-\beta^2} \ln \beta \left(1 - 2\nu + \dfrac{1}{\rho^2} \right) \right] + T_e(1-\nu) \right\}. \end{cases} \tag{10.44}$$

These relations describe the stress state and radial displacement in the long hollow cylindrical body with clamped ends, subject to a thermal load whose temperature variation with the radius follows Fourier's law.

The term T_e does not appear in the parentheses in the first two relations (10.44): this is because radial and tangential expansions are unconstrained except at the clamped ends, and the term T_e appearing in relation (10.43) makes no contribution to stresses σ_r and σ_t. In the third relation (10.44), on the other hand, a term in T_e appears between the braces, given that the axial expansions, unlike the other expansions, is constrained. A term in T_e also appears between the braces in the last relation (10.44), because it is assumed that the temperature at the outer radius differs from reference temperature T_0, which is in general assumed to be the assembly temperature. If T_e were to be assumed as the reference temperature,

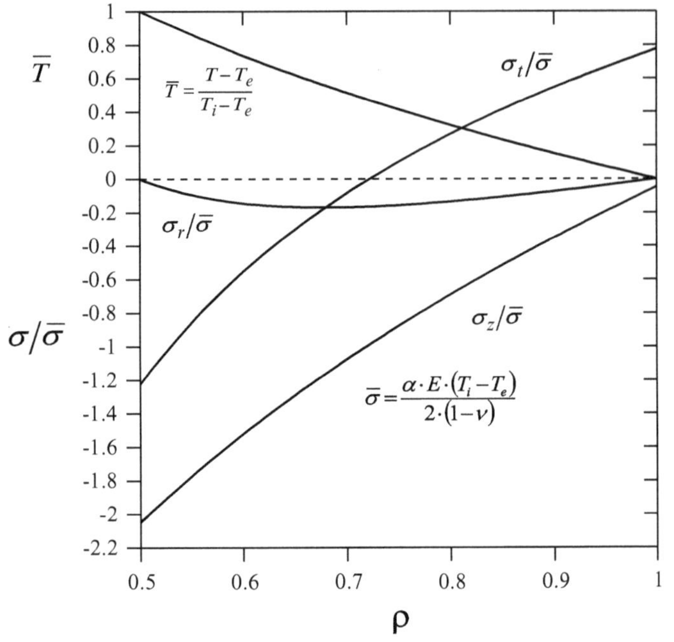

Fig. 10.3 Distribution curves of the three dimensionless principal stresses $\sigma_r/\bar{\sigma}$, $\sigma_t/\bar{\sigma}$e, $\sigma_z/\bar{\sigma}$ and dimensionless temperature $\bar{T}$ along the dimensionless radius of a hollow cylinder with clamped ends subjected to centrifugal heat flow. Reference temperature $T_0 = 20$ °C; outer temperature $T_e = 20$ °C

the associated term between braces would disappear. The value of T_e to be introduced in calculations is the difference between the temperature measured on the outer surface and reference temperature T_0.

Figure 10.3 shows distributions of temperatures and stresses in a steel cylindrical body ($E = 210$ GPa; $\nu = 0.3$; $\alpha = 12 \times 10^{-6}$ °C^{-1}), with $\beta = 0.5$ ($r_e = 2r_i$), in dimensionless form. Stresses are made dimensionless by dividing them by the factor $\bar{\sigma} = \alpha E\ (T_i - T_e)/2(1 - \nu)$ which, for steels, assumes a value of around 1.80 MPa for each °C of difference between the temperatures at the inner and outer radii. Temperature is also shown in dimensionless form, as the ratio $\bar{T} = (T - T_e)/(T_i - T_e)$, and thus assumes values of 1 and 0 at the inner radius and outer radius respectively. Its behaviour between these two extremes is logarithmic.

As is immediately clear from the figure, the most dangerous stress state as regards strength capacity is that at the inner radius, where stresses σ_z and σ_t, both compressive, reach their highest values, viz., 3.68 MPa and 2.20 MPa respectively for each °C of difference between the temperatures at the inner and outer radii. For a temperature differential $(T_i - T_e)$ of 100 °C, these two stresses would be 368 MPa and 220 MPa, or in other words sizeable fractions of yield stress σ_y. For steady-state heat flow, i.e., for static thermal load, and a solid consisting of AISI 1060 steel, whose yield stress is 480 MPa, this stress state would correspond to safety factors of

1.30, 1.30 and 1.50 respectively, determined using the σ_{max}, τ_{max} and maximum distortion energy failure criteria.

The end constraints thus result in a stress state that can be doubly dangerous, as it not only influences failure conditions, but can also trigger global and local buckling if the cylindrical body's walls are thin. It goes without saying that in designing long rotors (and in designing tube systems, as a tube is nothing more than a long, non-rotating cylindrical body), it is advisable to avoid axial constraints at the ends whenever possible.

Passing from the inner radius to the outer, both stresses σ_z and σ_t drop in absolute value. However, while longitudinal stress remains negative, reaching its minimum absolute value at the outer radius, the hoop stress changes sign for a value of ρ which can be readily determined from the second relation (10.44), becoming a tensile stress and, at the outer radius, assuming a value of 1.40 MPa for each °C of difference between the temperatures at the inner and outer radii. The radial stress, zero at inner and outer radii, is always negative and reaches its maximum value, which is in any case not very high and can be readily calculated, for a value of ρ that can once again be determined by establishing that the ratio $d\sigma_r/d\rho$ obtained from the first relation (10.44) is zero.

Figure 10.3 applies for centrifugal heat flow ($T_i > T_e$). For centripetal heat flow ($T_i < T_e$), the distribution of the three principal stresses along the dimensionless radius is obtained from that indicated in Fig. 10.3 as the mirror image of the curves shown therein, i.e., by rotating the plots for $\sigma_r/\bar{\sigma}$, $\sigma_t/\bar{\sigma}$ and $\sigma_z/\bar{\sigma}$ by 180° relative to the axis of the abscissa. The temperature distribution is obtained by rotating the $\bar{T} = (T - T_e)/(T_i - T_e)$ curve shown in Fig. 10.3, again by 180°, but around a vertical axis, so that the dimensionless temperature assumes values of 0 and 1 at the inner radius and outer radius respectively.

When the cylindrical body has thin walls, as in the case of a cylinder sleeve for an internal combustion engine or a compressor, the temperature distribution along the dimensionless radius may by assumed to be linear (Fig. 10.4), as will be demonstrated below, and corresponds to an equally linear distribution of hoop and longitudinal stresses. Radial stress is constantly zero. The maximum values of stress σ_t are found at the inner radius and outer radius, where they equal the factor $\bar{\sigma} = \alpha E(T_i - T_e)/2(1 - \nu)$, as they are respectively compressive and tensile stresses for centrifugal heat flow. Stress σ_z is zero at the outer radius and reaches a maximum value of $2 \cdot \bar{\sigma}$ at the inner radius. Figure 10.4 shows the distribution curves for the three principal stresses in a cylinder with a ratio $\beta = 0.91$ subjected to centrifugal heat flow; once again, stresses are made dimensionless by dividing them by the factor $\bar{\sigma}$. This value of β corresponds to a thickness/diameter ratio of 1/20, which by convention is the upper limit for a thin wall (for which β is thus greater than 0.91).

Stresses σ_t and σ_z are compressive at the inner radius and assume their maximum negative values of 1.80 MPa and 3.60 MPa respectively for each degree of temperature difference ($T_i - T_e$). Passing from the inner radius to the outer, hoop stress drops linearly in absolute value, changes sign to become a tensile stress, and at the outer radius assumes a value equal and opposite to that at the inner radius. Axial

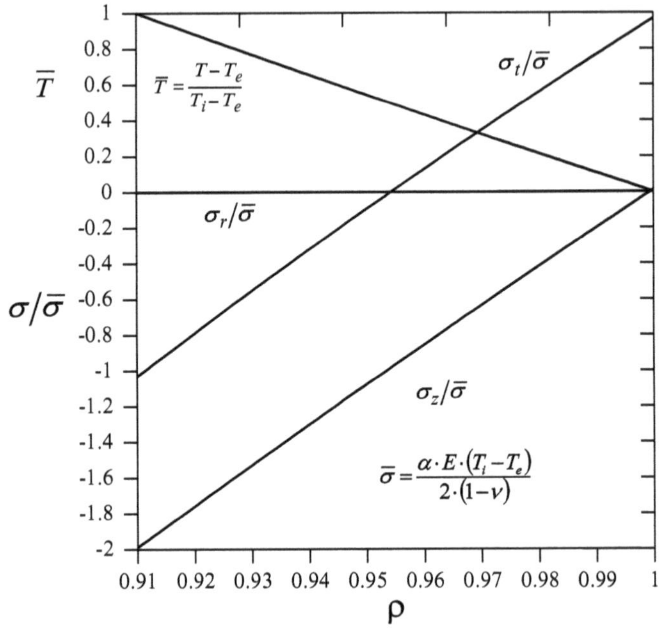

Fig. 10.4 Distribution curves of the three dimensionless principal stresses $\sigma_r/\bar{\sigma}$, $\sigma_t/\bar{\sigma}$ and $\sigma_z/\bar{\sigma}$ and dimensionless temperature $\bar{T}$ along the dimensionless radius of a thin-wall hollow cylinder with clamped ends subjected to centrifugal heat flow. Reference temperature $T_0 = 20$ °C; outer temperature $T_e = 20$ °C

stress also drops in absolute value, again following a linear distribution, remaining negative at all times until reaching zero at the outer radius. Figure 10.4 shows temperature in dimensionless form as the ratio $\bar{T} = (T - T_e)/(T_i - T_e)$; as can be seen, temperature assumes values of 1 and 0 at the inner radius and outer radius respectively.

Figure 10.4 applies for centrifugal heat flow ($T_i > T_e$). For centripetal heat flow ($T_i < T_e$), the distribution of the two non-zero principal stresses in the wall thickness is obtained from that indicated in Fig. 10.4 as the mirror image of the curves shown therein, i.e., by rotating the plots for $\sigma_t/\bar{\sigma}$ and $\sigma_z/\bar{\sigma}$ by 180° relative to the axis of the abscissa. The temperature distribution is obtained by rotating the $\bar{T} = (T - T_e)/(T_i - T_e)$ curve shown in Fig. 10.4, again by 180°, but around a vertical axis, so that the dimensionless temperature assumes values of 0 and 1 at the inner radius and outer radius respectively.

Given axisymmetry and compatibility, the curvature in the circumferential direction remains constant, so bending stresses are zero and the only stresses are membrane stresses. Near the ends of the cylinder, however, the structure's interactions with the constraint come into play. In particular, if there are clamped edges, bending effects arise which must be taken into account. The increases in the stress state can be quantified on the basis of Coates' theory [10]. Specifically, if the structure's material is steel, this increase is approximately 25 %.

The first two relations (10.44) can also be written in the following more compact form:

$$\begin{cases} \sigma_r = \dfrac{\alpha E}{2(1-\nu)}(T_i - T_e)H = \bar{\sigma}H \\ \sigma_t = \dfrac{\alpha E}{2(1-\nu)}(T_i - T_e)K = \bar{\sigma}K, \end{cases} \quad (10.45)$$

where $H = H(\beta,\rho)$ and $K = K(\beta,\rho)$ are functions of β and of ρ, given respectively by:

$$\begin{cases} H = \dfrac{1}{\ln\beta}\cdot\left(\dfrac{\beta^2}{1-\beta^2}\cdot\dfrac{1-\rho^2}{\rho^2}\cdot\ln\beta - \ln\rho\right) \\ K = \dfrac{1}{\ln\beta}\cdot\left(-1-\dfrac{\beta^2}{1-\beta^2}\cdot\dfrac{1+\rho^2}{\rho^2}\cdot\ln\beta - \ln\rho\right). \end{cases} \quad (10.46)$$

The behaviour of these functions, and thus of the correlated radial and hoop stresses, is shown versus the dimensionless radial coordinate $x = (r - r_i)/(r_e - r_i) = (\rho - \beta)/(1 - \beta)$ in Figs. 10.5 and 10.6.

As can be seen from these figures, the maximum values of H and K and, consequently, the maximum values of radial stress σ_r and hoop stress σ_t are heavily influenced by ratio β. For $\beta \geq 0.91$ (thin wall), the maximum values assumed by K are an order of magnitude higher than those assumed by H: when values of K are higher in absolute value than 1, H assumes a maximum value, again in absolute value, of around 0.023, which for steels corresponds to values of radial stress $\sigma_r = \overline{\sigma}H$ and hoop stress $\sigma_t = \overline{\sigma}K$ which are higher and lower by 0.04 MPa and 1 MPa respectively for a temperature difference of 1 °C. Accordingly, it can be maintained, with an error that is entirely negligible from the design standpoint, that radial stress is zero at any radius whatsoever. By way of example, we can compare the maximum values assumed by radial and hoop stresses in a steel cylindrical body, with $\beta = 0.91$, i.e., at the upper limit of thickness for a thin wall, for a temperature difference $(T_i - T_e) = 100$ °C. We will have $\sigma_{r,max} = \overline{\sigma}\cdot H_{max} = 180 \times 0.023 = 4.14$ MPa and $\sigma_{t,max} = \overline{\sigma}K_{max} = 180 \times 1.45 = 261$ MPa; there is thus a difference of almost two orders of magnitude between the hoop stress and the radial stress.

Both Figs. 10.5 and 10.6 apply for centrifugal heat flow $(T_i > T_e)$. For centripetal heat flow $(T_i < T_e)$, the distributions of H and K along the dimensionless radius are obtained from those indicated in Figs. 10.5 and 10.6 as the mirror image of the family of curves shown therein, i.e., by rotating the plots for H and K by 180° relative to the axis of the abscissa.

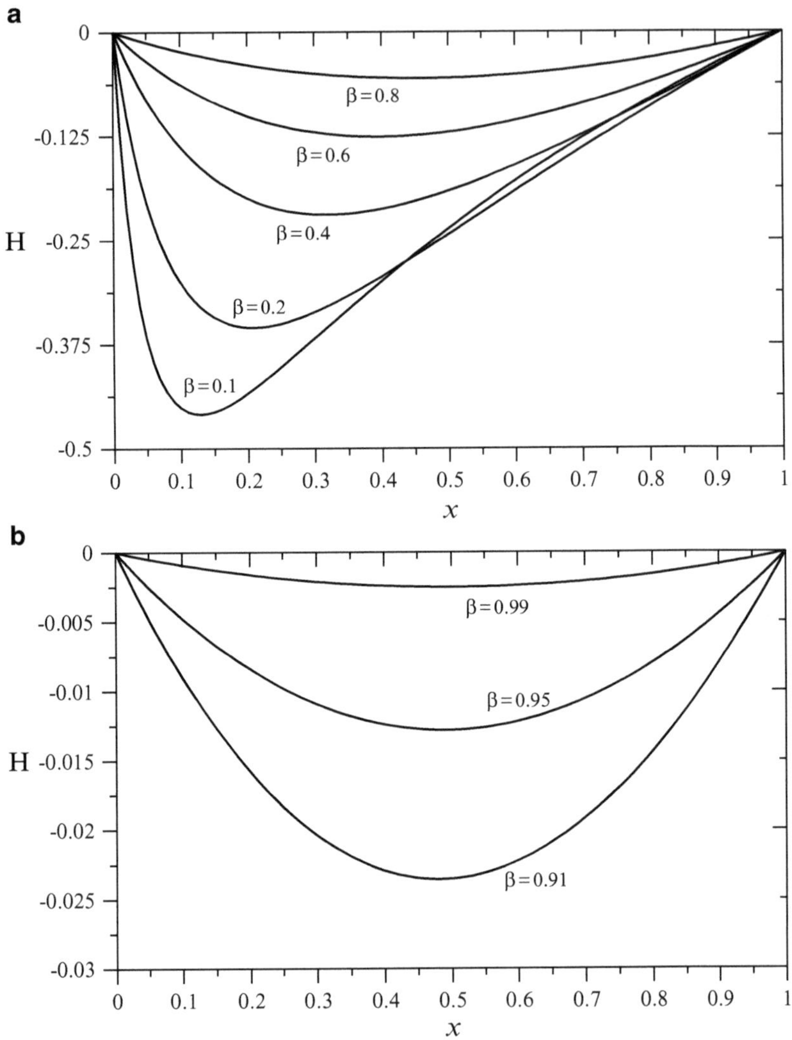

Fig. 10.5 Function H versus dimensionless radial coordinate $x = (r - r_i)/(r_e - r_i) = (\rho - \beta)/(1 - \beta)$ for different values of parameter β: (**a**) $0.1 < \beta < 0.8$; (**b**) $0.91 < \beta < 0.99$

10.2.2.1 Example

A hollow cylinder body made of unquenched AISI 1060 steel, with $\sigma_y = 480$ MPa, is subjected to steady-state centrifugal heat flow and has clamped ends. It has outside and inside diameters of 2 and 1 m respectively, and is 3 m long. Let $T_i = 120$ °C be the inner temperature, $T_e = 20$ °C the outer temperature, and $T_0 = 20$ °C the assembly temperature, taken as the reference temperature.

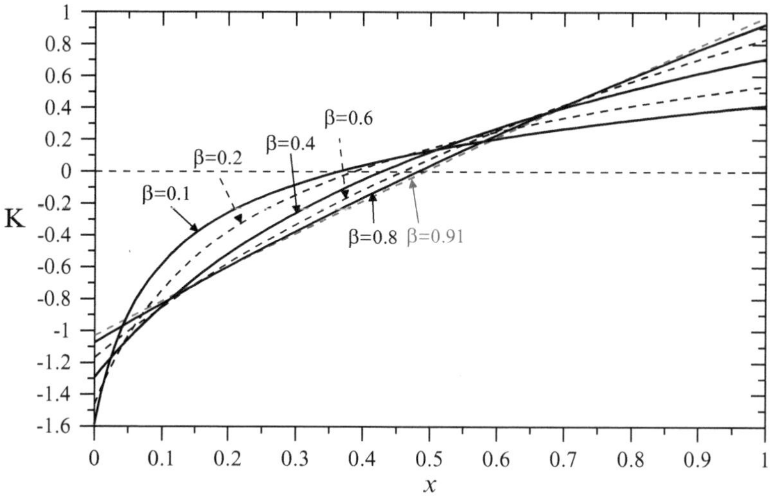

Fig. 10.6 Function K versus dimensionless radial coordinate $x = (r - r_i)/(r_e - r_i) = (\rho - \beta)/(1 - \beta)$ for different values of parameter β

We will calculate the stress state in the entire cylindrical solid, comparing the results with those obtained from a finite element model. The example will conclude with several considerations regarding design.

Further problem data are as follows: $\alpha = 12 \times 10^{-6}$ °C $^{-1}$; $\nu = 0.3$; $E = 210$ GPa.

Using the first three relations (10.44) which, as has been demonstrated, express the stress state resulting from heat flow in a hollow cylindrical body with clamped ends, we obtain the three dashed-line curves shown in Fig. 10.7a. As can be seen, the material is at greatest risk at the inner radius, where the principal stresses assume the following values: $\sigma_r = 0$; $\sigma_t = -220$ MPa; $\sigma_z = -368$ MPa.

With the three σ_{max}, τ_{max} and maximum distortion energy failure criteria, the safety factor for the material is 1.30, 1.30 and 1.50 respectively.

To confirm the theoretical results, a finite element numerical model (Fig. 10.7c) was developed which uses 8-node solid elements with one dof per node for thermal calculation and 8-node solid elements with three dofs per node for structural calculation. The constraints at the ends were simulated by using a zero longitudinal degree of freedom. Figure 10.7a also summarises the numerical results obtained with the model (solid-line curves). Axial stress σ_z along the wall thickness of the cylinder in question is shown in Fig. 10.7b.

Comparison of the theoretical and numerical results shows that the differences are entirely insignificant from the standpoint of design.

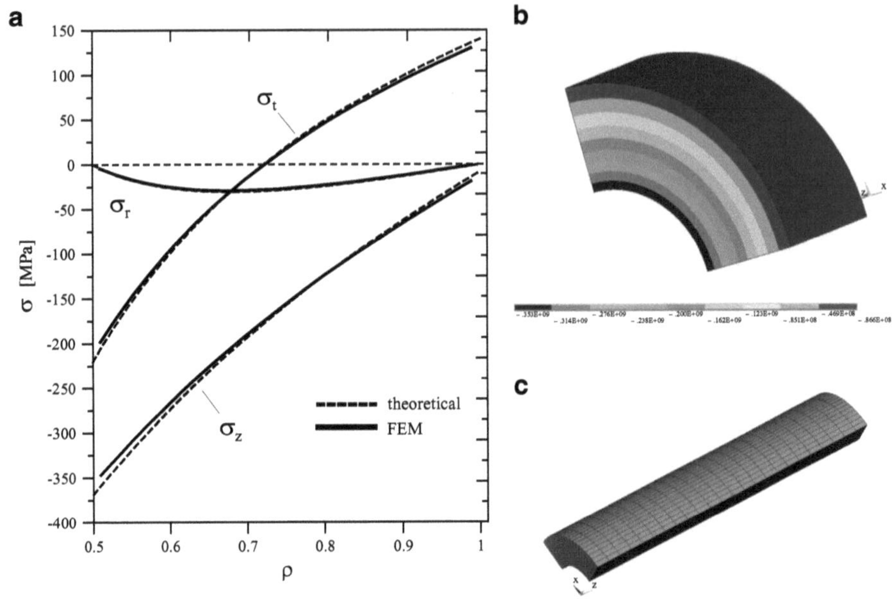

Fig. 10.7 (**a**) Distribution curves of the three principal stresses σ_r, σ_t and σ_z along the dimensionless radius of a hollow cylinder with clamped ends subjected to centrifugal heat flow. Reference temperature $T_0 = 20\ °\mathrm{C}$; outer temperature $T_e = 20\ °\mathrm{C}$; inner temperature $T_i = 120\ °\mathrm{C}$; cylinder inner radius $r_i = 0.5$ m and outer radius $r_e = 1$ m. (**b**) Stress σ_z along the wall thickness of a hollow cylinder with clamped ends. (**c**) FEM model: 8-node solid elements with one dof per node for thermal calculation; 8-node solid elements with three dofs per node for structural calculation

10.3 Circular Cylindrical Body of Finite Length with Free Ends Subjected to Centrifugal and Thermal Loads

We will now consider the somewhat more complex case of a circular cylindrical body with free ends whose length is finite but large compared to its cross section diameter, and subjected to both centrifugal and thermal loads. As there is no external axial force to balance the resultant of axial stress σ_z given by the third relation (10.29) and having the distribution shown in Fig. 10.8b, the latter cannot represent the real axial stress state. The effective axial stress, which we will call σ'_z, should have a distribution such that the resultant F'_z on any section normal to the axis and sufficiently far from the ends is zero, so as to satisfy the condition indicated above of global equilibrium in the direction of the rotational axis. Then, (10.5) must be fulfilled, and the integral of stress σ'_z with area, i.e. the net force across the section must be zero. From the elementary area of Fig. 10.8a, we have:

$$F'_z = \int_A \sigma'_z \cdot dA = 2\pi \cdot \int_{r_i}^{r_e} \sigma'_z \cdot r \cdot dr = 0. \tag{10.47}$$

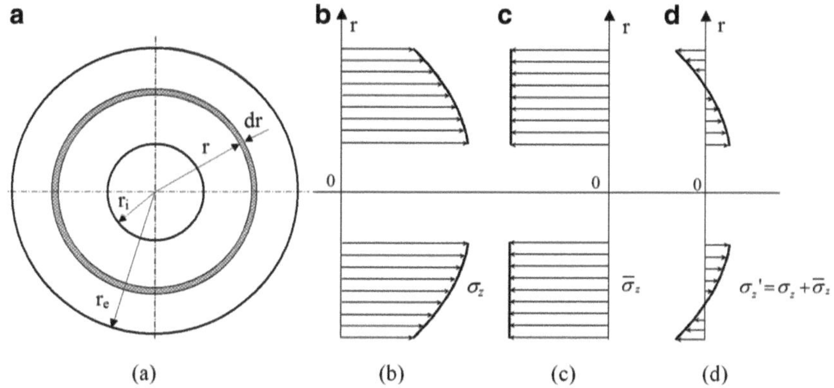

Fig. 10.8 Axial stresses in a cylinder of finite length only rotating: (**a**) element area of the cross section away from the ends; (**b**) axial stress σ_z with clamped ends; (**c**) uniform stress $\bar{\sigma}_z$; (**d**) axial stress $\sigma_z^{'}=\sigma_z + \bar{\sigma}_z$with free ends

To satisfy this equilibrium condition, a new system of stresses $\bar{\sigma}_r$, $\bar{\sigma}_t$ and $\bar{\sigma}_z$ must be superposed to the system of stresses σ_r, σ_t and σ_z, calculated earlier. This new system of stresses fulfilling all conditions must be the following: $\bar{\sigma}_r = \bar{\sigma}_t = 0$; $\bar{\sigma}_z =$ *const* (Fig. 10.8c). It is obvious that the radial and hoop stresses do not change with the superposition of these two stress systems, while the effective axial stress σ'_z will be given by the sum of stress σ_z, which is distributed on the generic cross section according to a nonlinear function of the radius and satisfies the plane strain state assumption ($\varepsilon_z = 0$), and of stress $\bar{\sigma}_z$, which is distributed uniformly on the same section, so that the latter can remain plane and perpendicular to the axis in compliance with the assumption (10.4) of a strain state characterised by a uniform axial translation of a generic cross section. In this case, this assumption enables us to write $\varepsilon'_z = \varepsilon_z + \bar{\varepsilon}_z = \bar{\varepsilon}_z =$ *const.*

To determine $\bar{\sigma}_z$, we can use condition (10.47), setting $\sigma'_z = \sigma_z + \bar{\sigma}_z$; by introducing variable ρ, condition (10.47) thus enables us to write:

$$\int_{\beta}^{1} (\sigma_z + \bar{\sigma}_z) \cdot \rho \cdot d\rho = 0. \tag{10.48}$$

To facilitate calculating stress $\bar{\sigma}_z$, we prefer here to use the following expression of axial stress σ_z which is also entirely general, instead of the third relation (10.29):

$$\sigma_z = -\frac{\alpha \cdot E \cdot T}{(1-\nu)} - \frac{\nu \cdot \sigma_0}{2 \cdot (1-\nu)} \cdot \rho^2 + 2 \cdot \nu \cdot \bar{A}. \tag{10.49}$$

This expression is obtained by substituting relations (10.24) in relation (10.14). Introducing this expression in relation (10.48) yields:

$$-\frac{\alpha \cdot E}{(1-\nu)} \cdot \int_{\beta}^{1} T \cdot \rho \cdot d\rho - \frac{\nu \cdot \sigma_0}{2 \cdot (1-\nu)} \cdot \int_{\beta}^{1} \rho^3 \cdot d\rho + [2 \cdot \nu \cdot \bar{A} + \bar{\sigma}_z] \cdot \int_{\beta}^{1} \rho \cdot d\rho = 0. \qquad (10.50)$$

Integrating this relation gives the following constant value of $\bar{\sigma}_z$:

$$\bar{\sigma}_z = \frac{2 \cdot \alpha \cdot E}{(1-\nu) \cdot (1-\beta^2)} \cdot \int_{\beta}^{1} T \cdot \rho \cdot d\rho + \frac{\nu \cdot \sigma_0}{4 \cdot (1-\nu)} \cdot (1+\beta^2) - 2 \cdot \nu \cdot \bar{A}. \quad (10.51)$$

Adding relation (10.51) and relation (10.28) gives the following expression of effective axial stress $\sigma_z^{'}$:

$$\begin{aligned} \sigma_z^{'} &= \sigma_z + \bar{\sigma}_z \\ &= \frac{E \cdot \alpha}{(1-\nu)} \cdot \left(-T + \frac{2}{(1-\beta^2)} \cdot \int_{\beta}^{1} T \cdot \rho \cdot d\rho \right) + \frac{\nu \cdot \sigma_0}{4 \cdot (1-\nu)} \cdot (1+\beta^2 - 2\rho^2) \end{aligned} \qquad (10.52)$$

which, for pure centrifugal load, can be specialised in:

$$\sigma_z^{'} = \frac{\nu}{4 \cdot (1-\nu)} \cdot \sigma_0 \cdot (1+\beta^2 - 2\rho^2). \qquad (10.53)$$

Figure 10.8d shows the distribution of effective axial stress σ'_z in a circular cylindrical body of finite length with free ends subjected only to centrifugal load, as the sum of the distributions of σ_z and $\bar{\sigma}_z$. Naturally, the effective axial stress σ'_z is correlated with a uniform axial strain ε'_z as well as a constant axial displacement w in the generic cross section of abscissa z. It is equally obvious that this expression of σ'_z does not hold close to the end edges, since at these free boundaries, there can be no reaction, and σ'_z is thus zero at all points (at any radius). It goes in any case without saying that, in accordance with Saint Venant's principle, relation (10.53) represents the actual axial stress state of a rotating circular cylindrical body with free ends starting from a distance from the edges which is of the same order of magnitude as the outer radius.

In conclusion, the three principal stresses which define the stress state in a circular cylindrical body of finite length with free ends, with no load at either the inner or outer radii and subjected to both centrifugal load and a temperature gradient along the radius according to any function $T = T(\rho)$ are as follows:

$$\begin{cases} \sigma_r = \dfrac{3-2\nu}{8\cdot(1-\nu)}\cdot\sigma_0\cdot\left(1+\beta^2-\dfrac{\beta^2}{\rho^2}-\rho^2\right)+\dfrac{\alpha\cdot E}{\rho^2\cdot(1-\nu)}\cdot\left[\dfrac{\rho^2-\beta^2}{1-\beta^2}\cdot\int\limits_\beta^1 T\cdot\rho\cdot d\rho-\int\limits_\beta^\rho T\cdot\rho\cdot d\rho\right] \\ \sigma_t = \dfrac{3-2\nu}{8\cdot(1-\nu)}\cdot\sigma_0\cdot\left(1+\beta^2+\dfrac{\beta^2}{\rho^2}-\dfrac{1+2\nu}{3-2\nu}\cdot\rho^2\right)+\dfrac{\alpha\cdot E}{\rho^2\cdot(1-\nu)}\cdot\left[\dfrac{\rho^2+\beta^2}{1-\beta^2}\cdot\int\limits_\beta^1 T\cdot\rho\cdot d\rho+\int\limits_\beta^\rho T\cdot\rho\cdot d\rho\right] \\ -\dfrac{\alpha\cdot E\cdot T}{(1-\nu)} \\ \sigma_z = \dfrac{\nu\cdot\sigma_0}{4\cdot(1-\nu)}\cdot(1+\beta^2-2\rho^2)+\dfrac{E\cdot\alpha}{(1-\nu)}\cdot\left(\dfrac{2}{(1-\beta^2)}\cdot\int\limits_\beta^1 T\cdot\rho\cdot d\rho-T\right). \end{cases} \tag{10.54}$$

Introducing relations (10.54) in relation (10.30) written in terms of dimensionless variable ρ yields the following relation expressing radial displacement $u = u(\rho)$:

$$u = \frac{r_e\cdot\rho}{E\cdot(1-\nu)}\cdot\left\{\frac{3-2\nu}{8}\sigma_0\left[(1+\beta^2)\cdot\frac{3-5\nu}{3-2\nu}+\frac{\beta^2}{\rho^2}\cdot(1+\nu)-\frac{(1-\nu-2\nu^2)}{(3-2\nu)}\cdot\rho^2\right]\right. \\ \left.+\frac{E\cdot\alpha\cdot(1+\nu)}{\rho^2}\cdot\left[\frac{(\rho^2+\beta^2)}{(1-\beta^2)}\cdot\int\limits_\beta^1 T\cdot\rho\cdot d\rho+\int\limits_\beta^\rho T\cdot\rho\cdot d\rho\right]\right\}. \tag{10.55}$$

As relation (10.4) gives $\varepsilon_z = dw/dz = const$ and, consequently, $w = \varepsilon_z\cdot z$, given that $\varepsilon_z = const$ and the integration constant, with $w = 0$ for $z = 0$, is zero, substituting relations (10.54) in the third Eq. (10.6) and multiplying by z gives the following relation expressing axial displacement $w = w(z)$ which, assuming a strain state characterised by a uniform axial translation of a generic cross section of the cylindrical body, is independent of variable ρ:

$$w = z\cdot\varepsilon_z = \frac{z}{E}\cdot\left[\sigma_z-\nu\cdot(\sigma_r+\sigma_t)+\alpha\cdot E\cdot T\right] = \\ = \frac{z}{E}\cdot\left\{-\frac{\nu\cdot\sigma_0}{2}\cdot(1+\beta^2)+\frac{2\cdot\alpha\cdot E}{1-\nu}\cdot\left[\frac{1+\nu}{1-\beta^2}\cdot\int\limits_\beta^1 T\cdot\rho\cdot d\rho-\nu\cdot T\right]\right\}. \tag{10.56}$$

Relations (10.54), (10.55) and (10.56) fully describe the stress state and displacements of the hollow cylindrical body with free ends considered here. If the same cylindrical body is also subjected to surface force distributions acting at the inner and outer radii as well as at the free ends, the overall stress state and radial displacements can once again be determined as the sum of the contributions made by the various stresses acting on the solid. Relations (10.54), (10.55) and (10.56) also apply if the cylindrical body is not hollow; in this case, it is sufficient to set $\beta = 0$. As has been repeatedly specified, the problem regarding singularity remains unchanged.

10.3.1 Circular Cylindrical Body of Finite Length with Free Ends Subjected to Centrifugal Load

If the temperature terms are omitted from relations (10.54), (10.55) and (10.56), we obtain the following relations expressing the stress state and radial and axial displacements in the hollow cylindrical body of finite length with free ends subjected only to centrifugal load:

$$\begin{cases} \sigma_r = \dfrac{(3-2\nu)}{8\cdot(1-\nu)}\cdot\sigma_0\cdot\left(1+\beta^2-\dfrac{\beta^2}{\rho^2}-\rho^2\right) \\ \sigma_t = \dfrac{(3-2\nu)}{(1-\nu)}\cdot\sigma_0\cdot\left(1+\beta^2+\dfrac{\beta^2}{\rho^2}-\dfrac{1+2\nu}{3-2\nu}\cdot\rho^2\right) \\ \sigma_z = \dfrac{\nu\cdot\sigma_0}{4\cdot(1-\nu)}\cdot\left(1+\beta^2-2\rho^2\right) \\ u = \dfrac{r_e\cdot\rho\cdot\sigma_0}{8\cdot E}\cdot\dfrac{(3-2\nu)}{(1-\nu)}\cdot\left[\left(1+\beta^2\right)\cdot\dfrac{3-5\nu}{3-2\nu}+\dfrac{\beta^2}{\rho^2}\cdot(1+\nu)-\dfrac{(1-\nu-2\nu^2)}{(3-2\nu)}\cdot\rho^2\right] \\ w = -\dfrac{z}{E}\cdot\dfrac{\nu\cdot\sigma_0}{2}\cdot\left(1+\beta^2\right). \end{cases} \tag{10.57}$$

The radial stress, which is zero at the inner and outer edges, is a tensile stress at all points, as is the hoop stress. Axial stress is zero for

$$\rho = \sqrt{(1+\beta^2)/2},$$

or in other words for

$$r = \sqrt{(r_i^2+r_e^2)/2},$$

as it is positive towards the inner radius and negative towards the outer radius.

Figure 10.9 shows the distribution of the three principal stresses σ_r, σ_t and σ_z along the dimensionless radius of a hollow cylinder only rotating. As can be seen from this figure and from a comparison with Fig. 10.2 for the long cylindrical body with clamped ends, the radial and hoop stresses do not change. The longitudinal stress, however, changes, assuming positive values towards the inner radius and negative values towards the outer radius. The maximum values of stresses σ_t and σ_z are found at the inner radius, where the values they assume are respectively:

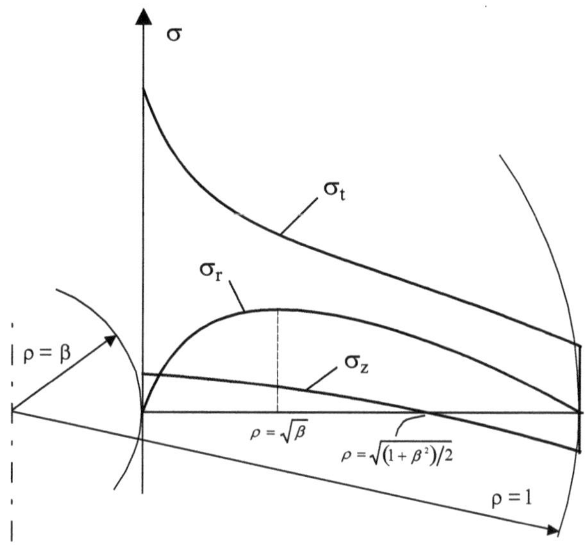

Fig. 10.9 Distribution curves of the three principal stresses σ_r, σ_t and σ_z along the dimensionless radius of a long hollow cylinder with free ends in simple rotation

$$\begin{cases} \sigma_{t,max} = \dfrac{(3-2\nu)}{4\cdot(1-\nu)}\cdot\sigma_0\cdot\left(1+\dfrac{1-2\nu}{3-2\nu}\beta^2\right) \\ \sigma_{z,max} = \dfrac{\nu\cdot\sigma_0}{4\cdot(1-\nu)}\cdot\left(1-\beta^2\right). \end{cases} \quad (10.58)$$

The minimum values of stresses σ_t and σ_z are found at the outer radius, where they are respectively:

$$\begin{cases} \sigma_{t,min} = \dfrac{(3-2\nu)}{4\cdot(1-\nu)}\cdot\sigma_0\cdot\left(\beta^2+\dfrac{1-2\nu}{3-2\nu}\right) \\ \sigma_{z,min} = -\dfrac{\nu\cdot\sigma_0}{4\cdot(1-\nu)}\cdot\left(1-\beta^2\right). \end{cases} \quad (10.59)$$

Radial stress, which as we have said is zero at the inner and outer radii, assumes its maximum value when the sum $[(\beta^2/\rho^2)+\rho^2]$ is minimum, i.e., when $\beta^2/\rho^2=\rho^2$, given that the product of the two addends is constant. The maximum radial stress, which we have for $\rho=\sqrt{\beta}$, or in other words for $r=\sqrt{r_i\cdot r_e}$, is thus:

$$\sigma_{r,max} = \frac{(3-2\nu)}{8\cdot(1-\nu)}\cdot\sigma_0\cdot(1-\beta)^2. \quad (10.60)$$

At the inner radius, the maximum shear stress failure criterion (whose results in this case coincide with those obtained using the maximum normal stress criterion) gives:

$$\tau_{\max} = \left(\frac{\sigma_t-\sigma_r}{2}\right)_{\rho=\beta} = (\sigma_t/2)_{\rho=\beta} = \frac{(3-2\nu)}{8\cdot(1-\nu)}\cdot\sigma_0\cdot\left(1+\frac{1-2\nu}{3-2\nu}\cdot\beta^2\right). \quad (10.61)$$

The maximum octahedral shear stress criterion, which is one of the most reliable criteria for predicting yield, and hence the onset of the plastic state, leads to the general relation (9.8). Specialised at the inner radius, where $\sigma_r = 0$, the latter relation gives:

$$\tau_{oct} = \frac{\sqrt{2}}{3} \cdot \left(\sigma_t^2 + \sigma_z^2 - \sigma_z \sigma_t\right)^{\frac{1}{2}}. \qquad (10.62)$$

At the inner edge where the material is at greatest risk, it is thus clear that the τ_{max} criterion indicates that the danger of failure depends on the higher principal stress, which is the hoop stress, whereas the τ_{oct} criterion makes this danger depend on both the non-zero principal stresses, i.e., stresses σ_t and σ_z.

Introducing relations (10.58) in (10.62) gives the following relation which expresses τ_{oct} on the inner edge:

$$\tau_{oct} = \frac{\sqrt{2}}{3} \cdot \frac{\sigma_0}{4 \cdot (1-\nu)} \cdot \left[9 + 7\nu^2 - 15\nu + \beta^4 \cdot \left(1 + 3\nu^2 - 3\nu\right) + \beta^2 \cdot \left(6 + 6\nu^2 - 14\nu\right)\right]^{\frac{1}{2}}. \qquad (10.63)$$

The latter relation, applied to cases $\nu = 0.3$ and $\nu = 0.5$ (note that $\nu = 0.5$ represents the fully plastic state), enables us to write:

$$(\tau_{oct})_{\nu=0.3} = 0.10 \cdot \sigma_0 \cdot \sqrt{\beta^4 + 6.4\beta^2 + 13.8}$$
$$(\tau_{oct})_{\nu=0.5} = 0.12 \cdot \sigma_0 \cdot \sqrt{\beta^4 + 2\beta^2 + 13}\,.$$

If the cylinder only rotating considered above is not hollow ($\beta = 0$), relations (10.57) are specialised in:

$$\begin{cases} \sigma_r = \dfrac{(3-2\nu)}{8 \cdot (1-\nu)} \cdot \sigma_0 \cdot \left(1 - \rho^2\right) \\ \sigma_t = \dfrac{(3-2\nu)}{8 \cdot (1-\nu)} \cdot \sigma_0 \cdot \left(1 - \dfrac{1+2\nu}{3-2\nu} \cdot \rho^2\right) \\ \sigma_z = \dfrac{\nu \cdot \sigma_0}{4 \cdot (1-\nu)} \cdot \left(1 - 2\rho^2\right) \\ u = \dfrac{r_e \cdot \rho \cdot \sigma_0}{8E \cdot (1-\nu)} \cdot \left[3 - 5\nu - \left(1 - \nu - 2\nu^2\right) \cdot \rho^2\right] \\ w = -\dfrac{z}{E} \cdot \dfrac{\nu \cdot \sigma_0}{2}. \end{cases} \qquad (10.64)$$

The radial stress, which is zero at the outer edge, is a tensile stress at all points, as is the hoop stress. Axial stress is zero for $\rho = 1/\sqrt{2}$, or in other words for $r = r_e/\sqrt{2}$, as it is positive towards the axis and negative towards the outer edge.

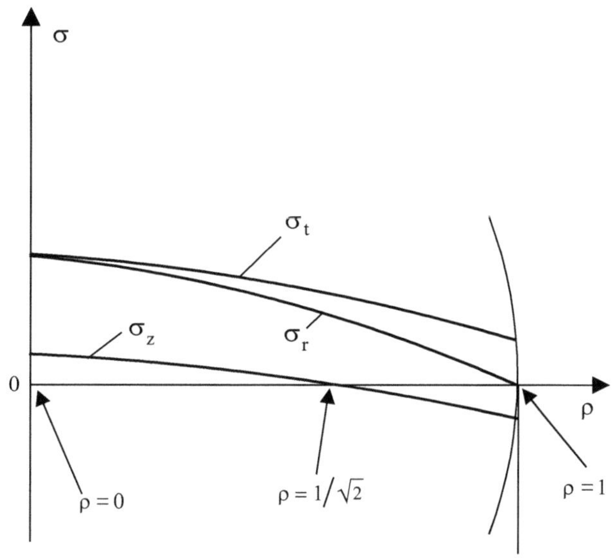

Fig. 10.10 Distribution curves of the three principal stresses σ_r, σ_t and σ_z along the dimensionless radius of a long solid cylinder, made of steel ($\nu = 0.3$) and only rotating

Figure 10.10 shows the distribution curves of the three principal stresses σ_r, σ_t and σ_z along the dimensionless radius of a long solid cylinder only rotating. As can be seen, the maximum values of these stresses are found on the axis, where they are:

$$\begin{cases} \sigma_{r,\max} = \sigma_{t,\max} = \dfrac{3-2\nu}{8\cdot(1-\nu)}\cdot\sigma_0 \\ \sigma_{z,\max} = \dfrac{\nu}{4\cdot(1-\nu)}\cdot\sigma_0 \end{cases} \quad (10.65)$$

The following minimum values of the three principal stresses are found on the outer edge, i.e., for $\rho = 1$:

$$\begin{cases} \sigma_{r,\min} = 0 \\ \sigma_{t,\min} = \dfrac{1-2\nu}{4\cdot(1-\nu)}\cdot\sigma_0 \\ \sigma_{z,\min} = -\dfrac{\nu}{4\cdot(1-\nu)}\cdot\sigma_0. \end{cases} \quad (10.66)$$

On the axis, the maximum shear stress failure criterion gives:

$$\tau_{max} = \left(\frac{\sigma_t - \sigma_z}{2}\right)_{\rho=0} = \frac{3-4\nu}{16\cdot(1-\nu)}\cdot\sigma_0. \quad (10.67)$$

Again on the axis, the maximum octahedral shear stress failure criterion expressed by relation (9.8) gives:

Table 10.1 Comparison of equivalent stresses at the inner radius, at yield point with the τ_{max} and τ_{oct} failure criteria for steel disks and cylinders subjected only to centrifugal load

	Hollow cylinder		Annular disk		Solid cylinder	Solid disk
Failure criterion	$\beta = 0.1$	$\beta \to 1$	$\beta = 0.1$	$\beta \to 1$	$\rho = 0$	$\rho = 0$
τ_{max}	$0.429 \cdot \sigma_0$	$0.500 \cdot \sigma_0$	$0.413 \cdot \sigma_0$	$0.500 \cdot \sigma_0$	$0.161 \cdot \sigma_0$	$0.206 \cdot \sigma_0$
τ_{oct}	$0.382 \cdot \sigma_0$	$0.471 \cdot \sigma_0$	$0.389 \cdot \sigma_0$	$0.471 \cdot \sigma_0$	$0.151 \cdot \sigma_0$	$0.194 \cdot \sigma_0$

$$\tau_{oct} = \frac{\sqrt{2} \cdot (3 - 4\nu)}{24 \cdot (1 - \nu)} \cdot \sigma_0. \tag{10.68}$$

From the design standpoint, a comparison of the yield strengths of a disk and a cylindrical body of finite length, all other conditions such as material, geometry, stresses and so forth remaining equal, is of considerable design interest. Table 10.1 summarises such a comparison based on the equivalent stress at the material's yield point, calculated at the inner radius by means of the τ_{max} and τ_{oct} failure criteria. The table applies to hollow cylinders and annular disks, as well as solid cylinders and disks, made of steel ($\nu = 0.3$) and subjected only to centrifugal load.

On the basis of the maximum shear stress criterion, we can conclude from this comparison between the annular disk and hollow cylinder of finite length subjected only to centrifugal load that, *ceteris paribus*, the equivalent stress is slightly higher for the cylinder ($0.429\sigma_0$ as against $0.413\sigma_0$, for $\beta = 0.1$), but the difference tends to lessen as β increases, and disappears for β tending to unity (for $\beta \to 1$, i.e., for structural elements with very small radial thickness, the equivalent stress for both cylinder and disk is $0.500\sigma_0$).

On the basis of the maximum octahedral shear stress failure criterion, the same comparison, again with all other conditions remaining equal, leads us to conclude that since the cylinder has a stress $\sigma_z > 0$, the equivalent stress is slightly higher for the disk ($0.389\sigma_0$ as against $0.382\sigma_0$, again for $\beta = 0.1$). Once again, the difference tends to lessen as β increases, and disappears for β tending to unity (for both structural elements with very small radial thickness, the equivalent stress is $0.471\sigma_0$).

The comparison between the solid disk and cylinder subjected only to centrifugal load carried out on the basis of the maximum shear stress failure criterion with all other conditions remaining equal leads us to conclude that the cylinder is intrinsically stronger than the disk. At its rotation axis, in fact, the equivalent stress is $0.161\sigma_0$, well over 20 % less than that ($0.206\sigma_0$) in the disk. The same comparison carried out on the basis of the maximum octahedral shear stress failure criterion confirms the results give by the τ_{max} criterion, except that the numerical values are necessarily lower because the stress state is triaxial.

10.3.1.1 Example

A hollow cylinder made of unquenched AISI 1060 steel, with $\sigma_y = 480$ MPa, rotates at a velocity of 14,000 rpm and has free ends. It is 3 m long and has an outside diameter of 250 mm and an inside diameter of 56 mm.

We will compare the stress state and displacements with those of the thin disk of identical cross section subjected to the same operating conditions discussed in the example given in Sect. 2.5.1 of Chap. 2.

Relations (10.57), bearing in mind that $\nu = 0.3$, $\beta = r_i/r_e = 0.224$ and $E = 210$ GPa, give:

$$\sigma_r = 112 \cdot \left(1.05 - \frac{0.05}{\rho^2} - \rho^2\right)$$
$$\sigma_t = 112 \cdot \left(1.05 + \frac{0.05}{\rho^2} - 0.667\rho^2\right)$$
$$\sigma_z = 28 \cdot \left(1.05 - 2\rho^2\right)$$
$$u = 6.68 \cdot 10^{-5} \cdot \rho \cdot \left(0.66 + \frac{6.5 \cdot 10^{-2}}{\rho^2} - 0.217\rho^2\right)$$
$$w = -2 \cdot 10^{-7} \cdot z\,.$$

Here as for the disk, radial stress is zero at the outer radius ($\rho = 1$) and at the inner radius ($\rho = \beta$), and is maximum for $\rho = \sqrt{\beta}$, or in other words for $\rho = 0.47$, where it is 67.7 MPa. Though it is a tensile stress, this value is slightly higher than that found in the comparable disk.

The values assumed by hoop stress at the inner and outer radii respectively are:

$$(\sigma_t)_{\rho=\beta} = 226 \text{ MPa}; \; (\sigma_t)_{\rho=1} = 48 \text{ MPa}.$$

These local hoop stress values are respectively slightly higher and slightly lower than those found in the comparable disk.

The values assumed by axial stress at the inner and outer radii respectively are:

$$(\sigma_z)_{\rho=\beta} = 27 \text{ MPa}; \; (\sigma_z)_{\rho=1} = -27 \text{ MPa}.$$

In the problem examined here, these values are not particularly significant unless other axial stresses deriving from other external longitudinal loads are superposed on them. It is then clear that stress σ_z is a tensile stress at the inner radius and a compressive stress at the outer radius.

Radial displacements at the inner and outer radii assume the following values:

$$(u)_{\rho=\beta} = 2.91 \times 10^{-5} \text{ m}; \; (u)_{\rho=1} = 3.38 \times 10^{-5} \text{ m}.$$

According to the maximum shear stress failure criterion, τ_{max} occurs at the inner radius and is:

$$\tau_{max} = \left(\frac{\sigma_t}{2}\right)_{\rho=\beta} = 113 \text{ MPa}.$$

If the maximum octahedral shear stress failure criterion is used, we obtain:

$$(\tau_{oct})_{\rho=\beta} = \frac{1}{3} \cdot \left[(\sigma_t - \sigma_r)^2 + (\sigma_t - \sigma_z)^2 + (\sigma_z - \sigma_r)^2\right]^{\frac{1}{2}} = 101\text{MPa}.$$

The safety factors s for the material according to the maximum shear stress criterion (which here coincides with the maximum normal stress criterion) and the maximum octahedral shear stress criterion are respectively:

$$s = \frac{\sigma_s/2}{\tau_{\max}} = 240/113 = 2.12$$

$$s = \frac{\sigma_s/2}{\tau_{oct}} = 240/101 = 2.38.$$

10.3.2 Circular Cylindrical Body of Finite Length with Free Ends Subjected to Thermal Load

If the terms associated with centrifugal load are omitted from relations (10.54), (10.55) and (10.56), we obtain the following relations expressing the stress state and radial and axial displacements in the hollow cylindrical body of finite length with free ends subjected only to thermal load:

$$\begin{cases} \sigma_r = \dfrac{\alpha \cdot E}{\rho^2 \cdot (1-v)} \cdot \left(\dfrac{\rho^2 - \beta^2}{1-\beta^2} \cdot \displaystyle\int_\beta^1 T \cdot \rho \cdot d\rho - \int_\beta^\rho T \cdot \rho \cdot d\rho \right) \\ \sigma_t = \dfrac{\alpha \cdot E}{\rho^2 \cdot (1-v)} \cdot \left(\dfrac{\rho^2 + \beta^2}{1-\beta^2} \cdot \displaystyle\int_\beta^1 T \cdot \rho \cdot d\rho + \int_\beta^\rho T \cdot \rho \cdot d\rho \right) - \dfrac{\alpha \cdot E \cdot T}{(1-v)} \\ \sigma_z = \dfrac{E \cdot \alpha}{(1-v)} \cdot \left(\dfrac{2}{(1-\beta^2)} \cdot \displaystyle\int_\beta^1 T \cdot \rho \cdot d\rho - T \right) \\ u = r_e \cdot \dfrac{\alpha \cdot (1+v)}{\rho \cdot (1-v)} \cdot \left\{ \dfrac{\rho^2 + \beta^2}{1-\beta^2} \cdot \displaystyle\int_\beta^1 T \cdot \rho \cdot d\rho + \int_\beta^\rho T \cdot \rho \cdot d\rho \right\} \\ w = z \cdot \dfrac{2\alpha}{1-v} \cdot \left(\dfrac{1+v}{1-\beta^2} \cdot \displaystyle\int_\beta^1 T \cdot \rho \cdot d\rho - v \cdot T \right). \end{cases} \tag{10.69}$$

The first two of these relations obviously coincide with the first two relations (10.29), both without the centrifugal load term. Here again, introducing relation (10.43) in relations (10.69) and performing integrations gives the following:

$$\begin{cases} \sigma_r = \dfrac{\alpha E}{2(1-\nu)} \dfrac{T_i - T_e}{\ln\beta} \left(\dfrac{\beta^2}{\rho^2} \dfrac{1-\rho^2}{1-\beta^2} \ln\beta - \ln\rho \right) \\ \sigma_t = \dfrac{\alpha E}{2(1-\nu)} \dfrac{T_i - T_e}{\ln\beta} \left(-1 - \dfrac{\beta^2}{\rho^2} \dfrac{1+\rho^2}{1-\beta^2} \ln\beta - \ln\rho \right) \\ \sigma_z = \dfrac{\alpha E}{2(1-\nu)} \dfrac{T_i - T_e}{\ln\beta} \left(-1 - \dfrac{2\beta^2}{1-\beta^2} \ln\beta - 2\ln\rho \right) \\ u = r_e \rho \alpha \dfrac{T_i - T_e}{2(1-\nu)\ln\beta} \left\{ -(1-\nu) + (1+\nu)\ln\rho - \dfrac{\beta^2}{1-\beta^2} \ln\beta \left[\dfrac{(1+\nu)}{\rho^2} + (1-3\nu) \right] \right\} \\ w = z\alpha \dfrac{T_i - T_e}{2\ln\beta} \left(1 + \dfrac{2\beta^2}{1-\beta^2} \ln\beta \right). \end{cases} \tag{10.70}$$

These relations describe the stress state and the radial and axial displacements in hollow cylindrical body of finite length with free ends, subject to a thermal load whose temperature variation with the radius follows Fourier's law.

The first two relations (10.70) coincide with the corresponding relations (10.44); from the standpoint of stresses σ_r and σ_t, there is thus no difference between the free cylindrical body and the cylindrical body constrained at its ends or indefinitely extended in the direction of its axis.

From the first three relations (10.70), we find that the longitudinal stress σ_z in the same solid is equal to the sum $(\sigma_r + \sigma_t)$. These relations also show that a constant temperature in this solid with free ends results in a stress state characterized by $\sigma_r = \sigma_t = \sigma_z = 0$; unlike the corresponding solid constrained at its ends and subjected to the same temperature distribution $T = const$, the longitudinal stress will be zero, as the solid is free to expand in the axial direction.

It should also be noted that here, in contrast to the situation for the solid constrained at its ends or indefinitely extended in the direction of its axis, the radial and longitudinal expansions are both unconstrained, and terms in T_e consequently do not appear in the last two relations (10.70). This conclusion is reached using relations (10.30) and (10.56), which express radial and longitudinal displacements respectively, and assuming that displacements are evaluated starting from the configuration for T_e appearing in relation (10.43) and taken as the reference (or assembly) temperature. If the latter differs from T_e, the additional terms $r_e \rho \alpha T_e$ and $z\alpha T_e$ would appear in the expressions $u = u(\rho)$ and $w = w(z)$ respectively, which are given by the last two relations (10.70), while the value of T_e to be introduced in calculations is the difference between the temperature measured on the outer surface and reference temperature T_0.

It can be concluded from the first relation (10.70) that the radial stress σ_r is zero at the inner radius ($\rho = \beta$) and at the outer radius ($\rho = 1$), while it retains the same sign for every other value of ρ, as it is negative for centrifugal heat flow ($T_i > T_e$) and positive for centripetal heat flow ($T_i < T_e$). The second and third relations (10.70) show that stresses σ_t and σ_z assume the same values at the inner and outer radii, where they are respectively:

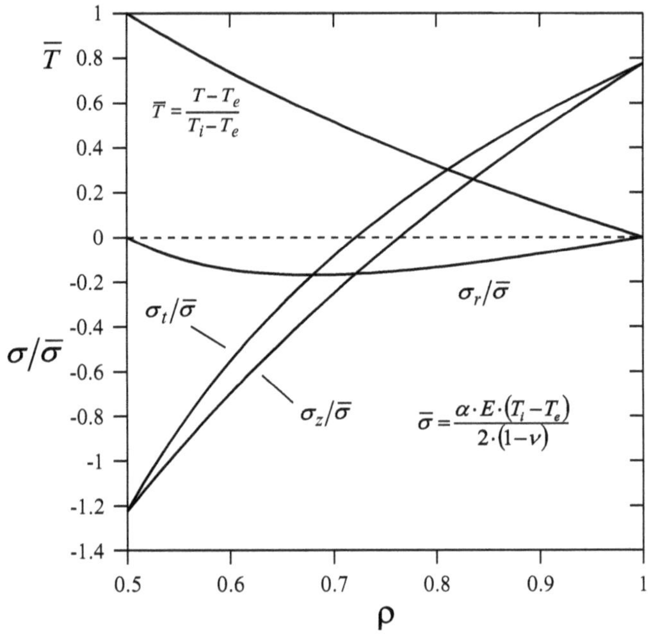

Fig. 10.11 Distribution curves of the three dimensionless principal stresses $\sigma_r/\bar{\sigma}$, $\sigma_t/\bar{\sigma}$ and $\sigma_z/\bar{\sigma}$ and dimensionless temperature $\bar{T}$ along the dimensionless radius of a hollow cylinder with free ends subjected to centrifugal heat flow

$$\begin{aligned}
(\sigma_t)_{\rho=\beta} = (\sigma_z)_{\rho=\beta} &= -\frac{\alpha \cdot E}{2 \cdot (1-\nu)} \cdot \frac{T_i - T_e}{\ln \beta} \cdot \left(1 + \frac{2 \cdot \ln \beta}{1-\beta^2}\right) = \\
&= \frac{\alpha \cdot E}{2 \cdot (1-\nu)} \cdot \frac{T_i - T_e}{\ln(r_e/r_i)} \cdot \left(1 - \frac{2 \cdot r_e^2}{r_e^2 - r_i^2} \cdot \ln \frac{r_e}{r_i}\right) \\
(\sigma_t)_{\rho=1} = (\sigma_z)_{\rho=1} &= -\frac{\alpha \cdot E}{2 \cdot (1-\nu)} \cdot \frac{T_i - T_e}{\ln \beta} \cdot \left(1 + \frac{2 \cdot \beta^2 \cdot \ln \beta}{1-\beta^2}\right) = \\
&= \frac{\alpha \cdot E}{2 \cdot (1-\nu)} \cdot \frac{T_i - T_e}{\ln(r_e/r_i)} \cdot \left(1 - \frac{2 \cdot r_i^2}{r_e^2 - r_i^2} \cdot \ln \frac{r_e}{r_i}\right).
\end{aligned} \tag{10.71}$$

These are compression stresses towards the inner radius and tensile stresses towards the outer radius for centrifugal heat flow, and the converse for centripetal heat flow. For all other values of ρ differing from $\rho = \beta$ and from $\rho = 1$, we have $\sigma_t \neq \sigma_z$.

Figure 10.11 shows distribution curves of temperatures and principal stresses in dimensionless form in a steel cylindrical body ($E = 210$ GPa; $\nu = 0.3$; $\alpha = 12 \times 10^{-6}\ ^\circ\text{C}^{-1}$), with $\beta = 1/2$ ($r_e = 2r_i$), free at the ends and subjected to centrifugal heat flow. Stresses and temperature are made dimensionless in the same way illustrated for the cylindrical body clamped at the ends.

As can be seen from the figure, the absolute maximum values of the stresses are those for longitudinal and hoop stresses at the inner radius, where the two values are equal, negative (i.e., are compressive), and equal to 2.20 MPa for each degree of difference between the temperatures at the inner and outer radii. For a temperature difference ($T_i - T_e$) of 100 °C, both of these stresses would be 220 MPa. In this case, for steady-state heat flow, i.e., for static stress, for the same AISI 1060 steel considered for the cylindrical body clamped at the ends, and again using the three σ_{max}, τ_{max} and maximum distortion energy failure criteria, this stress state would correspond to a safety factor of 2.18 for all three failure criteria. All other conditions remaining equal, the cylindrical body with free ends is much more resistant to thermal load than the clamped body.

Passing from the inner radius to the outer, these stresses drop in absolute value, change sign to become a tensile stress, and at the outer radius reach the same value of 1.34 MPa for each °C of difference between the temperatures at the inner and outer radii. For steel or metals in general, the designer's attention must be focused on the inner radius, where the material is at greatest risk. For reinforced concrete or masonry construction as used in certain smokestacks, this tensile stress state at the outer radius often causes more or less generalised fractures, as these materials have low tensile strength.

In the entire range of variability for ρ, the distribution curve of axial stress remains below that for hoop stress except at the ends, where the two values coincide.

Figure 10.11 applies for centrifugal heat flow ($T_i > T_e$). For centripetal heat flow ($T_i < T_e$), here as for the clamped cylindrical body, the distribution of the three principal stresses and the temperature distribution along the dimensionless radius are obtained from those indicated in Fig. 10.11 by rotating the curves for $\sigma_r/\bar{\sigma}$, $\sigma_t/\bar{\sigma}$ and $\sigma_z/\bar{\sigma}$and the $\bar{T} = (T - T_e)/(T_i - T_e)$ curve by 180° relative to the axis of the abscissa and to a vertical axis respectively.

When the cylindrical body has thin walls, as in the case of a cylinder sleeve for an internal combustion engine or a compressor, the temperature distribution along the dimensionless radius may be assumed to be linear, and corresponds to an equally linear distribution of hoop and longitudinal stresses whose curves overlap for the entire range of variability for ρ. Radial stress is constantly zero.

To see that this is in fact true, it is sufficient to consider relations (10.71), which express hoop and axial stresses at the inner and outer radii as a function of ratio r_e/r_i. These relations can be rewritten to introduce the cylindrical body's wall thickness s, as $r_e/r_i = (1 + s/r_i)$. If s is small compared to r_i, a power series expansion yields:

$$\ln\frac{r_e}{r_i} = \frac{s}{r_i}\cdot\left[1 - \frac{1}{2}\cdot\left(\frac{s}{r_i}\right) + \frac{1}{3}\cdot\left(\frac{s}{r_i}\right)^2 - \frac{1}{4}\cdot\left(\frac{s}{r_i}\right)^3 + \ldots\right] \tag{10.72}$$

Thus, by stopping the expansion at the third term in the square brackets, we find that the expression

$$\frac{1}{\ln(r_e/r_i)}\cdot\left(1 - \frac{2\cdot r_e^2}{r_e^2 - r_i^2}\cdot\ln\frac{r_e}{r_i}\right), \tag{10.73}$$

which appears in the second member of the first relation (10.71) can be reduced to the following

$$\frac{1}{\ln(r_e/r_i)}\cdot\left(1-\frac{2\cdot r_e^2}{r_e^2-r_i^2}\cdot\ln\frac{r_e}{r_i}\right)$$

$$=\frac{1}{\frac{s}{r_i}\cdot\left[1-\frac{1}{2}\cdot\left(\frac{s}{r_i}\right)+\frac{1}{3}\cdot\left(\frac{s}{r_i}\right)^2\right]}-\frac{2\cdot\left(1+\frac{s}{r_i}\right)^2}{\left(1+\frac{s}{r_i}\right)^2-1}=$$

$$=\frac{1}{\frac{s}{r_i}\cdot\left[1-\frac{1}{2}\cdot\left(\frac{s}{r_i}\right)+\frac{1}{3}\cdot\left(\frac{s}{r_i}\right)^2\right]}-\frac{\left(1+\frac{s}{r_i}\right)^2}{\frac{s}{r_i}\cdot\left(1+\frac{s}{2\cdot r_i}\right)}. \tag{10.74}$$

By expanding the two parentheses in the denominator of the last expression of this equality with the following series

$$(1+x)^a=\sum_{n=0}^{\infty}\binom{a}{n}\cdot x^n=1+a\cdot x+\frac{a\cdot(a-1)}{2!}x^2+\ldots, \tag{10.75}$$

with $a=-1$ (the two parentheses are in the denominator), stopping the expansion at the second degree terms, whereby:

$$(1+x)^{-1}=1-x+x^2+\ldots \tag{10.76}$$

and setting x equal to $-(s/2r_i-s^2/3r_i^2)$ in the square brackets of relation (10.74) and equal to $s/2r_i$ in the parentheses, we can then obtain the following from relation (10.74):

$$\frac{1}{\ln(r_e/r_i)}\cdot\left(1-\frac{2\cdot r_e^2}{r_e^2-r_i^2}\cdot\ln\frac{r_e}{r_i}\right)=\frac{r_i}{s}\cdot\left[\left(1+\frac{s}{2\cdot r_i}-\frac{s^2}{3\cdot r_i^2}\right)-\left(1+\frac{s}{r_i}\right)^2\cdot\left(1-\frac{s}{2\cdot r_i}-\frac{s^2}{4\cdot r_i^2}\right)\right]=$$

$$=-\left(1+\frac{s}{3\cdot r_i}\right). \tag{10.77}$$

In the series expansion of the square brackets appearing in relation (10.74), as in the elaboration of the square brackets appearing in relation (10.77), it should be noted that the powers $(s/r_i)^3$ and $(s/r_i)^4$ by respect to the linear and quadratic terms of this ratio were ignored.

Proceeding in the same way with the expression

$$\frac{1}{\ln r_e/r_i} \cdot \left(1 - \frac{2 \cdot r_i^2}{r_e^2 - r_i^2} \cdot \ln \frac{r_e}{r_i}\right), \tag{10.78}$$

which appears in the second relation (10.71), we obtain:

$$\frac{1}{\ln(r_e/r_i)} \cdot \left(1 - \frac{2 \cdot r_i^2}{r_e^2 - r_i^2} \cdot \ln \frac{r_e}{r_i}\right) = 1 - \frac{s}{3 \cdot r_i}. \tag{10.79}$$

Substituting expressions (10.77) and (10.79) in relations (10.71) gives:

$$\begin{aligned}
(\sigma_t)_{\rho=\beta} &= (\sigma_z)_{\rho=\beta} = -\frac{\alpha \cdot E \cdot (T_i - T_e)}{2 \cdot (1 - \nu)} \cdot \left(1 + \frac{s}{3 \cdot r_i}\right) \\
(\sigma_t)_{\rho=1} &= (\sigma_z)_{\rho=1} = \frac{\alpha \cdot E \cdot (T_i - T_e)}{2 \cdot (1 - \nu)} \cdot \left(1 - \frac{s}{3 \cdot r_i}\right).
\end{aligned} \tag{10.80}$$

If $s < < r_i$, i.e., in the case of a very thin wall, relations (10.80) can be further simplified to become:

$$\begin{aligned}
(\sigma_t)_{\rho=\beta} &= (\sigma_z)_{\rho=\beta} = -\frac{\alpha \cdot E \cdot (T_i - T_e)}{2 \cdot (1 - \nu)} \\
(\sigma_t)_{\rho=1} &= (\sigma_z)_{\rho=1} = \frac{\alpha \cdot E \cdot (T_i - T_e)}{2 \cdot (1 - \nu)}.
\end{aligned} \tag{10.81}$$

Consequently, the values assumed by the hoop and axial stresses at the inner radius, which equal the factor $\bar{\sigma}$, are equal and opposite to those assumed by the same stresses at the outer radius. With the same degree of approximation, we can say that temperature variation with the radius is linear, and thus

$$T = T_i - (T_i - T_e) \cdot \frac{r - r_i}{r_e - r_i} = T_e + (T_i - T_e) \cdot \frac{r_e - r}{r_e - r_i}. \tag{10.82}$$

Introducing this relation rewritten in terms of variable ρ in the first three relations (10.69) yields:

$$\begin{aligned}
\sigma_r &= 0 \\
\sigma_t &= \sigma_z = \frac{\alpha \cdot E \cdot (T_i - T_e)}{2 \cdot (1 - \nu)} \cdot \left(1 - 2 \cdot \frac{1 - \rho}{1 - \beta}\right).
\end{aligned} \tag{10.83}$$

For $\rho = \beta$ and for $\rho = 1$, the second relation (10.83) yields relations (10.81).

Figure 10.12 shows the distribution curves of the three principal stresses, which here again are made dimensionless by dividing them by the factor $\bar{\sigma}$, in a cylindrical body with a ratio $\beta = 0.91$ subjected to centrifugal heat flow. This value of β corresponds to a thickness/diameter ratio of 1/20, which by convention is the upper

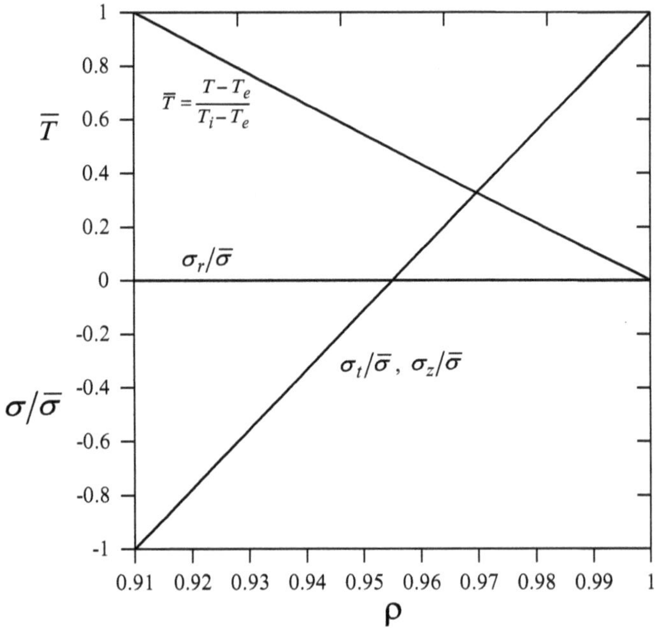

Fig. 10.12 Distribution curves of the three dimensionless principal stresses $\sigma_r/\bar{\sigma}$, $\sigma_t/\bar{\sigma}$ and $\sigma_z/\bar{\sigma}$ and dimensionless temperature $\bar{T}$ along the dimensionless radius of a thin-wall hollow cylinder with free ends subjected to centrifugal heat flow

limit for a thin wall (for which β is thus greater than 0.91). In this condition (where the thickness is small, but not very small), the maximum values for stresses σ_t and σ_z at the inner and outer radii are only approximately equal to the factor $\bar{\sigma} = \alpha E (T_i - T_e)/2(1 - \nu)$, but the differences are not significant for design purposes.

Stresses σ_t and σ_z are compressive at the inner radius and assume their maximum negative value of around 1.80 MPa for each degree of temperature difference $(T_i - T_e)$. Passing from the inner radius to the outer, these stresses drop linearly in absolute value, change sign to become tensile stresses, and at the outer radius assume a value equal and opposite to that at the inner radius. Figure 10.12 shows temperature in dimensionless form as the ratio $\bar{T} = (T - T_e)/(T_i - T_e)$.

These stress values correspond to the radial and hoop bending stresses found at the upper face and lower face of a circular plate clamped at the outer radius and subject to a temperature gradient along the thickness, with upper face temperature T_i and lower face temperature T_e. It is thus clear that the stresses in the thin-wall cylindrical body and those in the plate are comparable only as regards the numerical values and their behaviour in the direction of the temperature gradient, while the stress field is different, as it is characterised by σ_t and σ_z in the cylindrical body and by σ_r and σ_t in the plate similar to a disk.

Figure 10.12 applies for centrifugal heat flow $(T_i > T_e)$. For centripetal heat flow $(T_i < T_e)$, the distribution of the two non-zero principal stresses along the

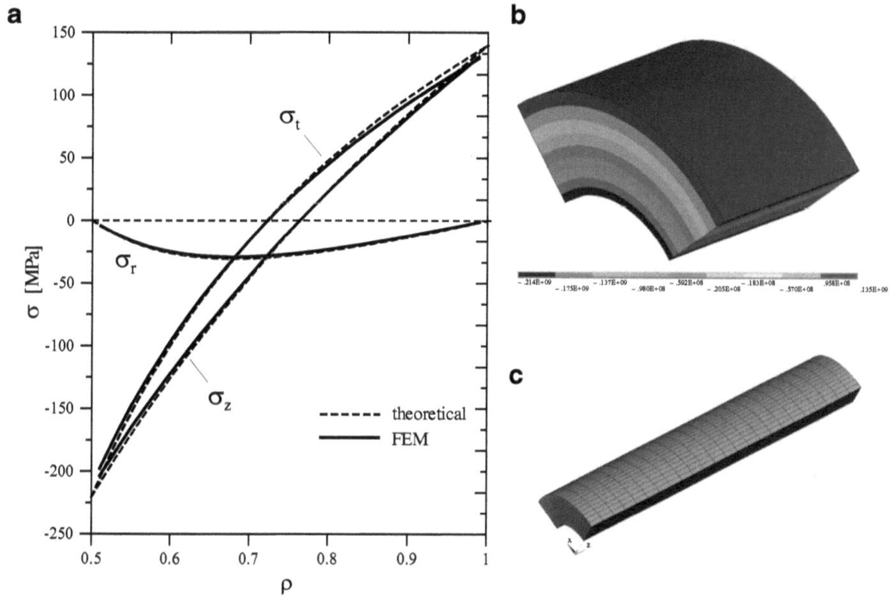

Fig. 10.13 (**a**) Distribution curves of the three principal stresses σ_r, σ_t and σ_z along the dimensionless radius of a hollow cylinder with free ends subjected to centrifugal heat flow. Reference temperature $T_0 = 20\ °\text{C}$; outer temperature $T_e = 20\ °\text{C}$; inner temperature $T_i = 120\ °\text{C}$; cylinder inner radius $r_i = 0.5$ m and outer radius $r_e = 1$ m. (**b**) Stress σ_z along the wall thickness of a hollow cylinder with free ends. (**c**) FEM model: 8-node solid elements with one dof per node (temperature) for thermal calculation; 8-node solid elements with three dofs per node for structural calculation

dimensionless radius is obtained from that indicated in Fig. 10.12 as the mirror image of the curves shown therein, i.e., by rotating the plots for $\sigma_t/\bar{\sigma}$ and $\sigma_z/\bar{\sigma}$ by 180° relative to the axis of the abscissa. The temperature distribution is obtained by rotating the $\bar{T} = (T - T_e)/(T_i - T_e)$ curve shown in Fig. 10.12 again by 180°, but around a vertical axis, so that the dimensionless temperature assumes values of 0 and 1 at the inner radius and outer radius respectively.

10.3.2.1 Example

We will now consider the same hollow cylindrical rotor discussed in the example in Sect. 10.2.2.1, but this time with free ends. Once again, we will calculate the stress state in the entire cylindrical body, comparing the results with those obtained from a finite element model, and concluding with several considerations regarding design. In addition, we will compare the results obtained here with those for the cylindrical body with clamped ends.

The problem date are those given in the example indicated above.

Using the first three relations (10.70) which express the stress state resulting from heat flow in a hollow cylindrical body of finite length with free ends, we obtain the three dashed-line curves shown in Fig. 10.13a. As can be seen, the material is at greatest risk at the inner radius, where the principal stresses assume the following values: $\sigma_r = 0$; $\sigma_t = -220$ MPa; $\sigma_z = -220$ MPa.

The three σ_{max}, τ_{max} and maximum distortion energy failure criteria lead to the same value for the material's safety factor, viz., 2.18.

Numerical analysis will be carried out with the same finite element model used for the cylindrical body with clamped ends, with the obvious difference that in this case, the longitudinal degree of freedom will not be zero at the ends. Figure 10.13a summarises the numerical results obtained with the model (solid-line curves), while axial stress σ_z along the dimensionless radius of the cylindrical body in question, which here coincides with the hoop stress, is shown in Fig. 10.13b.

In this case as in the other example, comparison of the theoretical and numerical results shows that the differences are entirely insignificant from the design standpoint.

Comparing these values for the three stresses σ_r, σ_t and σ_z at the inner radius from those found in Sect. 10.2.2.1, which apply to the same cylindrical rotor with clamped ends, we see that the values for radial stress (zero in both cases) and hoop stress (−220 MPa in both cases) are unchanged, while the axial stress in the constrained cylindrical body is more than 67 % above that in the free cylindrical body (−368 MPa as against −220 MPa).

10.4 Edge Effect in a Circular Cylindrical Body of Finite Length with Free Ends Subjected to Thermal Load

What we have said so far for the cylindrical body of finite length with free ends and subjected to thermal load holds as long as it can be assumed that there is a strain state characterised by a uniform axial translation of a generic cross section ($\varepsilon_z = \varepsilon_0 = const$). For a body of finite length, this assumption is respected at a sufficient distance from the ends, but as this distance is reduced, the stress state should vary given that there should be a plane stress state at the ends ($\sigma_z = 0$).

The stress and strain state at the free ends can be rigorously analysed by using the relations governing heat flow and those for thermoelasticity simultaneously. With the rather rough assumption that the finite length does not have an appreciable influence on temperature distribution along the radius and that the temperature varies according to a linear function along the radius, an approximate solution for the thin wall can be reached which consists of calculating the local stress state as the superposition of that calculated above assuming a strain state characterised by a uniform axial translation of a generic cross section and that determined by a localised bending moment which generates stresses σ_z equal and opposite to these stresses at the free ends. In this way, we will have $\sigma_z = 0$ at the free ends. This

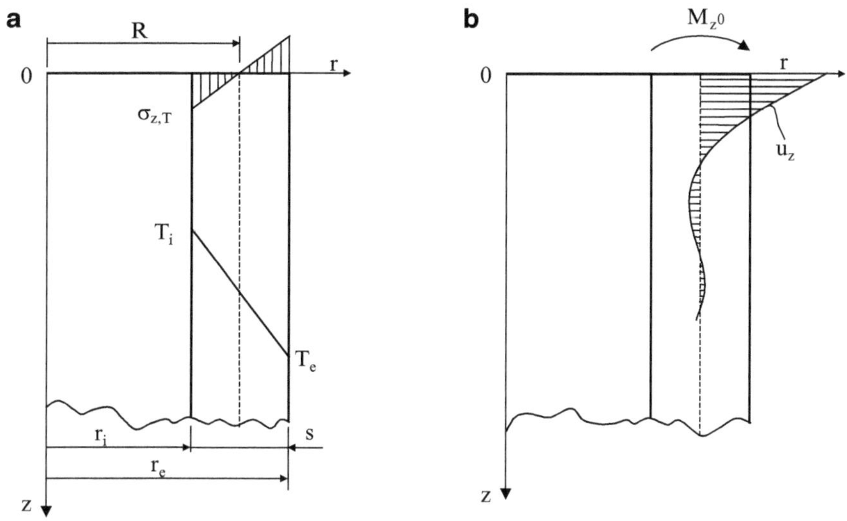

Fig. 10.14 (**a**) Distribution curves of $T = T(r)$ and the correlated axial stress σ_z; (**b**) deflection $u = u(z)$ resulting from localised bending moment M_{z0}

approach, though it respects the equilibrium conditions, leads to results that are highly approximate.

Bearing relations (10.81) and thin cylindrical shell theory in mind, it is obvious that the longitudinal bending moment M_{z0} per unit length of the edge, to be applied to the free ends of the cylindrical body (for example the end having $z = 0$) must satisfy the relation $\sigma_z = 6M_{z0}/s^2$, so that

$$M_{z0} = \frac{\alpha \cdot E \cdot (T_i - T_e) \cdot s^2}{12 \cdot (1 - \nu)}. \tag{10.84}$$

For centrifugal heat flow, it is also obvious that M_z must have a direction such that the fibers at the inner radius will be in tension and those at the outer radius will be compressed (this is the direction assumed as positive), with the converse being true for centripetal heat flow.

From the general bending theory of cylindrical shells [74] we have:

$$\begin{aligned} M_z &= D \cdot \frac{d^2u}{dz^2} = M_{z0} \cdot e^{-\beta \cdot z} \cdot (\cos \beta z + \sin \beta z) \\ M_t &= \nu \cdot M_z = \nu \cdot M_{z0} \cdot e^{-\beta \cdot z} \cdot (\cos \beta z + \sin \beta z) \\ u &= \frac{M_{z0}}{2\beta^2 D} \cdot e^{-\beta \cdot z} \cdot (\cos \beta z - \sin \beta z), \end{aligned} \tag{10.85}$$

where, with the meaning of the other symbols remaining the same, M_t is the anticlastic tangential bending moment for unit length, $D = E \cdot s^3 / 12(1 - \nu^2)$ is the flexural rigidity of the cylindrical shell, $\beta = \sqrt[4]{3(1-\nu^2)/R^2 s^2}$ and $R = (r_i + r_e)/2$ is the average radius. It should be noted that β here does not have the same meaning as the ($\beta = r_i/r_e$) used so far. In the edge effects of thin shell bending theory, β denotes a magnitude linked to the thickness, the flexural rigidity and the elastic properties of the shell material.

Figure 10.14a shows the distribution curves of $T = T(r)$ along the dimensionless radius for centrifugal heat flow and of the corresponding axial stress σ_z on the end surface with axial coordinate $z = 0$. Figure 10.14b, on the other hand, shows the distribution curve of radial displacement $u = u(z)$ and thus for the deflection resulting from applying the localised bending moment M_{z0}, uniformly distributed along the edge $z = 0$.

At the inner edge, the axial and hoop stresses due to the localised bending moment will be:

$$\sigma_{z,M} = \frac{6 \cdot M_z}{s^2} = \frac{6 \cdot M_{z0}}{s^2} \cdot e^{-\beta \cdot z} \cdot (\cos\beta z + \sin\beta z)$$
$$\sigma_{t,M} = \frac{E}{R} \cdot u + \frac{6 \cdot M_t}{s^2} = M_{z0} \cdot e^{-\beta \cdot z} \cdot \left[\frac{E}{2RD\beta^2}(\cos\beta z - \sin\beta z) + \frac{6\nu}{s^2}(\cos\beta z + \sin\beta z)\right] \tag{10.86}$$

At the outer edge, these stresses will be:

$$\sigma_{z,M} = -\frac{6 \cdot M_z}{s^2} = -\frac{6 \cdot M_{z0}}{s^2} \cdot e^{-\beta \cdot z} \cdot (\cos\beta z + \sin\beta z)$$
$$\sigma_{t,M} = \frac{E}{R} \cdot u - \frac{6 \cdot M_t}{s^2} = M_{z0} \cdot e^{-\beta \cdot z} \cdot \left[\frac{E}{2RD\beta^2}(\cos\beta z - \sin\beta z) - \frac{6\nu}{s^2}(\cos\beta z + \sin\beta z)\right]. \tag{10.87}$$

At the end of the cylindrical body ($z = 0$), the overall axial and hoop stresses, which are the sum of those due to thermal load and those caused by the localised bending moment, will thus be as follows at the inner edge:

$$\sigma_z = 0$$
$$\sigma_t = \sigma_{t,T} + \sigma_{t,M} = -\frac{\alpha \cdot E \cdot (T_i - T_e)}{2 \cdot (1-\nu)} \cdot \left(1 - \sqrt{\frac{1-\nu^2}{3}} - \nu\right) = \sigma_{t,T} \cdot \left(1 - \sqrt{\frac{1-\nu^2}{3}} - \nu\right) \tag{10.88}$$

At the outer edge, they will be:

$$\sigma_z = 0$$
$$\sigma_t = \sigma_{t,T} \cdot \left(1 + \sqrt{\frac{1-\nu^2}{3}} - \nu\right). \tag{10.89}$$

For centrifugal heat flow, it can thus be concluded that hoop stress decreases at the inner radius and increases at the outer radius. For $\nu = 0.3$ (steels), the increase in σ_t is around 25% over the value with a plane strain state, i.e., at a sufficient distance from the edge $z = 0$. The edge stresses, in any case, drop rapidly as the distance z from the end increases. At a distance equal to R ($z = R$), for $R/s = 10$ (the upper limit for a thin wall), we have $e^{-\beta z} = 0.017$; in other words, the effect of the localised bending moment has practically disappeared.

10.5 Solid Cylindrical Body of Finite Length and with Free Ends, Subjected to Transient Thermal Load

In Sects. 10.2.2 and 10.3.2, we analysed the stress and strain states in cylindrical bodies of finite length with clamped ends (as was discussed earlier, cylindrical bodies indefinitely extended in the direction of their axis can be treated in the same way) and with free ends, both subjected only to steady-state thermal load. In Sect. 10.2.2 it was also mentioned that in design, there are cases in which steady-state heat flow conditions are not respected. One of these cases is that of the cylindrical rod of finite length with free ends. In this case, the thermal transient must always be considered, as the entire mass will be at the same temperature after the transient has ended and there will be no stresses due to thermal loading.

We will consider a solid cylindrical body (solid cylindrical rod) with free ends, initially at reference temperature T_0, and we will suppose that, starting from instant $t = 0$, the temperature at the outer radius ($\rho = 1$) is brought to and maintained at $T_e \neq T_0$. After a theoretically infinite period of time, the temperature at all points of the cylindrical rod's mass should be $T = T_e$.

At the generic instant of time t starting from the initial time $t = 0$, temperature $T = T(\rho, t)$ is given by solving the following partial differential equation of first order:

$$2\pi\cdot\rho\cdot k\cdot\frac{\partial T}{\partial\rho}\cdot dt-2\pi\cdot(\rho+d\rho)\cdot k\cdot\frac{\partial}{\partial\rho}\left(T+\frac{\partial T}{\partial\rho}d\rho\right)\cdot dt-2\pi\cdot r_e^2\cdot\gamma\cdot c\cdot\frac{\partial T}{\partial t}\cdot\rho\cdot d\rho\cdot dt=0 \tag{10.90}$$

For the annular element of the cylinder having unitary axial length with inner radius ρ and outer radius $\rho + d\rho$, and bearing in mind that in accordance with Fourier's law, the heat flow moving through the generic cylindrical surface of radius ρ is proportional to the temperature gradient, this equation expresses the thermal balance between the heat entering the cylinder of radius ρ, the heat exiting from the cylinder of radius $\rho + d\rho$ and the heat accumulated in the element, expressed respectively by the three monomials appearing in equation (10.90). In this equation, with the meaning of the other symbols remaining the same, c denotes the specific heat of the rod material.

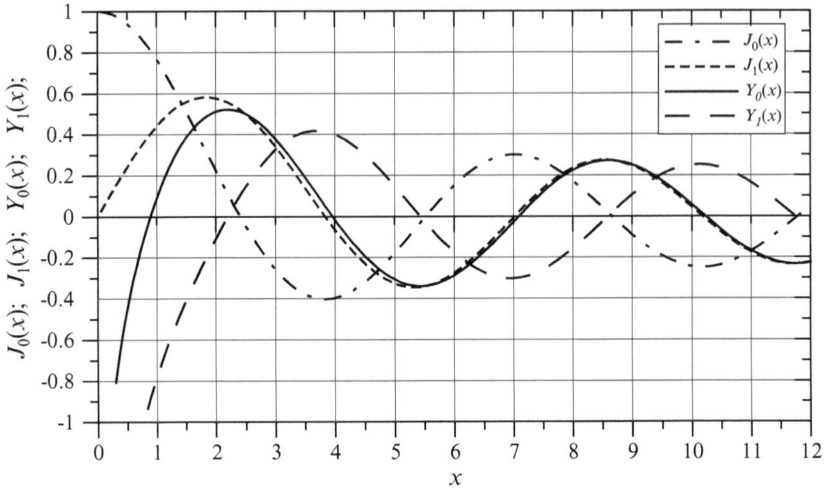

Fig. 10.15 Bessel functions of the first kind, $J_0(x)$ of order zero and $J_1(x)$ of order one, and Bessel functions of the second kind (or Weber functions), $Y_0(x)$ of order zero and $Y_1(x)$ of order one

Expanding (10.90) and ignoring the higher-order infinitesimals in the first member yields the following partial differential equation of second order:

$$k \cdot \rho \cdot \frac{\partial^2 T}{\partial \rho^2} + k \cdot \frac{\partial T}{\partial \rho} + c \cdot \gamma \cdot r_e^2 \cdot \rho \cdot \frac{\partial T}{\partial t} = 0. \tag{10.91}$$

The solution of this latter equation, which satisfies the boundary conditions and the assigned initial conditions, is given by the series expansion:

$$T = (T_e - T_0) \cdot \sum_{n=1}^{\infty} A_n \cdot J_0(\beta_n \rho) \cdot e^{-\frac{k \cdot \beta_n^2}{\gamma \cdot c} \cdot \frac{t}{r_e^2}}, \tag{10.92}$$

where J_0 and J_1 are the Bessel functions of the first kind, respectively of order zero and of order one, β_n are the zeros of the equation $J_0(x) = 0$ and $A_n = 2/[\beta_n J_1(\beta_n)]$. In this connection, it should be borne in mind that the Bessel functions of the first kind $J_{\pm n}(x)$, with n being a positive or negative integer and, in particular, zero, are the solutions of the second order differential equation:

$$x^2 \frac{d^2 w}{dx^2} + x \frac{dw}{dx} + \left(x^2 - n^2\right) w = 0, \tag{10.93}$$

which is the Bessel differential equation. It should also be borne in mind that, for every $n = p$, equation (10.93) admits other solutions: these are the Bessel functions of the second kind $Y_n(x)$, also called Weber functions, and of the third kind $H_n^{(1)}(x)$ and $H_n^{(2)}(x)$, also called Hankel functions [2, 66].

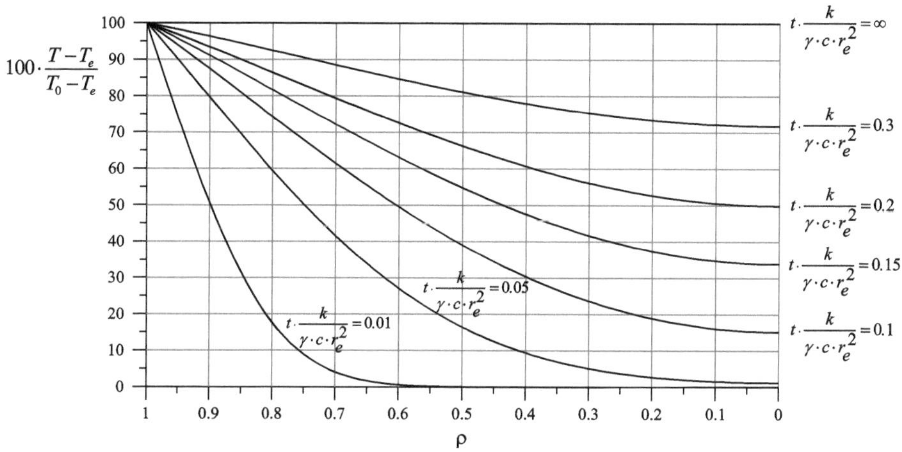

Fig. 10.16 Temperature distributions with time and along the radius in a steel solid cylinder

From Fig. 10.15, which shows the curves for the Bessel functions $J_0(x)$ and $J_1(x)$ as well as for the Weber functions $Y_0(x)$ and $Y_1(x)$, we can see that the first three zeros of equation $J_0(x)$ are $\beta_1 = 2.4048$, $\beta_2 = 5.5201$ and $\beta_3 = 8.6537$, while the first three zeros of equation $J_1(x)$ are $\beta'_1 = 3.8317$, $\beta'_2 = 7.0156$ and $\beta'_3 = 10.1735$.

Figure 10.16 shows temperature distribution along the radius in a steel solid cylinder at different instants of time $tk/\gamma c r_e^2$ starting from the initial instant $t = 0$; temperature is given in dimensionless form, as the percentage ratio $100 \cdot [(T - T_e)/(T_0 - T_e)]$.

Once function $T = T(\rho, t)$ at a given instant of time t is known, familiar numerical analysis methods can be used to calculate the integrals appearing in relations (10.69) written for $\beta = r_i = 0$ and thus determine the rod's stress and strain state at that instant. It is also possible to employ an analytical approach based on series expansions together with some of the relations linking the Bessel functions, their derivatives and their integrals. We thus arrive at the following relations:

$$
\begin{aligned}
\sigma_r &= \frac{2 \cdot \alpha \cdot E \cdot (T_0 - T_e)}{(1-\nu)} \cdot \sum_{n=1}^{\infty} e^{-\frac{k \cdot \beta_n^2 \cdot t}{\gamma \cdot c \cdot r_e^2}} \cdot \left[\frac{1}{\beta_n^2} - \frac{1}{\beta_n^2 \cdot \rho} \cdot \frac{J_1(\beta_n \cdot \rho)}{J_1(\beta_n)} \right] \\
\sigma_t &= \frac{2 \cdot \alpha \cdot E \cdot (T_0 - T_e)}{(1-\nu)} \cdot \sum_{n=1}^{\infty} e^{-\frac{k \cdot \beta_n^2 \cdot t}{\gamma \cdot c \cdot r_e^2}} \cdot \left[\frac{1}{\beta_n^2} + \frac{1}{\beta_n^2 \cdot \rho} \cdot \frac{J_1(\beta_n \cdot \rho)}{J_1(\beta_n)} - \frac{J_0(\beta_n \cdot \rho)}{\beta_n \cdot J_1(\beta_n)} \right] \\
\sigma_z &= \frac{2 \cdot \alpha \cdot E \cdot (T_0 - T_e)}{(1-\nu)} \cdot \sum_{n=1}^{\infty} e^{-\frac{k \cdot \beta_n^2 \cdot t}{\gamma \cdot c \cdot r_e^2}} \cdot \left[\frac{2}{\beta_n^2} - \frac{J_0(\beta_n \cdot \rho)}{\beta_n \cdot J_1(\beta_n)} \right].
\end{aligned}
\tag{10.94}
$$

When t is very short, the temperature on the outer surface and that inside the cylinder can still be considered to be T_e and T_0 respectively. Relations (10.69) written for $\beta = r_i = 0$, with t tending to zero, yield:

$$\begin{aligned} \sigma_r &= 0 \\ \sigma_t = \sigma_z &= \frac{\alpha \cdot E \cdot (T_0 - T_e)}{(1 - \nu)}. \end{aligned} \tag{10.95}$$

We arrive at the same results using the series expansions (10.94), given that

$$\sum_{n=1}^{\infty} \frac{1}{\beta_n^2} = \frac{1}{4}. \tag{10.96}$$

The second relation (10.95) gives the maximum values of the stresses characterizing the stress state in a cylindrical rod subjected to heat flow along the radius. This stress is compressive during heating and tensile during cooling, and equals that needed to annul thermal expansion. To reduce this stress state in shafts and rotors, it is necessary to heat or cool gradually, or to increase the heating time in proportion to the square of the diameter.

Chapter 11
Stress Analysis in Rotating Disks Loaded Beyond Yielding: Non-Hardening Materials

11.1 Basic Hypotheses and Their Limitations in Disks Made of Non-Hardening Materials

Spinning tests on disks demonstrate unequivocally that failure occurs as a result of rotational speeds corresponding to maximum nominal elastic stresses that are not only much above the material's yield strength, but also exceed its failure strength. If the equivalent stress calculated with a yield criterion locally exceeds the material's elastic limit, the stress field will be profoundly changed. In particular, there will be a more uniform redistribution of stresses, together with a reduction in their concentrations at notches (holes, fillets, geometric discontinuities, etc.). In addition, once centrifugal loading stops, the disk stressed beyond yielding will have residual stresses that in the following load cycles will be superposed on the stresses resulting from applied loads. As these internal stresses are generally compressive, the designer can use them to advantage in reducing the maximum tensile stresses due to centrifugal loading.

Analyzing the stress state in a disk loaded beyond yielding until the material reaches the fully plastic state is of considerable importance for highly stressed disks, for carrying out spinning tests and interpreting their results, and for reducing stress peaks by creating an appropriate state of coaction in the material. This analysis, conducted with the usual assumptions of axisymmetry and a plane stress state so that the problem is mono-dimensional, leads to closed form solutions only for the constant-thickness disk, and even then only when rough simplifications are made concerning the material's stress-strain curve.

Here, it will be assumed that this $\sigma = \sigma(\varepsilon)$ curve, plotted from tensile tests on a standard specimen, is as shown in Fig. 11.1, which corresponds to linear elastic-perfectly plastic behaviour with no strain hardening, as occurs in ductile metals and alloys; the material shows linear elastic behaviour up to the yield stress σ_y (in this case, the proportional limit, the elastic limit and the yield point coincide), followed by perfectly plastic behaviour up to failure. It will also be assumed that, regardless of the material's condition (i.e., virgin or otherwise), the $\sigma = \sigma(\varepsilon)$ curve for a

V. Vullo and F. Vivio, *Rotors: Stress Analysis and Design*,
Mechanical Engineering Series, DOI 10.1007/978-88-470-2562-2_11,
© Springer-Verlag Italia 2013

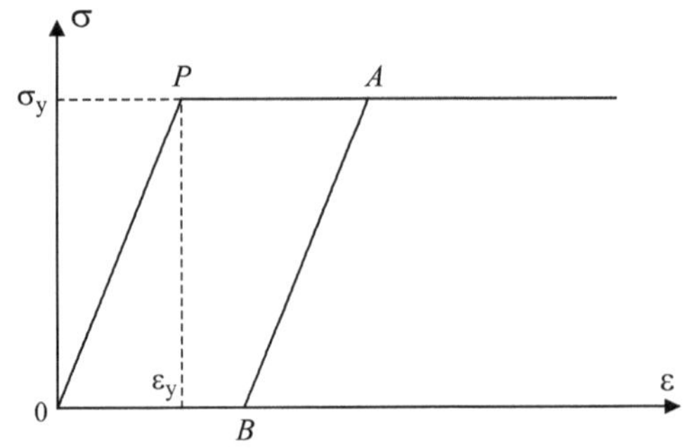

Fig. 11.1 $\sigma = \sigma(\varepsilon)$ curve for a linear elastic-perfectly plastic material

specimen subjected to a compression test is the mirror image of the origin of the tensile test curve. Here and in the rest of the text, in other words, we will ignore the Bauschinger effect, whereby there will be a reduction in the compressive (or tensile) yield strength of a specimen whose material has undergone prior non-zero plastic tensile (or compressive) strain; in other words, the lowering of the yield stress when deformation in one direction is followed by deformation in the opposite direction is called Bauschinger effect.

It is then assumed that, starting from a generic point *A* (Fig. 11.1) located beyond *P* on the horizontal straight-line of the $\sigma = \sigma(\varepsilon)$ curve and reducing the load, the representative point will move along path *AB* essentially parallel to the straight-line portion *OP*. Consequently, when the load reaches zero, there will be a permanent strain given by *OB* on the abscissa. Finally, it will be assumed that, starting from point *B* and increasing load again, the representative point will move along path *BA* in the opposite direction to unloading path *AB*. In other words, it will be assumed that the hysteresis cycle associated with the energy losses per unit volume occurring in the unloading and loading cycle will have zero area. A further increase in load beyond point *A* will result in a movement towards the right along the horizontal segment of the $\sigma = \sigma(\varepsilon)$ curve. Obviously, this model of the material's behaviour is highly approximate. A model which is closer to reality will be described in Chap. 12 below.

In a disk stressed symmetrically around its axis and loaded uniformly along the latter (in this case, as a plane stress state is assumed, we will have $\sigma_z = 0$), the radial, tangential and axial directions will be the principal directions of both stresses and strains. From the design standpoint, most interest focuses on cases where deformations are small (small deformations) and those where small deformations cannot be neglected but must be taken into consideration.

Below, it will be assumed that the strains are small, and that the associated strain state is thus described by conventional (Lagrangian or engineering) strains ε_r, ε_t and ε_z. In cases involving large deformations, it will clearly be necessary to introduce the true (logarithmic or natural) strains $\bar{\varepsilon}_r = \ln(1 + \varepsilon_r)$, $\bar{\varepsilon}_t = \ln(1 + \varepsilon_t)$ and $\bar{\varepsilon}_z = \ln(1 + \varepsilon_z)$. The assumption that deformations are small is fully verified for materials that show well defined yielding in the $\sigma = \sigma(\varepsilon)$ curve. In such cases, there is a gradual passage from an elastic strain state to the plastic strain state, with a plastic-elastic interface that, from the moment permanent deformation first appears, shifts to mark the boundary between two zones, one plastic and one elastic. Because of the two zones' continuity and contiguity, strains in the plastic region must be of comparable magnitude with those in the elastic region, or in other words must be small because those of the elastic zone are small.

11.2 Yield Criteria

For a rigorous treatment, in fact, it is necessary to consider a yield criterion. If deformations are small and the material is homogeneous and isotropic, shows ductile behaviour and is in the virgin state (zero Bauschinger effect), the yield theory that best interprets experimental data is that of maximum distortion energy (the Von Mises criterion) which, for a biaxial stress field ($\sigma_z = 0$), can be expressed by (9.4) with $\sigma_e = \sigma_y$, which thus yields the relation:

$$\sigma_r^2 + \sigma_t^2 - \sigma_r \cdot \sigma_t = \sigma_y^2 = const. \tag{11.1}$$

In plane (σ_r, σ_t), this is the equation of an ellipse (the ellipse of plasticity), with semi-major and semi-minor diameters equal to $\sigma_y \cdot \sqrt{2}$ and $\sigma_y \cdot \sqrt{2/3}$, positioned so that they bisect dihedra (σ_r, σ_t) and (σ_r, $-\sigma_t$) respectively. This ellipse is represented in Fig. 11.2 in plane (σ_r/σ_y, σ_t/σ_y), as it is often convenient to introduce stresses that are dimensionless relative to yield stress σ_y.

As can be seen from the curves in Fig. 2.6, both stresses σ_r and σ_t in a disk subjected to centrifugal load are tensile. Given that the hoop stress at any radius always assumes values above those for radial stress, or in other words the ratio σ_t/σ_r is above unity at all times, the arc of ellipse of interest to us here is that between the line bisecting dihedron (σ_r, σ_t) and the axis of stresses σ_t. This arc can be substituted, with a degree of approximation that is acceptable from the design standpoint, by the horizontal straight-line of ordinate $\sigma_t = k = const.$ The value of this constant can be assumed to be equal to the average of the yield stress σ_y in pure tension (ordinate of point B intersecting the ellipse with the axis of the ordinates) and the maximum possible values of the ordinate of the ellipse, which is $2 \cdot \sigma_y/\sqrt{3}$. Consequently:

$$\sigma_t = k = \frac{\sigma_y}{2} \cdot \left(1 + 2\Big/\sqrt{3}\right) = 1.077 \cdot \sigma_y; \tag{11.2}$$

With this approximation, which enables us to arrive at a closed form solution, onset of yielding will take place when the hoop stress (and the equivalent stress σ_e) reaches $1.077\sigma_y$.

If we use the maximum shear stress yield criterion (the Tresca criterion), which can be expressed by (9.3) with $\sigma_e = \sigma_y$, we obtain the relation:

$$\sigma_t = \sigma_y. \tag{11.3}$$

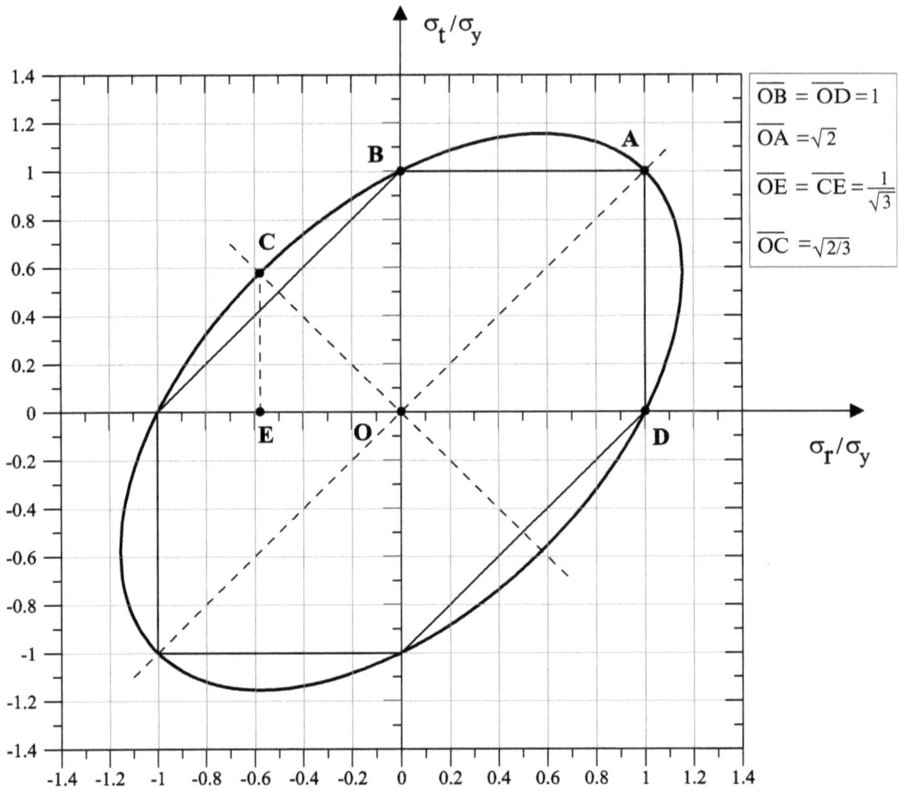

Fig. 11.2 Ellipse and hexagon of plasticity in plane (σ_r/σ_y, σ_t/σ_y)

With the same approach outlined above, we substitute the arc of ellipse *AB* in plane (σ_r, σ_t) with the horizontal straight-line $\overline{AB}$ and, more generally, the ellipse of plasticity with the hexagon of plasticity, likewise represented in plane (σ_r/σ_y, σ_t/σ_y) in Fig. 11.2. In this case, radial stress does not contribute to yielding, which will begin when hoop stress reaches the value of yield stress σ_y.

At the inner radius ($\rho = \beta$), where $\sigma_r = 0$, these two yield criteria lead to the same result: onset of yielding will take place at the inner radius when hoop stress σ_t equals yield stress σ_y. If, on the other hand, we consider the immediate neighbourhood of the disk's inner radius ($\rho \neq \beta$, but adjacent to it), the maximum distortion energy criterion and the maximum shear stress criteria lead to results that differ by approximately 8 %. This difference, in the light of the other simplifying assumptions that are introduced, is of little significance from the design standpoint, and justifies the considerations made above.

Consequently, and in view of the τ_{max} criterion's greater formal simplicity compared to the maximum distortion energy criterion, it will be assumed that no plastic deformation takes place in the disk until the hoop stress at the inner radius reaches $\sigma_t = \sigma_y$. This result is also reached by using the maximum normal stress criterion, though the latter provides a good interpretation of the experimental results for brittle materials, but not for those with ductile behaviour. It goes without saying that at the inner radius, where $\sigma_r = \sigma_z = 0$, the three criteria considered above all lead to the same result.

11.3 Annular Disk

We will consider a rotating annular disk of constant thickness and consisting of a material whose $\sigma = \sigma(\varepsilon)$ curve is as represented in Fig. 11.1. As Fig. 2.6 shows, the most dangerous stresses are at the inner radius, where hoop stress reaches its absolute maximum value and radial stress is zero. As the radius increases, the hoop stress drops while radial stress rises. At the inner edge and for a large zone adjacent to it, however, the radial stress remains small by comparison with the hoop stress. As will be illustrated below, its contribution to material yielding can be regarded as negligible.

11.3.1 Elastic Limit Angular Velocity

The rotational velocity ω_y at the start of plastic deformation at the inner radius (ω_y is the elastic limit angular velocity or angular velocity at initial yielding, that is the largest angular velocity for entirely elastic behaviour) can be found from the second (2.34), setting $\sigma_t = \sigma_y$ and $\rho = \beta$; we thus obtain:

$$\omega_y = \sqrt{\frac{\sigma_y}{\gamma \cdot r_e^2} \cdot \frac{4}{(3+\nu) + \beta^2 \cdot (1-\nu)}}. \tag{11.4}$$

This velocity is the maximum rotational speed for an annular disk of known material that can be permitted if the onset of plastic deformation at the inner radius is to be avoided. With the assumed behaviour of the $\sigma = \sigma(\varepsilon)$ curve, (11.4) applies up to yield; if the latter is not reached, the service stress σ (below σ_y) will reappear at the root. With the three yield criteria considered above, this service stress coincides with hoop stress σ_t at the inner edge. The stress state at the onset of plastic flow is represented by a diagram such as that shown in Fig. 2.6, where we have $\sigma_t = \sigma_0 \cdot [(3 + \nu) + \beta^2 \cdot (1 - \nu)]/4 = \sigma_y$ at the inner radius.

Curve (I) in Fig. 11.3a shows angular velocity ω_y versus outer radius r_e in a constant thickness annular steel disk with material having $\sigma_y = 480$ MPa and $\nu = 0.3$. For a given material, the curves corresponding to different values of β exhibit hyperbolic behaviour of the type $\omega_y = C/r_e$, with C being a dimensional constant that depends on β ($C \cong 273$ rad·m/s for β varying from approximately 0 to

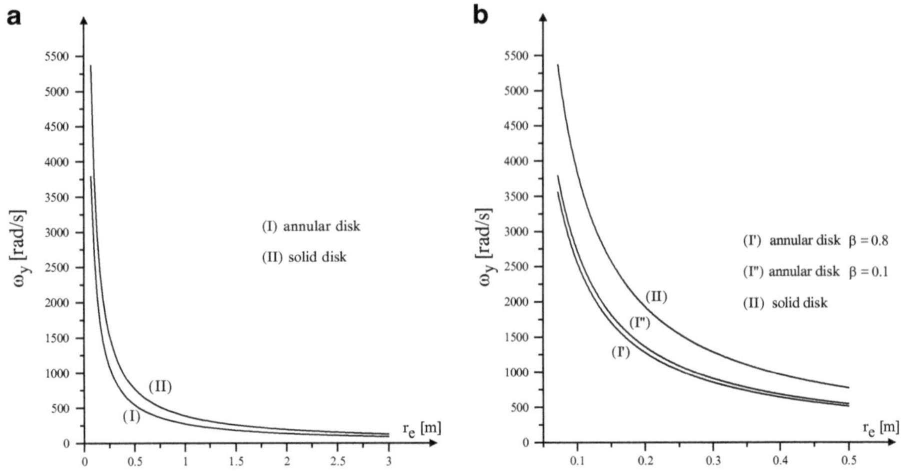

Fig. 11.3 (**a**) Elastic limit angular velocity ω_y versus outer radius in annular steel disks and in a solid disk; (**b**) detail of the zone adjacent to the origin of the axes

0.2, $C \cong 266$ rad·m/s for $\beta = 0.5$, $C \cong 252$ rad·m/s for $\beta = 0.9$, $C = 248$ rad·m/s for $\beta \to 1$). As these curves overlap in practice, curve (I) indicates what happens in a annular disk consisting of steel with known mechanical properties, with $0 < \beta < 1$. The detail in Fig. 11.3b shows that when values of β are highly dissimilar (curves I′ and I″), the differences are not significant.

On the other hand, for a radial thin walled disk rotating uniformly around its axis, the plastic angular velocity (its intensity is the largest angular velocity that the radial thin-walled disk can sustain without plastic collapse and the entire disk is in fully plastic state) which actually coincide with the elastic limit angular velocity, is given by:

$$\omega_0 = \sqrt{\frac{\sigma_s}{\gamma \cdot r_e^2}}\,. \tag{11.5}$$

This equation is derived from the relation $\sigma_0 = \gamma \cdot \omega^2\, r_e{}^2$ for $\sigma_0 = \sigma_y$; for stresses beyond the yield point, ω_0 is assumed as the reference angular velocity. Relation (11.5) provides the value of ω_0 for a ring of assigned material and geometry. As it can be solved for r_e, however, this relation can be applied to a given material and an assigned rotational velocity to determine the ring radius at which, assuming that the material's behaviour is linear elastic-perfectly plastic, the fully plastic state will be reached and the ring will thus explode. To express the idea more clearly, this radius r_e could be called the "*self-supporting radius*", in the sense that it is the limit radius below which there is no plastic flow for a given material and rotational velocity.

From the design standpoint, it can be seen from relation (11.5) that the ratio σ_y/γ of yield strength to material density is an extremely important parameter when selecting the material for turbine and compressor disks or developing new materials for these applications.

Bearing relation (11.5) in mind, (11.4) can be rewritten in the following form:

$$\omega_y = \omega_0 \cdot \sqrt{\frac{4}{(3+\nu) + \beta^2 \cdot (1-\nu)}}, \tag{11.6}$$

which shows that for a given material and for an assigned outer radius, ω_y is a function of ν and of β and that, for a steel ($\nu = 0.3$), β has little influence on elastic angular velocity ω_y; we pass from $\omega_y \cong 1.10 \cdot \omega_0$ for $\beta = 0.1$ (disks with a small hole) to $\omega_y \cong 1.02 \cdot \omega_0$ for $\beta = 0.9$ (disks with a large hole, i.e., thin walled disks).

Beyond the velocity given by (11.4) or (11.6), two zones can be identified in the disk, with a plastic-elastic interface consisting of the cylindrical surface with unknown elastic-plastic junction radius r_p: the first or inner zone located between the surface of radius r_i and the plastic-elastic interface, where the material's behaviour is plastic, and the second or outer zone located between said interface and radius r_e, where the material still exhibits linear elastic behaviour. The latter zone can be treated as a rotating disk of inner radius r_p, where it is subject to a radial stress σ_{rp} to be determined, arising from the interaction between the two zones.

11.3.2 Elastic-Plastic Analysis of Rotating Annular Disks

The stress state in the inner zone, where plastic deformation occurs, can be calculated using the boundary conditions together with the equilibrium equation given by the first of equations (1.10) and one of the three maximum-distortion-energy, maximum-shear-stress and maximum-normal-stress yield criteria, with the limitations discussed above.

As a basic hypothesis, it is assumed that the features of the elastic solution, whereby stresses σ_r and σ_t are both positive and the ratio σ_t/σ_r is always greater than 1, are also maintained for the plastic solution, subsequently verifying whether this assumption is reliable. With this assumption, and with the stress-strain curve shown in Fig. 11.1, we can use a horizontal straight-line (see the foregoing discussion of the ellipse of plasticity, as regards the assumption $\sigma_t = k = const$, and of the hexagon of plasticity) to represent the equation expressing the yield condition for any of the three criteria. Consequently, we can say that hoop stress σ_t will remain constant and equal to σ_y.

In these conditions, the compatibility equation written either in form (1.17) or in form (1.23) is satisfied; accordingly, once the hoop stress depending on the condition supplied by the yield criterion is known, the radial stress can be determined from simple equilibrium considerations. Bearing in mind that $h = const$ and that $\sigma_t = \sigma_y$, the first (1.10) becomes:

$$\frac{d}{dr}(\sigma_r \cdot r) - \sigma_y + \gamma \cdot \omega^2 \cdot r^2 = 0, \tag{11.7}$$

This is a first order differential equation with separable variables in $(\sigma_r \cdot r)$. Introducing the dimensionless variable ρ, (11.7) becomes:

$$d(\sigma_r \cdot \rho) = \left(\sigma_y - \gamma \cdot \omega^2 \cdot r_e^2 \cdot \rho^2\right) \cdot d\rho. \tag{11.8}$$

Integrated, this gives the relation:

$$\sigma_r = \sigma_y - \sigma_0 \cdot \frac{\rho^2}{3} + \frac{C}{\rho}, \tag{11.9}$$

where C is the integration constant, which can be determined by imposing that radial stress assumes the known value σ_{ri} at the inner radius. As $\sigma_{ri} = 0$ for a disk in only rotating, imposing this boundary condition yields:

$$C = \frac{1}{3} \cdot \left(-3 \cdot \beta \cdot \sigma_y + \sigma_0 \cdot \beta^3\right). \tag{11.10}$$

Substituting the constant thus found in relation (11.9) and bearing in mind that $\sigma_t = \sigma_y = const$, we find that the equations describing the stress state in the inner zone exhibiting plastic behaviour are:

$$\begin{cases} \sigma_r = \sigma_y \cdot \left(1 - \dfrac{\beta}{\rho}\right) + \sigma_0 \cdot \dfrac{\beta^3 - \rho^3}{3\rho} \\ \sigma_t = \sigma_y. \end{cases} \tag{11.11}$$

At the interface between the two zones, or in other words at elastic-plastic junction dimensionless radius $\rho = \rho_p$, which is still unknown, the radial stress assumes the value σ_{rp} given by the following relation:

$$\sigma_{rp} = \sigma_y \cdot \left(1 - \frac{\beta}{\rho_p}\right) + \sigma_0 \cdot \frac{\beta^3 - \rho_p^3}{3 \cdot \rho_p}. \tag{11.12}$$

As mentioned earlier, the outer zone where the material still shows elastic behaviour can be treated as a rotating disk having inner radius r_p, and subject at that radius to a radial stress σ_{rp}. Thus, taking into account the first two (2.34), which apply for an only rotating annular disk, and the first two (2.19), which apply to a non-rotating annular disk loaded at the outer radius, using the method of superposition and noting that in this case $\beta = \rho_p$, we can conclude that the stress state in the outer zone is given by the relations:

$$\begin{cases} \sigma_r = \sigma_0 \cdot \dfrac{3+\nu}{8} \cdot \left(1 + \rho_p^2 - \dfrac{\rho_p^2}{\rho^2} - \rho^2\right) + \sigma_{rp} \cdot \dfrac{\rho_p^2}{\rho^2} \cdot \dfrac{1-\rho^2}{1-\rho_p^2} \\ \sigma_t = \sigma_0 \cdot \dfrac{3+\nu}{8} \cdot \left(1 + \rho_p^2 + \dfrac{\rho_p^2}{\rho^2} - \dfrac{1+3\nu}{3+\nu} \cdot \rho^2\right) - \sigma_{rp} \cdot \dfrac{\rho_p^2}{\rho^2} \cdot \dfrac{1+\rho^2}{1-\rho_p^2}. \end{cases} \tag{11.13}$$

Equation (11.13) can also be derived from general relations (2.28), calculating the two integration constants A and B from the first two relations by means of the two boundary conditions:

$$\begin{cases} \sigma_r = \sigma_{rp} \text{ for } \quad \rho = \rho_p \\ \sigma_r = 0 \quad \text{for } \ \rho = 1. \end{cases} \tag{11.14}$$

and substituting the constants thus found in them.

Dimensionless radius ρ_p which defines the position of the elastic-plastic junction can be determined from the second (11.13), establishing that, because of continuity, hoop stress at this interface assumes the value of the material's yield stress. With this condition, we have:

$$\sigma_y = \sigma_0 \cdot \frac{(3+\nu) + \rho_p^2 \cdot (1-\nu)}{4} - \sigma_{rp} \cdot \frac{1+\rho_p^2}{1-\rho_p^2}; \tag{11.15}$$

substituting the value of σ_{rp} given by (11.12) in this relation then yields:

$$\sigma_y = \sigma_0 \cdot \frac{(3+\nu)+\rho_p^2 \cdot (1-\nu)}{4} - \left[\sigma_y \cdot \left(1 - \frac{\beta}{\rho_p}\right) + \sigma_0 \cdot \frac{\beta^3 - \rho_p^3}{3 \cdot \rho_p}\right] \cdot \frac{1+\rho_p^2}{1-\rho_p^2}. \tag{11.16}$$

This relation can then be solved for ρ_p, obtaining the following fifth degree equation:

$$\rho_p^5 \cdot \left(\frac{1}{3} + \nu\right) - 2 \cdot \rho_p^3 \cdot \left(\frac{1}{3} + \nu\right) + 4\beta \cdot \rho_p^2 \cdot \left(\frac{\sigma_y}{\sigma_0} - \frac{\beta^2}{3}\right) + \rho_p \cdot \left(3 + \nu - \frac{8 \cdot \sigma_y}{\sigma_0}\right)$$
$$+ 4 \cdot \beta \cdot \left(\frac{\sigma_y}{\sigma_0} - \frac{\beta^2}{3}\right) = 0. \tag{11.17}$$

This equation can be solved numerically with relative ease.

However, it is simpler to solve relation (11.16) for angular velocity ω; in this case, we obtain the value of ω for which the interface between the plastic and elastic zones is located in a given position. Noting that $\sigma_0 = \gamma \cdot \omega^2 \cdot {r_e}^2$, and bearing (11.4) in mind, relation (11.16) gives:

$$\omega(\rho_p) = \omega_0 \cdot \sqrt{\frac{1 + \dfrac{1+\rho_p^2}{1-\rho_p^2} \cdot \dfrac{\rho_p - \beta}{\rho_p}}{\dfrac{(3+\nu)}{4} \cdot \left(1 + \dfrac{1-\nu}{3+\nu} \cdot \rho_p^2\right) + \dfrac{1+\rho_p^2}{1-\rho_p^2} \cdot \dfrac{\rho_p^3 - \beta^3}{3 \cdot \rho_p}}}. \tag{11.18}$$

Setting $\rho_p = \beta$ in relation (11.18) yields (11.6), i.e., yields the velocity at which the plastic flow state begins at the inner edge of the disk. In this case, the elastic-plastic junction coincides with the inner radius. It is also of interest to calculate the angular velocity which brings this interface to the outer edge. This velocity is determined from relation (11.18) with $\rho_p = 1$, which gives:

$$\omega_p = \omega_0 \cdot \sqrt{\frac{3}{1 + \beta + \beta^2}}. \tag{11.19}$$

This relation defines the angular velocity at which the entire disk is in the fully plastic state, which given the particular shape of the stress-strain curve, results in disk collapse. For a given material and an assigned outer radius, this relation indicates that ω_p, unlike ω_y, does not depend on ν and that it increases significantly as β is reduced, going from $\omega_p \cong 1.05{\cdot}\omega_0$, for $\beta = 0.9$ (thin-walled disk), to $\omega_p \cong 1.64{\cdot}\omega_0$ for $\beta = 0.1$ (disks with very large radial thickness).

Figure 11.4 shows the dimensionless distributions of radial and hoop stresses in the plastic zone and the elastic zone of a rotating steel ($\nu = 0.3$) disk of constant thickness, having $\beta = 1/6$ and $\rho_p = 2/3$ and thus, as determined from relation (11.18), operating at a velocity $\omega = 1.52 \cdot\omega_0$. From this figure, where the σ_t/σ_y and σ_r/σ_y curves have general validity, it can be seen that the basic assumption for which $\sigma_t/\sigma_r > 1$ holds true and the solution obtained can be considered correct.

11.3.3 Residual Stresses

As mentioned earlier, the precompression generated in a non-rotating disk prestressed by overspinning beyond yielding is of considerable design interest. This precompression can be used to advantage at the design stage in order to reduce the maximum stresses, which are generally tensile, that result from centrifugal loading. Assuming that when a disk that has been loaded by overspeeding beyond yielding is decelerated, the material's behaviour is linear, or in other words, with reference to Fig. 11.1, follows straight-line *AB* of the $\sigma = \sigma(\varepsilon)$ curve parallel to straight-line *OP* (in this connection, see Chap 12 below), and that angular

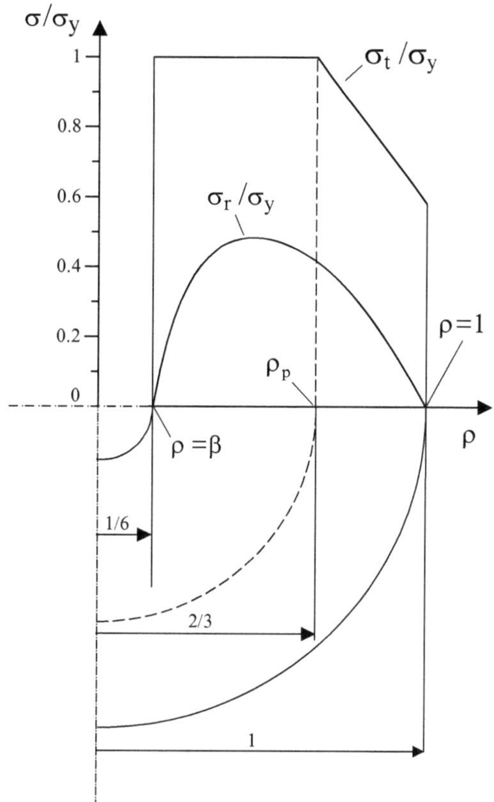

Fig. 11.4 Distribution curves of dimensionless principal stresses σ_r/σ_y and σ_t/σ_y in the plastic and elastic zones of a constant-thickness annular disk of given geometry and rotating at an assigned angular velocity

acceleration is sufficiently low that its effects can be neglected, the residual stress state in the stationary disk can be calculated by subtracting the stresses due to centrifugal loading given by the first two relations (2.34) from the stresses in the plastic zone and the elastic zone expressed by (11.11) and (11.13). We thus obtain the following for the inner zone, i.e., for $\beta \leq \rho \leq \rho_p$:

$$\begin{cases} \sigma_r = \dfrac{\sigma_0}{3} \cdot \left[\dfrac{\beta^3}{\rho} - \rho^2 - \dfrac{3 \cdot (3+\nu)}{8} \cdot \left(1 + \beta^2 - \dfrac{\beta^2}{\rho^2} - \rho^2\right)\right] + \sigma_y \cdot \left(1 - \dfrac{\beta}{\rho}\right) \\ \sigma_t = \sigma_y - \sigma_0 \cdot \dfrac{3+\nu}{8} \cdot \left(1 + \beta^2 + \dfrac{\beta^2}{\rho^2} - \dfrac{1+3\nu}{3+\nu} \cdot \rho^2\right). \end{cases} \tag{11.20}$$

For the outer zone, i.e., for $\rho_p \leq \rho \leq 1$, we obtain:

$$\begin{cases} \sigma_r = \sigma_0 \cdot \dfrac{3+\nu}{8} \cdot \left(\rho_p^2 - \beta^2\right) \cdot \left(1 - \dfrac{1}{\rho^2}\right) + \sigma_{rp} \cdot \dfrac{\rho_p^2}{\rho^2} \cdot \dfrac{1-\rho^2}{1-\rho_p^2} \\ \sigma_t = \sigma_0 \cdot \dfrac{3+\nu}{8} \cdot \left(\rho_p^2 - \beta^2\right) \cdot \left(1 + \dfrac{1}{\rho^2}\right) - \sigma_{rp} \cdot \dfrac{\rho_p^2}{\rho^2} \cdot \dfrac{1+\rho^2}{1-\rho_p^2}, \end{cases} \tag{11.21}$$

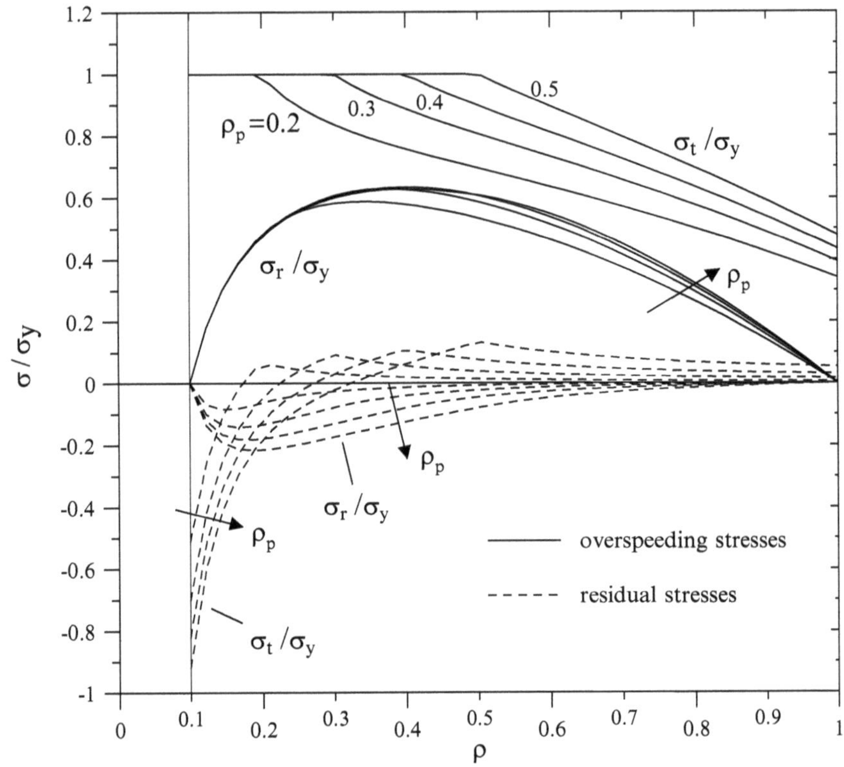

Fig. 11.5 Distribution curves of residual stresses (*dashed-line curves*) and overspeeding stresses (*solid-line curves*) in constant-thickness disks having $\beta = 0.1$, yielded at different angular velocities $\omega(\rho_p)$

where σ_0 is the reference stress corresponding to the angular velocity $\omega(\rho_p)$ of the disk in prestressing equipment, and ρ_p, which is a function of σ_0 and thus of ω, is the radius of the elastic-plastic junction given by the solution of (11.17). Note that no new yielding occurs upon the release of $\omega(\rho_p)$ and that any re-application of angular velocity values up to $\omega(\rho_p)$ will not cause any additional yielding.

Figure 11.5 shows distribution curves of dimensionless residual stresses (dashed-line curves) in constant-thickness disks with ratio $\beta = 0.1$, operated and yielded at different angular velocities $\omega(\rho_p)$ corresponding to different values for ρ_p. The figure also shows the overspeeding stresses (i.e., the stresses resulting from a rotational velocity $\omega(\rho_p)$ exceeding the elastic limit angular velocity ω_0) corresponding to the velocity $\omega(\rho_p)$ at which the elastic-plastic junction radius is ρ_p (solid-line curves), in the two plastic and elastic zones. Stresses are shown in dimensionless form, as ratio σ/σ_y.

Rewritten for the condition in which the disk is in the fully plastic state, i.e., for $\omega = \omega_p$ defined by relation (11.19), (11.11) becomes:

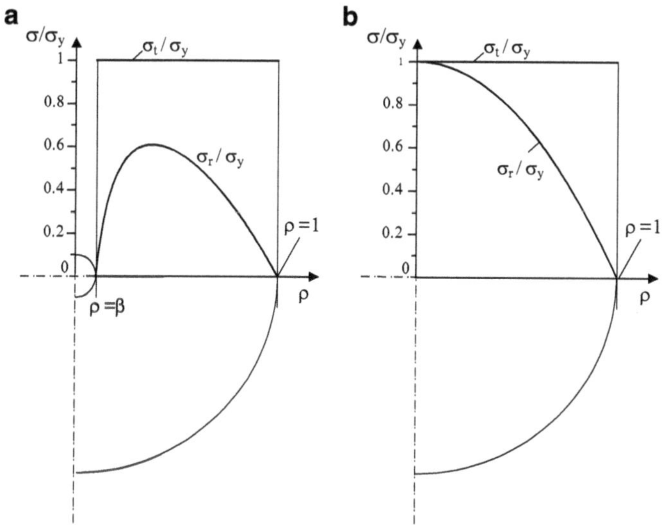

Fig. 11.6 (**a**) Distribution curves of dimensionless stresses $\sigma_r\ /\ \sigma_y$ and $\sigma_t\ /\ \sigma_y$ in a rotating constant-thickness annular disk with $\beta = 0.1$, in the fully plastic state; (**b**) distribution curves of dimensionless stresses $\sigma_r\ /\ \sigma_y$ and $\sigma_t\ /\ \sigma_y$ in a rotating constant-thickness solid disk in the fully plastic state

$$\begin{cases} \sigma_r = \sigma_y \cdot \left[\left(1 - \frac{\beta}{\rho}\right) + \frac{\beta^3 - \rho^3}{\left(1 + \beta + \beta^2\right) \cdot \rho}\right] \\ \sigma_t = \sigma_y. \end{cases} \tag{11.22}$$

Figure 11.6a shows distribution curves of dimensionless radial and hoop stresses in a constant-thickness annular disk ($\beta = 0.1$) in the fully plastic state, i.e., operating at the peripheral velocity $\omega_p{}^2 = 3 \cdot \sigma_y / \gamma \cdot r_e{}^2 (1 + \beta + \beta^2)$.

11.3.4 Example

We will consider the steel disk with small longitudinal thickness examined in the example given in Sect. 2.5.1 of Chap. 2 (Fig. 2.8), and will determine the rotational velocity which brings the elastic-plastic interface up to radius $r_p = 52$ mm, i.e., to approximately one quarter of the radial thickness starting from the inner radius. In addition, we will evaluate the residual stress state at the inner and outer radii, discussing whether it has beneficial effects on the disk's strength.

From the problem data, we have: $\sigma_y = 480$ MPa; $\nu = 0.3$; $r_i = 28$ mm; $r_e = 125$ mm; $\beta = 0.224$; $\rho_p = 52/125 = 0.416$; $\gamma = 7{,}800$ kg/m^3.

From relation (11.18), we find that ω ($\rho_p = 0.416$) = 2655.3 rad/s, or in other words that the rotational velocity that brings the elastic-plastic interface up to radius $\rho_p = 0.416$ is approximately 25,400 rpm.

This angular velocity corresponds to the reference stress $\sigma_0 = \gamma\,\omega^2 r_e^{\,2} = 859.3$ MPa to be introduced in (11.20) and (11.21) to obtain the residual stress states in the plastic and elastic zones respectively. But as our analysis here is restricted to the residual stresses at the inner edge and outer edge, where $\sigma_r = 0$ for $\rho = \beta$ and for $\rho = 1$, we need only calculate σ_t at the inner and outer radii.

From the second (11.20), we find that the residual hoop stress at the inner radius is compressive and equal to $(\sigma_t)_{\rho=\beta} = -236.5$ MPa.

The residual hoop stress at the outer radius, on the other hand, is calculated using the second (11.21) after determining radial stress σ_{rp} at the elastic-plastic junction radius. With the problem data, relation (11.12) gives $\sigma_{rp} = 179.7$ MPa and we thus find from the second (11.21) that the residual hoop stress at the outer radius is tensile and equal to $(\sigma_t)_{\rho=1} = 11.9$ MPa.

The residual stress state thus calculated at the inner radius can only be beneficial as regards the material's strength, as this high negative residual hoop stress will contribute to no small extent to reducing the peak hoop stress resulting from centrifugal loading under actual operating conditions. In other words, the effective hoop stress will be the algebraic sum of the hoop stress due to centrifugal load at operating speed (which is positive) and the calculated residual hoop stress, which is negative.

If the operating speed is 14,000 rpm (as in the example in Sect. 2.5.1 of Chap. 2), it is obvious that it is not necessary to bring the disk to the plastic state, as in this case the hoop stress at the inner radius is 218 MPa. As indicated earlier, in stationary conditions, or in other words with a load that can be regarded as static, this corresponds to a safety factor of 2.20, by respect to yield point.

But if the operating speed is such that stress σ_t at the inner radius reaches the value of yield stress σ_y (in this case, (11.6) gives $\omega_y = 2173.4$ rad/s, equal to approximately 20,754 rpm), the effect of precompression would be extremely useful, as the effective hoop stress at the inner radius would be $\sigma_t = 480 - 236.5 = 243.5$ MPa.

11.4 Solid Disk

11.4.1 Elastic Limit Angular Velocity

For a solid disk, yield begins at the axis ($\rho = 0$), at the following angular velocity

$$\omega_y = \omega_0 \cdot \sqrt{\frac{8}{3+\nu}}. \tag{11.23}$$

This velocity is obtained on the basis of a yield criterion (the three criteria considered for the annular disk are also used here, and once again lead to the same result when applied at the rotational axis, i.e., for $\rho = 0$), by substituting stresses

σ_t and σ_r given by the first two (2.70) and calculated for $\rho = 0$ in the relation expressing the yield criterion and then solving for ω. With the assumed behaviour of the $\sigma = \sigma(\varepsilon)$ curve (Fig. 11.1), (11.23), like (11.4) or the equivalent (11.6), applies up to yield, and thus makes it possible to determine the rotational velocity corresponding to a predetermined service stress σ for the material, after substituting this stress for σ_y in the root appearing in the second member of equation (11.5). Here again, the stress state at the onset of plastic flow is represented by a diagram such as that shown in Fig. 2.15a, where at the axis of revolution we have $\sigma_r = \sigma_t = \sigma_0\,(3 + \nu)/8 = \sigma_y$.

Curve (II) in Fig. 11.3a shows velocity ω_y versus outer radius r_e in a constant thickness solid steel disk with material having $\sigma_y = 480$ MPa and $\nu = 0.3$. Comparing curve (II) with curves (I) – as will be recalled, (I) represents a plurality of curves corresponding to $0 < \beta < 1$ – clearly shows that, *ceteris paribus*, the solid disk is considerably stronger than an annular disk. In particular, (11.23) shows that for a given material and for an assigned outer radius, ω_y depends on Poisson's ratio ν and that, for a steel ($\nu = 0.3$), $\omega_y \cong 1.56\omega_0$. In other words, the elastic limit angular velocity for a solid steel disk is 56 % higher than that for a radial thin-walled disk with the same r_e.

11.4.2 *Elastic-Plastic Analysis of Rotating Solid Disks*

Beyond yielding, the analysis to be performed for the solid disk is identical to that for the annular disk, except that the constant C in relation (11.9) must be zero to prevent radial stress from becoming infinite for $\rho = 0$; the same result is reached by establishing that $\sigma_r = \sigma_t = \sigma_y$ at the axis as a boundary condition. Accordingly, the equations describing the stress state in the inner zone of the solid disk exhibiting plastic behaviour are:

$$\begin{cases} \sigma_r = \sigma_y - \sigma_0 \cdot \dfrac{\rho^2}{3} \\ \sigma_t = \sigma_y. \end{cases} \qquad (11.24)$$

These equations can also be derived from (11.11) by setting $\beta = 0$ in the latter.

In this case, the interface between the two elastic and plastic zones, or in other words at the unknown dimensionless elastic-plastic junction radius $\rho = \rho_p$, the radial stress assumes the value σ_{rp} given by the following relation:

$$\sigma_{rp} = \sigma_y - \sigma_0 \cdot \frac{\rho_p^2}{3}, \qquad (11.25)$$

which can also be derived from (11.12) for $\beta = 0$.

For the outer zone where the solid disk's material shows elastic behaviour, nothing changes with respect to the treatment described for the annular disk.

Consequently, relations (11.13) also apply for the solid disk, and describe the stress state in this zone. Likewise, relation (11.15) which defines the value of the hoop stress at the interface between the elastic and plastic zones also continues to apply, with the obvious variation that in this case stress σ_{rp} is given by relation (11.25), which thus yields:

$$\sigma_y = \sigma_0 \cdot \frac{(3+\nu) + \rho_p^2 \cdot (1-\nu)}{4} - \left(\sigma_s - \sigma_0 \cdot \frac{\rho_p^2}{3}\right) \cdot \frac{1+\rho_p^2}{1-\rho_p^2}. \tag{11.26}$$

Once again, this can also be derived from (11.16) for $\beta = 0$.

Solving this relation for ρ_p gives the following fourth degree equation:

$$\rho_p^4 - 2 \cdot \rho_p^2 + \frac{3 \cdot \sigma_0 \cdot (3+\nu) - 24 \cdot \sigma_y}{\sigma_0 \cdot (1+3\nu)} = 0, \tag{11.27}$$

which can be solved with relative ease.

Here again, however, it is more usual to solve relation (11.26) for ω, thus obtaining the angular velocity for which the interface between the plastic and elastic zones is located in a given position defined by ρ_p. Proceeding in this way, we obtain:

$$\omega(\rho_p) = \sqrt{\frac{\sigma_y}{\gamma \cdot r_e^2}} \cdot \sqrt{\frac{2}{\frac{(3+\nu)\cdot\left(1-\rho_p^2\right)}{4} \cdot \left(1 + \frac{1-\nu}{3+\nu} \cdot \rho_p^2\right) + \frac{\rho_p^2}{3} \cdot \left(1+\rho_p^2\right)}}. \tag{11.28}$$

Setting $\rho_p = 0$ in relation (11.28) yields (11.23), i.e., we thus obtain the elastic limit angular velocity at the disk's axis, which was already found with another approach. For $\rho_p = 1$, on the other hand, we obtain the following plastic angular velocity for the solid disk:

$$\omega_p = \omega_0 \cdot \sqrt{3}. \tag{11.29}$$

We can also arrive at this value by starting from relation (11.9), writing it with $C = 0$ and establishing that the radial stress at the outer radius is zero. Comparing (11.23) and (11.29) shows that the angular velocity at which a solid steel disk is in the fully plastic state is slightly (i.e., approximately 11 %) higher than the velocity at the onset of plastic flow.

11.4.3 Residual Stresses

Here again, with the same assumptions made for the annular disk, the residual stress state in the solid disk can be calculated by subtracting the stresses due to centrifugal loading given by the first two relations (2.70) from the stresses in the plastic zone

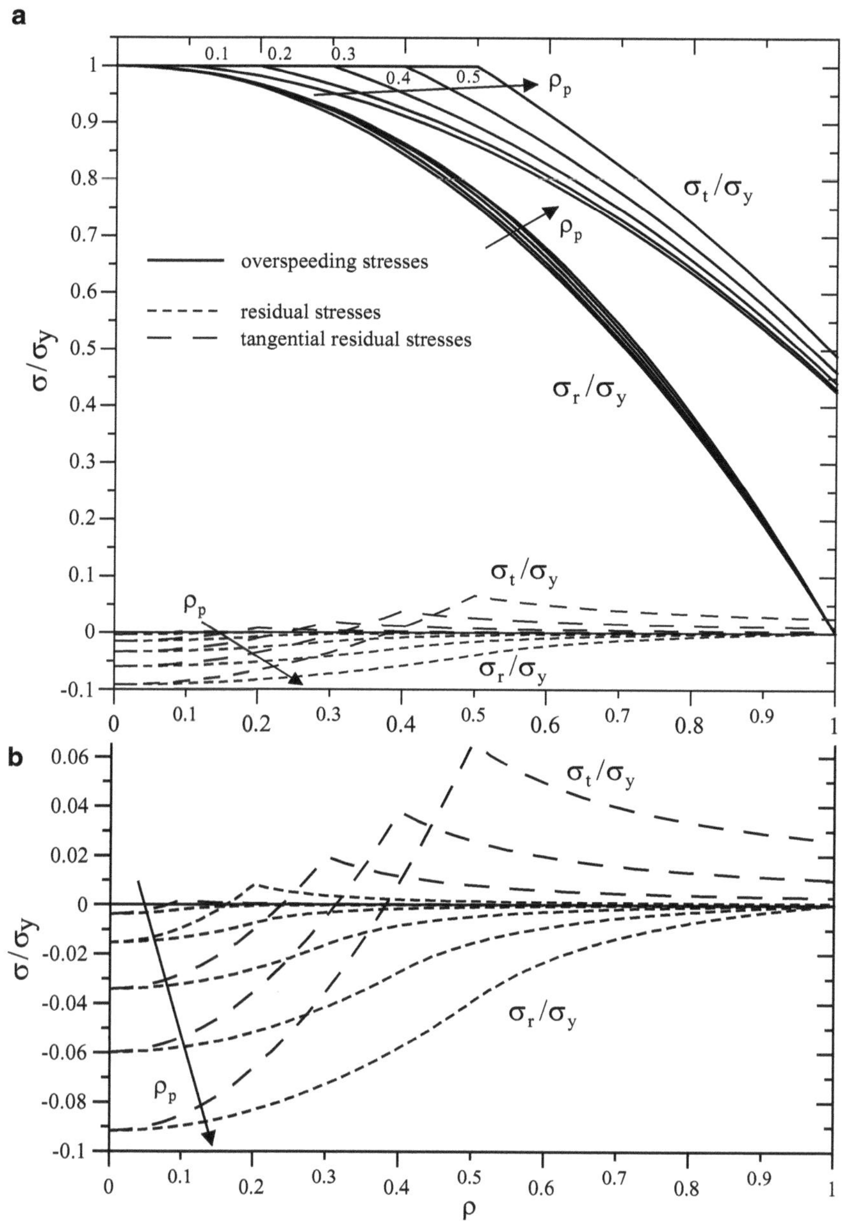

Fig. 11.7 (**a**) Distribution curves of dimensionless residual stresses (*dashed-line* and *thin solid-line* curves) and dimensionless overspeeding stresses (*solid-line curves*) in constant-thickness solid disks, yielded at different velocities ω (ρ_p); (**b**) detail of residual stresses

and the elastic zone expressed by relations (11.24) and (11.13). We thus obtain the following for the inner zone, i.e., for $0 \leq \rho \leq \rho_p$:

$$\begin{cases} \sigma_r = \sigma_y - \dfrac{\sigma_0}{3} \cdot \left[\rho^2 + \dfrac{3 \cdot (3+v)}{8} \cdot (1 - \rho^2) \right] \\ \sigma_t = \sigma_y - \dfrac{3+v}{8} \cdot \sigma_0 \cdot \left(1 - \dfrac{1+3v}{3+v} \cdot \rho^2 \right). \end{cases} \tag{11.30}$$

For the outer zone, i.e., for $\rho_p \leq \rho \leq 1$, we obtain:

$$\begin{cases} \sigma_r = \sigma_0 \cdot \dfrac{3+v}{8} \cdot \rho_p^2 \cdot \left(1 - \dfrac{1}{\rho^2} \right) + \sigma_{rp} \cdot \dfrac{\rho_p^2}{\rho^2} \cdot \dfrac{1 - \rho^2}{1 - \rho_p^2} \\ \sigma_t = \sigma_0 \cdot \dfrac{3+v}{8} \cdot \rho_p^2 \cdot \left(1 + \dfrac{1}{\rho^2} \right) - \sigma_{rp} \cdot \dfrac{\rho_p^2}{\rho^2} \cdot \dfrac{1 + \rho^2}{1 - \rho_p^2} \end{cases} . \tag{11.31}$$

In relations (11.30) and (11.31) σ_0 is the reference stress corresponding to the angular velocity $\omega(\rho_p)$ of the disk in prestressing equipment, and ρ_p, which is a function of σ_0 and thus of ω, is the radius of the elastic-plastic junction given by (11.27).

Figure 11.7a shows distribution curves of dimensionless residual stresses (dashed-line and thin solid-line curves) in constant-thickness solid disks, operated and yielded at different velocities ω (ρ_p). The figure also shows the overspeeding stresses corresponding to the velocity $\omega(\rho_p)$ at which the elastic-plastic junction radius is ρ_p (solid-line curves), in the two plastic and elastic zones. Stresses are shown in dimensionless form, as ratio σ/σ_y. By contrast with the situation in annular disks, where residual hoop stresses reach very high values, equal to a significant fraction of the yield stress, here they are very modest, and at most reach 10 % of σ_y (see detail in Fig. 11.7b).

Rewritten for the condition in which the disk is in the fully plastic state, i.e., for $\omega = \omega_p$ defined by relation (11.29), equations (11.24) become:

$$\begin{cases} \sigma_r = \sigma_y \cdot (1 - \rho^2) \\ \sigma_t = \sigma_y. \end{cases} \tag{11.32}$$

Figure 11.6b shows distribution curves of dimensionless radial and hoop stresses in a constant-thickness rotating disk in the fully plastic state, i.e., operating at the peripheral velocity $\omega_p{}^2 = 3\sigma_y/\gamma r_e{}^2$.

11.5 Bursting Angular Speed

Relations (11.19) and (11.29), which apply to the annular disk and the solid disk respectively, give the rotational speeds needed to cause the entire thickness becomes plastic. These angular velocities represent the lower limit of bursting speed. The upper limit of the same bursting speed is obtained from relations (11.19) and (11.29), by substituting the material ultimate strength σ_u for the yield strength σ_y. On the basis of limit analysis, the conventional bursting speed is the arithmetic average of these two speeds. For the annular disk and solid disk respectively, we thus have:

$$\omega_{esp} = \frac{\sqrt{\sigma_y} + \sqrt{\sigma_u}}{2} \cdot \sqrt{\frac{3}{\gamma \cdot r_e^2 \cdot \left(1 + \beta + \beta^2\right)}} \tag{11.33}$$

$$\omega_{esp} = \frac{\sqrt{\sigma_y} + \sqrt{\sigma_u}}{2} \cdot \sqrt{\frac{3}{\gamma \cdot r_e^2}}. \tag{11.34}$$

Chapter 12
Stress Analysis in Rotating Disks Loaded Beyond Yielding: Hardening Materials

12.1 General

Previous chapter addressed the problem of a disk in the partially or fully plastic states, assuming that the material's behaviour is elastic-perfectly plastic, or in other words that the $\sigma = \sigma(\varepsilon)$ curve plotted from tensile tests on a standard specimen is as shown in Fig. 11.1. On the basis of this assumption, which sees the material as exhibiting linear elastic behaviour up to yield and ideally plastic, i.e., non-hardening, behaviour thereafter, curves representing the principal stresses, residual stresses and effective stresses under service conditions were plotted which show discontinuities in the stress gradient along the radius, and thus cusps, at the interface between the innermost zone where the material's behaviour is plastic, and the outermost zone showing elastic behaviour.

As a number of researchers have found, the fact that the distribution curves for the stresses of interest here (the principal stresses from centrifugal loading, the residual stresses remaining when a disk that has been prestressed by overspeeding is decelerated, and the principal stresses associated with operating speed) are cusped is clearly a physical absurdity: in accordance with Linnaeus' dictum that *natura non facit saltus*, it has been demonstrated experimentally that the transition between these two zones is gradual, and takes place in a band of annular cross section between the two zones.

The treatment presented in the previous chapter is thus highly approximate, as it is based on drastic simplifying assumptions whose purpose is merely to permit an initial approach to the problem and, though they can provide information that is useful for design purposes, cannot claim to interpret the actual behaviour of the materials normally used for turbine disks. To overcome the limitations of this treatment and the physical absurdity deriving from it, and to reflect experimental findings more accurately, this chapter will tackle the problem in much more general terms by introducing a constitutive law that can describe the material's actual behaviour as faithfully as possible, approximating any experimental $\sigma = \sigma(\varepsilon)$ curve point by point.

V. Vullo and F. Vivio, *Rotors: Stress Analysis and Design*,
Mechanical Engineering Series, DOI 10.1007/978-88-470-2562-2_12,

© Springer-Verlag Italia 2013

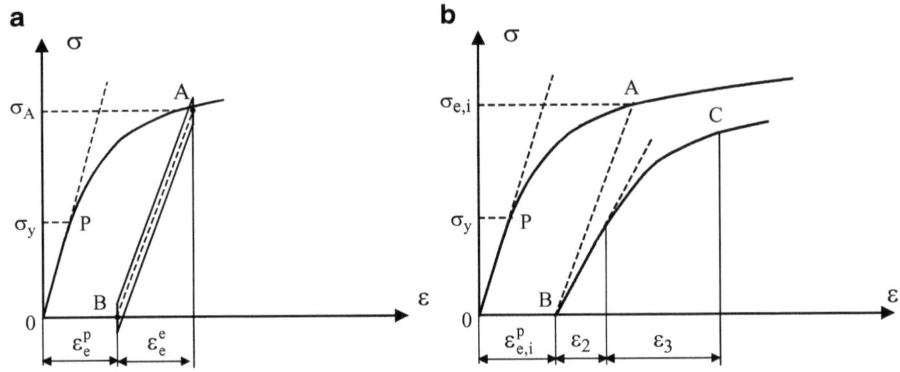

Fig. 12.1 (**a**) $\sigma = \sigma(\varepsilon)$ curve for a virgin material at a given temperature (curve *OPA*) and $\sigma = \sigma(\varepsilon)$ curve after one loading-unloading cycle at stress σ_A corresponding to point *A* at the same temperature (cyclic curve *ABA*); (**b**) $\sigma = \sigma(\varepsilon)$ curve for a virgin material at a predetermined temperature (curve OA) and $\sigma = \sigma(\varepsilon)$ curve after the first loading-unloading cycle, with stress beyond yielding at a higher temperature

For these reasons, the assumption that the material's behaviour is elastic-perfectly plastic beyond the yield point is very rough. Most materials now used in disk construction exhibit strain hardening phenomena in addition to plastic flow. To ensure that the treatment is no less general, but continuing to use the small displacement assumption, we will consider any $\sigma = \sigma(\varepsilon)$ curve (NB: Eulerian or true or actual stress, and Lagrangian or engineering strain) of the type shown in Fig. 12.1a, plotted at a given temperature from a uniaxial tensile test on a standard cylindrical specimen consisting of virgin material. In this curve, point *P*, corresponding to the unit yield stress σ_y (here again, it will be assumed that the proportional limit, the elastic limit and the yield point coincide), divides the curve portion showing elastic behaviour from that showing plastic behaviour. For the materials of interest here, the stress σ and its derivative $d\sigma/d\varepsilon$ are functions that monotonically increase and decrease respectively with ε, as these materials are strain-hardening. We will thus have $\sigma\cdot(d\sigma^2/d\varepsilon^2) < 0$ for any ε, while it is assumed that the constitutive relation $\sigma = f(\varepsilon)$ applies to each cross section of the specimen during the entire loading process.

Without going into the details of the location of point *P*, which will differ according to the convention used (in this connection, the reader is referred to specialised texts on the mechanics of materials), there will no longer be a one-to-one correspondence between stress and strain beyond this point. In other words, if we start from a generic point *A* beyond *P* and reduce the load, the representative point will move along path *AB*, which is essentially parallel to the linear elastic portion of the curve. Consequently, when the load reaches zero, there will be a permanent plastic deformation ε^p, while the recoverable elastic deformation will be ε^e. If load is increased again starting from point *B*, the representative point will move along a path near to but not coinciding with *BA*, resulting in a small hysteresis cycle correlated with the energy losses per unit volume occurring in the unloading and reloading process. Once point *A* is reached, further deformation will not occur unless load is increased: this is the strain hardening phenomenon. It is thus clear that stress in the plastic field depends on the total load operation and/or on the material's overall load history.

As is usual in plasticity theory, it will here be assumed that deformation is isothermal and independent of loading speed and, consequently, independent of time and distinct from phenomena such as creep and relaxation. So, we are within the boundaries of classical plasticity, where the stress-strain law is independent of the rate of deformation but is dependent on the history of deformation. Nevertheless, as specified in Sect. 11.1, it will be assumed that the Bauschinger effect is absent.

To take the material's actual behaviour as represented by the $\sigma = \sigma(\varepsilon)$ curve into account when determining the stress and strain state of rotating disks loaded beyond yielding, it will obviously be necessary to use more general analytic treatments or numerical solutions.

From the theoretical standpoint, a method which is more general than those discussed above was introduced by Odqvist [56]. This method makes it possible to analyse the stress state in a disk loaded beyond yielding and consisting of a ductile material that strain-hardens according to a function $\tau_{oct,y} = f\,(\gamma_{oct,y})$, which links octahedral shear yield stress $\tau_{oct,y}$ to the corresponding octahedral shear strain $\gamma_{oct,y}$. Odqvist addressed the case of a rotating disk whose geometry of revolution and thickness vary according to a linear function (conical disk), subject to slight creep, assuming that $\tau_{oct,S}$ is linked to the yield octahedral shear strain rate $\dot{\gamma}_{oct,S}$ by the relation $\tau_{oct,S} = C\cdot\dot{\gamma}^{m}_{oct,S}$, where C and m are constants depending on the material. For the treatment of this problem, which is beyond our scope here, the reader is referred to specialised publications.

In the following section, we will present two general methods that can be used to determine the stress state resulting from overspeeding in rotating disks made of strain-hardening material: (1) the first, which applies to disks whose thickness varies according to any function $h = h(r)$, is a finite difference numerical method, specifically Millenson-Manson's method, in the formulation that takes the material's non-linear behaviour due to plastic flow and creep into account; (2) the second, which applies to disks whose thickness varies non-linearly according to a power of a linear function, is an original analytical method that leads to general closed form relations, though the later can only be integrated numerically.

12.2 Millenson-Manson's Method

The assumptions underlying Millenson-Manson's method for non-linear behaviour are the same as those illustrated for linear behaviour in Sect. 8.2 of Chap. 8. The equilibrium equation given by the first of (8.20) or (8.21) does not change. For isotropic materials, the compatibility equation is derived by substituting the following constitutive laws in (1.17):

$$\begin{cases} \varepsilon_r = \dfrac{du}{dr} = \dfrac{1}{E}(\sigma_r - \nu \cdot \sigma_t) + \alpha \cdot T + \Delta_r + \delta_r \\ \varepsilon_t = \dfrac{u}{r} = \dfrac{1}{E}(\sigma_t - \nu \cdot \sigma_r) + \alpha \cdot T + \Delta_t + \delta_t. \end{cases} \quad (12.1)$$

These laws represent the stress-strain link for stresses beyond yielding due to plastic flow and creep, and are thus an extension of (1.25) in this specific field; here, Δ_r and Δ_t are the plastic increments of strain in the radial and tangential directions, while δ_r and δ_t are the analogous creep increments, again in the radial and tangential directions.

In this case, compatibility equation (1.24) includes further terms and, written in Millenson-Manson's formulation but introducing dimensionless variable ρ, becomes:

$$\frac{d}{d\rho}\cdot\left(\frac{\sigma_t}{E}-\frac{\nu\cdot\sigma_r}{E}+\alpha\cdot T+\Delta_t+\delta_t\right)-\frac{1+\nu}{E\cdot\rho}\cdot(\sigma_r-\sigma_t)-\frac{\Delta_r-\Delta_t}{\rho}-\frac{\delta_r-\delta_t}{\rho}=0 \quad (12.2)$$

The equilibrium equation given by the first of (8.20) or (8.21) and the compatibility equation derived above and rewritten with finite differences lead to the same system of (8.23), where the first nine constants are still those given by (8.24), specialised for an isotropic material ($E_{ti} = E_{ri} = E_{i,}$; $\mu_i = E_{ti}/E_{ri} = 1$; $\alpha_t = \alpha_r = \alpha$; $\nu_{tr} = \nu_{rt} = \nu$). The only constant that differs is H'_i, which must be substituted by the constant H'_{pi} defined by the following relation:

$$\begin{aligned} H'_{pi} = {} & -2\cdot[(\alpha_i\cdot T_i-\alpha_{i-1}\cdot T_{i-1})+(\Delta_{ti}-\Delta_{ti-1})+(\delta_{ti}-\delta_{ti-1})]+ \\ & +(\rho_i-\rho_{i-1})\cdot\left(\frac{\Delta_{ri}-\Delta_{ti}}{\rho_i}+\frac{\Delta_{ri-1}-\Delta_{ti-1}}{\rho_{i-1}}+\frac{\delta_{ri}-\delta_{ti}}{\rho_i}+\frac{\delta_{ri-1}-\delta_{ti-1}}{\rho_{i-1}}\right). \end{aligned} \quad (12.3)$$

With no creep, this relation becomes

$$\begin{aligned} H'_{pi} = {} & -2\cdot[(\alpha_i\cdot T_i-\alpha_{i-1}\cdot T_{i-1})+(\Delta_{ti}-\Delta_{ti-1})]+(\rho_i-\rho_{i-1}) \\ & \times\left(\frac{\Delta_{ri}-\Delta_{ti}}{\rho_i}+\frac{\Delta_{ri-1}-\Delta_{ti-1}}{\rho_{i-1}}\right), \end{aligned} \quad (12.4)$$

which can also be written in the following form:

$$H'_{pi} = H'_i - 2\cdot(\Delta_{ti}-\Delta_{ti-1})+(\rho_i-\rho_{i-1})\cdot\left(\frac{\Delta_{ri}-\Delta_{ti}}{\rho_i}+\frac{\Delta_{ri-1}-\Delta_{ti-1}}{\rho_{i-1}}\right) \quad (12.4)$$

12.2.1 Effect of Plastic Flow

To analyse the rotor's stress state for plastic flow alone, it is first necessary to have the $\sigma = \sigma(\varepsilon)$ curve beyond yielding.

It should also be borne in mind that for a triaxial stress state, the plastic strains in the radial, tangential and axial directions can be expressed in the following form:

$$\begin{cases} \Delta_r = \dfrac{R}{3} \cdot [(\sigma_r - \sigma_t) + (\sigma_r - \sigma_z)] \\ \Delta_t = \dfrac{R}{3} \cdot [(\sigma_t - \sigma_r) + (\sigma_t - \sigma_z)] \\ \Delta_z = \dfrac{R}{3} \cdot [(\sigma_z - \sigma_r) + (\sigma_z - \sigma_t)] \end{cases} \tag{12.5}$$

with

$$R = \frac{3 \cdot \varepsilon_e^p}{2 \cdot \sigma_e} \tag{12.6}$$

where ε_e^p is the equivalent plastic strain, which Millenson and Manson assume to be equal to that determined by means of a simple tensile test (Fig. 12.1a), and σ_e is the ideal or equivalent Von Mises stress, i.e.:

$$\sigma_e = \frac{1}{\sqrt{2}} \cdot \left[(\sigma_r - \sigma_t)^2 + (\sigma_t - \sigma_z)^2 + (\sigma_z - \sigma_r)^2\right]^{\frac{1}{2}}. \tag{12.7}$$

For the plane stress state ($\sigma_z = 0$), the first two (12.5), (12.6) and (12.7), written for the generic *i-th* radius of the $(n + 1)$ radii introduced in the finite difference procedure (see Sect. 8.2), respectively become:

$$\begin{cases} \Delta_{ri} = \dfrac{R}{3} \cdot \left(2 \cdot \sigma_{r,i} - \sigma_{t,i}\right) \\ \Delta_{ti} = \dfrac{R}{3} \cdot \left(2 \cdot \sigma_{t,i} - \sigma_{r,i}\right) \end{cases} \tag{12.8}$$

$$R = \frac{3 \cdot \varepsilon_{e,i}^p}{2 \cdot \sigma_{e,i}} \tag{12.9}$$

$$\sigma_{e,i} = \sqrt{\sigma_{r,i}^2 + \sigma_{t,i}^2 - \sigma_{r,i} \cdot \sigma_{t,i}} \tag{12.10}$$

Substituting (12.9) in (12.8) yields:

$$\begin{cases} \Delta_{ri} = \dfrac{\varepsilon_{e,i}^{p}}{2 \cdot \sigma_{e,i}} \cdot \left(2 \cdot \sigma_{r,i} - \sigma_{t,i}\right) \\ \Delta_{ti} = \dfrac{\varepsilon_{e,i}^{p}}{2 \cdot \sigma_{e,i}} \cdot \left(2 \cdot \sigma_{t,i} - \sigma_{r,i}\right). \end{cases} \tag{12.11}$$

Given the non-linear nature of the problem, calculation necessarily involves an iterative procedure. We begin by calculating the stresses without taking the material's plasticity into account. If the equivalent stress at any radii exceeds the proportional elastic limit as specified above, the plastic strain increments Δ_r and Δ_t at these radii are calculated using any yield criterion. If, for instance, the maximum distortion energy criterion is applied, we will use (12.7) for the triaxial stress state, or the relation deriving from it for $\sigma_z = 0$, if the stress state is biaxial.

Once the stress-strain curve is known (Fig. 12.1a shows a typical example for a material used in disk construction), plastic strain $\varepsilon_{e,i}^{p}$ can be found using a graphic procedure. If the analytical expression of the $\sigma_e = \sigma_e(\varepsilon_e)$ curve (this, as will be recalled, is the Eulerian stress vs. Langrangian strain curve, and not the Langrangian or engineering $\sigma = \sigma(\varepsilon)$ curve) is known, this plastic strain can also be determined mathematically. In this connection, it should be borne in mind that once the $\sigma_e = \sigma_e(\varepsilon_e)$ curve to failure has been determined experimentally, it can be described by a polynomial relation of the following type:

$$\varepsilon_e = \sum_{j=1}^{n} a_j \cdot \sigma_e^{j} \tag{12.12}$$

where σ_e is the equivalent stress, ε_e is the total equivalent strain and a_j, with $j = (1, 2, \ldots, n)$, are constants; obviously, it is not necessary to model the entire curve, but only the elastic part and the plastic part characterised by uniform plastic flow, as the remaining plastic part, which goes from the onset of instable plastic flow (necking down or local instability point of the $\sigma = \sigma(\varepsilon)$ curve) up to failure, is of no interest to the structural designer.

In the linear elastic field (where stress is proportional to strain), the equivalent strain is a linear function of the equivalent stress, and (12.12) thus reduces to Hooke's law $\varepsilon_e = \sigma_e/E$, which holds true as long as $\varepsilon_e \leq \varepsilon_p$, where ε_p is the strain corresponding to the proportional limit. Where it is assumed that proportional limit, elastic limit and yield point coincide, Hooke's law will hold true as long as $\varepsilon_e \leq \varepsilon_y$, where ε_y is the strain corresponding to the yield point.

Beyond yielding ($\varepsilon_e > \varepsilon_y$), in the plastic field with uniform plastic flow characterised by constant-volume strain, the link between the plastic component ε_e^p of total equivalent strain (note that $\varepsilon_e = \varepsilon_e^p + \varepsilon_e^e$, or in other words, the total equivalent strain is the sum of the equivalent plastic strain and the equivalent elastic strain) for the materials currently used in disk construction is well represented by the following relations:

$$\varepsilon_e^p = \left(\frac{\sigma_e - \sigma_y}{k}\right)^{1/n} \tag{12.13}$$

$$\varepsilon_e^p = \left(\frac{\sigma_y}{k_1}\right)^{n_1} \tag{12.14}$$

$$\varepsilon_e^p = \frac{1}{\eta} \cdot \left[\left(\frac{\sigma_e}{\sigma_y}\right)^m - 1\right] \tag{12.15}$$

$$\varepsilon_e^p = a_3 \cdot \sigma_e^3 + a_5 \cdot \sigma_e^5 \tag{12.16}$$

presented by Ludwik [45], Ramberg and Osgood [61], Swift [71] and You et al. [83] respectively. In these relations, where the meaning of the other symbols has already been indicated, k and n are the coefficient and exponent of Ludwik's law, k_1 and n_1 are the coefficient and exponent of Ramberg and Osgood's law, η and m are the strain hardening factor and exponent of Swift's law, and a_3 and a_5 are factors in the function proposed by You et al. In these functions, k, n, k_1, n_1, η, m, a_3 and a_5 are factors that depend on the type of material.

In any case, once plastic strain $\varepsilon_{e,i}^p$ has been determined, either using the graphic procedure (Fig. 12.1a) or analytically through one of the relations (12.13, 12.14, 12.15, and 12.16), relations (12.11) can be used to calculate the plastic increments of strain in the radial and tangential directions at the radii affected by plastic behaviour, and the new values of constant H'_{pi} given by (12.4) will be determined for these radii, and the stress state is calculated again. This procedure is repeated several times until the stress state calculated with the last iteration differs from that determined in the previous iteration by a small amount specified at the outset.

The residual stress state in a non-rotating disk previously prestressed by overspinning can be determined by subtracting the stress state with the material exhibiting elastic behaviour calculated in the first iteration from that calculated in the last iteration.

12.2.2 *Effect of Previous Plastic History*

The procedure described above, however, can be applied only if the disk has never been subjected to centrifugal loads that generate stresses beyond the yield point in it. In cases where the disk has undergone centrifugal loading with $\omega_y \leq \omega \leq \omega_p$, in a given moment of its history, the constitutive equations to be used are as follows:

$$\begin{cases} \varepsilon_r = \dfrac{du}{dr} = \dfrac{1}{E}(\sigma_r - \nu \cdot \sigma_t) + \alpha \cdot T + [\varDelta_r] + \varDelta_r + \delta_r \\ \varepsilon_t = \dfrac{u}{r} = \dfrac{1}{E}(\sigma_t - \nu \cdot \sigma_r) + \alpha \cdot T + [\varDelta_t] + \varDelta_t + \delta_t, \end{cases} \tag{12.17}$$

where $[\delta_r]$ and $[\delta_t]$ are the radial and tangential strains that already existed in the disk material before centrifugal loading generated the stress state characterised by stresses σ_r and σ_t. These prior strains are to be considered constant for the purposes of the method. $\varDelta_r$ and $\varDelta_t$ are the plastic components of strain correlated with stresses σ_r and σ_t.

In this case, compatibility equation (1.24) includes further terms and becomes:

$$\frac{d}{d\rho}\cdot\left(\frac{\sigma_t}{E}-\frac{\nu\cdot\sigma_r}{E}+\alpha\cdot T+\varDelta_t+\delta_t\right)-\frac{1+\nu}{E\cdot\rho}\cdot(\sigma_r-\sigma_t)-\frac{[\varDelta_r]-[\varDelta_t]}{\rho}-\frac{\varDelta_r-\varDelta_t}{\rho}-\frac{\delta_r-\delta_t}{\rho}=0, \tag{12.18}$$

while the following additional term will appear in the second member of relation (12.3) and (12.4):

$$A = (\rho_i - \rho_{i-1}) \cdot \left(\frac{[\varDelta_{ri}] - [\varDelta_{ti}]}{\rho_i} + \frac{[\varDelta_{ri-1}] - [\varDelta_{ti-1}]}{\rho_{i-1}}\right). \tag{12.19}$$

Where the preexisting plastic strains in the rotor were produced by a single cycle of loading beyond yield applied to the virgin material, the above additional term will be identical to that appearing in relations (12.3) and (12.4). If the pre-existing plastic strains result from n successive cycles of loading beyond yield, the additional term appearing in the second member of relations (12.3) and (12.4) will be the algebraic sum of the n terms of the type in (12.19). As the structure of relations (12.3) and (12.4) shows, this additional term must be added algebraically to the temperature term.

The calculation procedure outlined by Millenson and Manson is in fact equivalent to assuming that at a given load: (1) the prior plastic strain state was generated along loading-unloading path *OAB* on the virgin material's $\sigma = \sigma(\varepsilon)$ curve for a specified operating temperature (Fig. 12.1b); (2) as load and temperature vary, there will be a new $\sigma = \sigma(\varepsilon)$ curve whose origin is B; (3) with the subsequent load cycle, this second curve will be followed along the path from B up to C beyond yielding, so that the total strain correlated with this point is the sum of three strains: the residual strain $\varepsilon_1 = \varepsilon_{e,i}^p$ due to the first load cycle, the elastic component ε_2 of strain from the second load cycle, and the plastic component ε_3 of strain, also from the second load cycle. In general, it should be noted that, for a given temperature, the new $\sigma = \sigma(\varepsilon)$ curve for a material prestressed beyond yielding differs from that determined for the same material in the virgin state. In the absence of experimental data, however, the $\sigma = \sigma(\varepsilon)$ curve assumed must necessarily be that for the virgin material at the temperature of interest.

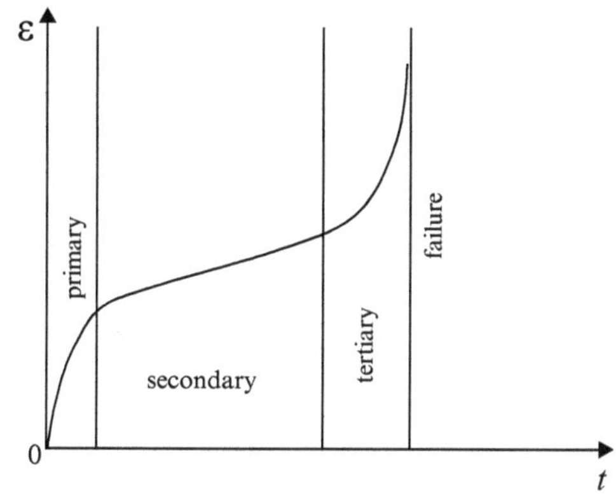

Fig. 12.2 Typical creep curve $\varepsilon = \varepsilon(t)$ from constant-temperature uniaxial creep test, and constant load

12.2.3 *Effect of Creep*

To analyse the stress state in a rotor subject to creep, we must have a relation linking creep deformation to stress. Although many problems of design significance would require a precise characterization of the material's creep behaviour with varying load and temperature, it must be acknowledged that the only experimental date available in this specific sector are those provided by uniaxial tensile tests run under constant temperature and load. Figure 12.2 shows a typical $\varepsilon = \varepsilon(t)$curve (where ε denotes the total strain and t is time) from a uniaxial creep test. This curve highlights the three known stages: the primary creep stage where deformation proceeds at a decreasing rate; the secondary stage, with a constant creep rate, and the tertiary stage, where creep proceeds at an increasing rate until failure occurs.

With a methodological approach similar to that outlined for calculating the two components of strain in plastic flow, Millenson and Manson arrive at the following relations:

$$\begin{cases} \delta_{ri} = \dfrac{\Gamma_i}{2 \cdot \sigma_{e,i}} \cdot \left(2 \cdot \sigma_{r,i} - \sigma_{t,i}\right) \\ \delta_{ti} = \dfrac{\Gamma_i}{2 \cdot \sigma_{e,i}} \cdot \left(2 \cdot \sigma_{t,i} - \sigma_{r,i}\right), \end{cases} \tag{12.20}$$

where Γ_i represents the creep total strain occurring in time Δt under a uniaxial equivalent stress $\sigma_{e,i}$. It is assumed that for sufficiently small values of time Δt the creep may be considered as occurring instantaneously at the end of the time period.

During the secondary stage of creep, there is a characteristic creep rate c_i corresponding to the stress $\sigma_{e,i}$ at temperature T_i, and Γ_i is given by the relation:

$$\Gamma_i = c_i \cdot \Delta t. \tag{12.21}$$

Rate c_i is the value usually published in papers and specialised texts on creep. During the primary and tertiary creep stages, the creep rate is also a function of time, which is thus more complex than relation (12.21). However, this does not complicate the computation, which proceeds in steps that are entirely similar to those described for plastic flow. Thus, once creep deformations have been determined, new values at the radii where creep occurs can be computed for the constant $H'_{p,i}$ given by relation (12.3) from which the plastic flow terms have been omitted, and hence by relation:

$$H'_{pi} = -2 \cdot [(\alpha_i \cdot T_i - \alpha_{i-1} \cdot T_{i-1}) + (\delta_{ti} - \delta_{ti-1})] + (\rho_i - \rho_{i-1}) \cdot \left(\frac{\delta_{ri} - \delta_{ti}}{\rho_i} + \frac{\delta_{ri-1} - \delta_{ti-1}}{\rho_{i-1}}\right). \quad (12.22)$$

We can then proceed with a new calculation of the stress state.

If the calculated values of stresses $\sigma_{r,i}$ and $\sigma_{t,i}$ differ by more than a predetermined small amount from the values of these stresses before creep occurred, a shorter time interval should be selected and the calculation repeated for each successive time interval required to equal the total time during which creep occurs. For each time interval, the effect of creep that occurred at previous time intervals is considered in a manner similar to that employed for previous plastic flow as described above. In other words, the following additional term will appear in the second member of relation (12.22):

$$B = (\rho_i - \rho_{i-1}) \cdot \left(\frac{[\delta_{ri}] - [\delta_{ti}]}{\rho_i} + \frac{[\delta_{ri-1}] - [\delta_{ti-1}]}{\rho_{i-1}}\right), \quad (12.23)$$

Where $[\delta_r]$ $[\delta_t]$ are the radial and tangential creep strains that already existed in the disk material.

The effects of this term must be added for all time intervals preceding that considered in the current calculation step.

Where plastic flow and creep operate simultaneously, their cumulative effects can obviously be computed using the more general relation (12.3), with the additions indicated above.

More complicated solutions based on finite element methods can be used for complex problems where the limitations of the one-dimensional approach are all too evident.

12.3 Analytical Method

The assumptions underlying the analytical method proposed here are those described at the Sect. 12.1 for stress beyond yielding, with the additional assumption of an isotropically strain-hardening material, and in Chap. 1 as regards stresses in the elastic field. However, the basic relations for the methods are rewritten below for the reader's convenience.

We will consider a rotating disk having thickness varying according to $h = h(r)$ and density $\gamma = \gamma(r)$ variable along radius r, subjected to thermal load and characterised by a temperature distribution $T = T(r)$, also varying along the radius r. A plane stress state is assumed. In these conditions, the equilibrium equation in the radial direction is given by the first (1.10), here reported:

$$\frac{d}{dr}(\sigma_r hr) - \sigma_t h + \gamma\omega^2 r^2 h = 0 \tag{12.24}$$

The first two geometric relations (1.14) continue to hold, and are given here:

$$\varepsilon_r = \frac{du}{dr} \text{ and } \varepsilon_t = \frac{u}{r} \tag{12.25}$$

Normal strain components ε_r and ε_t are only dependent from radial displacement u, and they are independent of thickness and material density. However, as we are dealing here with stresses beyond yielding, ε_r and ε_t represent total radial strain and total hoop strain respectively. Since relations (12.25) still apply, it follows that the compatibility equation in terms of strain given by (1.17) also continues to hold.

The total radial and tangential strains can be written as the sum of the elastic and plastic components (creep will not be considered here), which we designate with the apexes e and p respectively:

$$\begin{cases} \varepsilon_r = \dfrac{du}{dr} = \varepsilon_r^e + \varepsilon_r^p \\ \varepsilon_t = \dfrac{u}{r} = \varepsilon_t^e + \varepsilon_t^p. \end{cases} \tag{12.26}$$

The elastic components of total radial strain ε_r^e and tangential strain ε_t^e are related to radial and hoop stresses by relation (1.25):

$$\begin{cases} \varepsilon_r^e = \dfrac{1}{E}(\sigma_r - \nu\sigma_t) + \alpha T \\ \varepsilon_t^e = \dfrac{1}{E}(\sigma_t - \nu\sigma_r) + \alpha T, \end{cases} \tag{12.27}$$

As we know from the theory of plasticity, the plastic components ε_r^p and ε_t^p of radial and tangential strains are given by relations:

$$\begin{cases} \varepsilon_r^p = \dfrac{\varepsilon_e^p}{\sigma_e}\left(\sigma_r - \dfrac{1}{2}\sigma_t\right) = \dfrac{\varepsilon_e^p}{2\sigma_e}(2\sigma_r - \sigma_t) \\ \varepsilon_t^p = \dfrac{\varepsilon_e^p}{\sigma_e}\left(\sigma_t - \dfrac{1}{2}\sigma_r\right) = \dfrac{\varepsilon_e^p}{2\sigma_e}(2\sigma_t - \sigma_r), \end{cases} \tag{12.28}$$

where ε_e^p is the equivalent plastic strain, whereas σ_e is the equivalent stress.

According to Von Mises' criterion, the equivalent stress with the plane stress hypothesis ($\sigma_z = 0$) is given by relation (9.4):

$$\sigma_e = \sqrt{\sigma_r^2 + \sigma_t^2 - \sigma_r \sigma_t}. \tag{12.29}$$

To determine general differential equations governing the stress state in both the plastic and elastic range, here we choose the second of the two alternative procedures introduced in Sect. 1.3 of Chap. 1. In this second procedure, we first obtain the expression of stress σ_r as a function of σ_t from the equilibrium equation, and then substitute the resulting relation in the compatibility equation. Introducing the stress function $\phi = \sigma_r hr$ and considering (12.24) yields the following relations expressing stresses σ_r and σ_t as functions of ϕ:

$$\begin{cases} \sigma_r = \dfrac{\phi}{hr} \\ \sigma_t = \dfrac{1}{h}\dfrac{d\phi}{dr} + \gamma\omega^2 r^2. \end{cases} \tag{12.30}$$

By substituting the latter in (12.27) and the expressions thus obtained in (12.26), we obtain the following relations:

$$\begin{cases} \varepsilon_r = \dfrac{1}{E}\left(\dfrac{\phi}{hr} - \dfrac{\nu}{h}\phi' - \nu\gamma\omega^2 r^2\right) + \alpha T + \varepsilon_r^p \\ \varepsilon_t = \dfrac{1}{E}\left(\dfrac{\phi'}{h} - \dfrac{\nu}{hr}\phi + \gamma\omega^2 r^2\right) + \alpha T + \varepsilon_t^p, \end{cases} \tag{12.31}$$

where $\phi' = d\phi/dr$.

Developing the compatibility equation (1.17) and introducing (12.31) as well as the first derivative with respect to r of the second of the latter equations in the relation thus developed yields the following second-order non-homogeneous differential equation:

$$\begin{aligned} &\phi'' + \left(\frac{1}{r} - \frac{h'}{h}\right)\phi' - \left(\frac{1}{r^2} - \frac{\nu h'}{rh}\right)\phi + h\omega^2 r[(3+\nu)\gamma + \gamma' r] \\ &+ \alpha E h T' + \frac{Eh}{r}\left[r\frac{d\varepsilon_t^p}{dr} + \left(\varepsilon_t^p - \varepsilon_r^p\right)\right] = 0, \end{aligned} \tag{12.32}$$

where $\phi'' = d^2\phi/dr^2$, $h' = dh/dr$, $\gamma' = d\gamma/dr$ and $T' = dT/dr$.

If density is constant and there is no temperature gradient, terms γ' and T' disappear from differential Equation 12.32; obviously, the latter will be valid as long as the temperature is such that creep phenomena cannot occur.

Continuing to discuss the problem in its most general terms, relations (12.28) must be substituted together with the derivative with respect to r of the second relation (12.28) in (12.32) in order to obtain final governing differential equations.

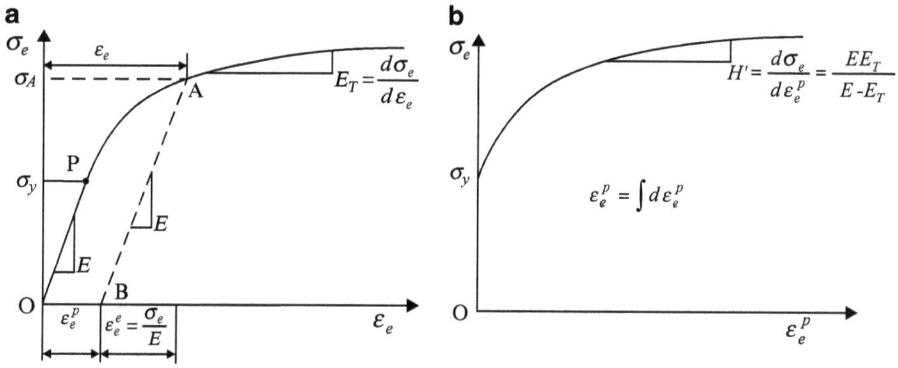

Fig. 12.3 (**a**) True stress-true strain curve $\sigma_e = \sigma_e(\varepsilon_e^p)$ from uniaxial tensile test (E elastic modulus, E_T tangent modulus); (**b**) curves $\sigma_e = \sigma_e(\varepsilon_e^p)$ obtained from the first by removing elastic strains ε_e^e, i.e., by imposing $\varepsilon_e^p = (\varepsilon_e - \varepsilon_e^e)$

Before proceeding, moreover, a hardening material model must be defined by introducing a function $\sigma_e = \sigma_e(\varepsilon_e^p)$ (actual equivalent stress - actual equivalent plastic strain) that interprets its actual behaviour beyond yielding. In this respect, a new correlation $\varepsilon_e^p = \varepsilon_e^p(\sigma_e)$ is proposed and used here that is more general than those found in the literature, which are based on the laws proposed by Ludwik [45], Ramberg and Osgood [61], Swift [71] and You et al. [83] mentioned above and given by relations (12.13), (12.14), (12.15) and (12.16) respectively.

To this end, the experimental true stress-true strain curve featured in Fig. 12.3a should be considered. In this curve, as in Fig. 12.1a, the elastic range covered by Hooke's law terminates at the point P where yielding starts (it is assumed that the proportional limit, elastic limit and yield stress σ_y are all identical). At any point A beyond P, we will have $\sigma_e = \sigma_A > \sigma_y$ and $\varepsilon_e = \varepsilon_e^p + \varepsilon_e^e$, i.e., equivalent stress is the ordinate of the generic point A, whereas the corresponding equivalent strain is the sum of elastic component ε_e^e, completely recovered on release of load, and of plastic component ε_e^p, due to which the material is hardened; in this connection, it is assumed that hardening is isotropic.

The $\sigma_e = \sigma_e(\varepsilon_e^p)$ curve featured in Fig. 12.3b is now drafted from the above experimental curve. This curve starts at coordinates ($\varepsilon_e^p = 0;\ \sigma_e = \sigma_y$) and, at a generic point, features a slope

$$H' = \frac{EE_T}{(E - E_T)} \tag{12.33}$$

which is obtained considering

$$d\sigma_e = E_T d\varepsilon_e;\ d\sigma_e = H' d\varepsilon_e^p;\ d\varepsilon_e^p = d\varepsilon_e - d\sigma_e/E \tag{12.34}$$

and replacing the first two in the third.

Function $\varepsilon_e^p = \varepsilon_e^p(\sigma_e)$ corresponding to the curve in Fig. 12.3b is simulated by means of the following n-grade polynomial

$$\varepsilon_e^p = a_1\sigma_e + a_2\sigma_e^2 + \ldots + a_n\sigma_e^n = \sum\nolimits_{i=1}^n a_i\sigma_e^i \tag{12.35}$$

where a_i, with $i = 1, 2, \ldots, n$, are numerical factors which may be easily determined once curve (b) is obtained from the experimental curve (a) in Fig. 12.3. For $\sigma_e = \sigma_y$, i.e., at the start of yielding, we must have $\varepsilon_e^e = \sigma_e/E$ and $\varepsilon_e^p = 0$ and, therefore, all coefficients in the series (12.35) must be such as to fulfil the above condition of initial yielding. On the basis of (12.35), we thus have:

$$\frac{\varepsilon_e^p}{\sigma_e} = \frac{\sum_{i=1}^n a_i\sigma_e^i}{\sigma_e} = a_1 + a_2\sigma_e + \ldots + a_n\sigma_e^{n-1} = \sum\nolimits_{i=1}^n a_i\sigma_e^{i-1} \tag{12.36}$$

whereby relations (12.28) become:

$$\begin{cases} \varepsilon_r^p = \dfrac{1}{2}(2\sigma_r - \sigma_t)\sum_{i=1}^n a_i\sigma_e^{i-1} \\ \varepsilon_t^p = \dfrac{1}{2}(2\sigma_t - \sigma_r)\sum_{i=1}^n a_i\sigma_e^{i-1}. \end{cases} \tag{12.37}$$

These relations can then be used to obtain the following:

$$\varepsilon_t^p - \varepsilon_r^p = \frac{3}{2}(\sigma_t - \sigma_r)\sum\nolimits_{i=1}^n a_i\sigma_e^{i-1} \tag{12.38}$$

and

$$\frac{d\varepsilon_t^p}{dr} = \frac{1}{2}(2\sigma_t - \sigma_r)\sum\nolimits_{i=1}^n \left[(i-1)a_i\sigma_e^{i-2}\frac{d\sigma_e}{dr}\right] + \frac{1}{2}\left(2\frac{d\sigma_t}{dr} - \frac{d\sigma_r}{dr}\right)\sum\nolimits_{i=1}^n a_i\sigma_e^{i-1}. \tag{12.39}$$

Also, differentiating equation (12.29) with respect to r gives:

$$\frac{d\sigma_e}{dr} = \frac{1}{2\sigma_e}\left[\frac{d\sigma_r}{dr}(2\sigma_r - \sigma_t) + \frac{d\sigma_t}{dr}(2\sigma_t - \sigma_r)\right], \tag{12.40}$$

which, introduced in (12.39), makes it possible to obtain:

$$\begin{aligned} \frac{d\varepsilon_t^p}{dr} = \frac{(2\sigma_t - \sigma_r)}{4}\left\{\left[\frac{d\sigma_r}{dr}(2\sigma_r - \sigma_t) + \frac{d\sigma_t}{dr}(2\sigma_t - \sigma_r)\right]\sum_{i=1}^n\left[(i-1)a_i\sigma_e^{i-3}\right]\right\} \\ + \frac{1}{2}\left(2\frac{d\sigma_t}{dr} - \frac{d\sigma_r}{dr}\right)\sum_{i=1}^n a_i\sigma_e^{i-1}. \end{aligned} \tag{12.41}$$

By introducing (12.38) and (12.41) in (12.32) and considering (1.26), we obtain:

$$\phi''+\left(\frac{1}{r}-\frac{h'}{h}\right)\phi'-\left(\frac{1}{r^2}-\frac{\nu h'}{rh}\right)\phi+h\omega^2 r[(3+\nu)\gamma+\gamma' r]+\alpha EhT'+$$
$$+\frac{Eh}{4}(2\sigma_t-\sigma_r)(2\sigma_r-\sigma_t)\left(\frac{\phi'}{hr}-\frac{\phi h'}{h^2 r}-\frac{\phi}{hr^2}\right)\sum_{i=1}^{n}[(i-1)a_i\sigma_e^{i-3}]+$$
$$+\frac{Eh(2\sigma_t-\sigma_r)^2}{4}\left(\frac{\phi''}{h}-\frac{\phi' h'}{h^2}+\gamma'\omega^2 r^2+2\gamma\omega^2 r\right)+$$
$$+\frac{Eh}{2}\left[2\left(\frac{\phi''}{h}-\frac{\phi' h'}{h^2}+\gamma'\omega^2 r^2+2\gamma\omega^2 r\right)-\left(\frac{\phi'}{hr}-\frac{\phi h'}{h^2 r}-\frac{\phi}{hr^2}\right)\right]$$
$$\times\sum_{i=1}^{n}a_i\sigma_e^{i-1}+\frac{3Eh}{2r}(\sigma_t-\sigma_r)\sum_{i=1}^{n}a_i\sigma_e^{i-1}=0. \tag{12.42}$$

Developing and rearranging this relation gives the following second-order non-homogeneous differential equation governing the plastic field in a rotating disk having variable thickness and density along its radius, subject to thermal load, and made of a material hardening according to the polynomial relation given by (12.35):

$$\phi''\left\{1+E\left[\frac{(2\sigma_r-\sigma_t)^2}{4}\sum_{i=1}^{n}(i-1)a_i\sigma_e^{i-3}+\sum_{i=1}^{n}a_i\sigma_e^{i-1}\right]\right\}+\phi'\left\{\left(\frac{1}{r}-\frac{h'}{h}\right)+\right.$$
$$\left.\frac{E}{4}\left[\left(\frac{(2\sigma_r-\sigma_t)(2\sigma_t-\sigma_r)}{r}-\frac{h'}{h}(2\sigma_t-\sigma_r)^2\right)\sum_{i=1}^{n}[(i-1)a_i\sigma_e^{i-3}]-2\left(\frac{1}{r}+2\frac{h'}{h}\right)\sum_{i=1}^{n}a_i\sigma_e^{i-1}\right]\right\}+$$
$$-\phi\left\{\left(\frac{1}{r^2}-\frac{\nu h'}{rh}\right)+\frac{E}{4}\left(\frac{1}{r^2}+\frac{h'}{rh}\right)\left[(2\sigma_r-\sigma_t)(2\sigma_t-\sigma_r)\sum_{i=1}^{n}[(i-1)a_i\sigma_e^{i-3}]-2\sum_{i=1}^{n}a_i\sigma_e^{i-1}\right]\right\}+$$
$$+\gamma\omega^2 hr\left\{(3+\nu)+E\left[\frac{(2\sigma_t-\sigma_r)^2}{2}\sum_{i=1}^{n}[(i-1)a_i\sigma_e^{i-3}]+2\sum_{i=1}^{n}a_i\sigma_e^{i-1}\right]\right\}+$$
$$+\gamma'\omega^2 hr^2\left\{1+E\left[\frac{(2\sigma_t-\sigma_r)^2}{4}\sum_{i=1}^{n}[(i-1)a_i\sigma_e^{i-3}]+\sum_{i=1}^{n}a_i\sigma_e^{i-1}\right]\right\}$$
$$+\alpha EhT'+\frac{3}{2}\frac{Eh}{r}(\sigma_t-\sigma_r)\sum_{i=1}^{n}a_i\sigma_e^{i-1}=0. \tag{12.43}$$

This relation is simplified in the elastic field, given that all terms with summations disappear, as they are related to stress beyond initial yielding; in this specific case, the relation becomes:

$$\phi''+\left(\frac{1}{r}-\frac{h'}{h}\right)\phi'-\left(\frac{1}{r^2}-\frac{\nu h'}{rh}\right)\phi+h\omega^2 r[(3+\nu)\gamma+\gamma' r]+\alpha EhT'=0. \tag{12.44}$$

At the elastic-plastic interface separating the inner plastic region from the outer elastic region, there must be a continuity of radial stress σ_r and, therefore, of stress function ϕ; in other words, values σ_r and ϕ, calculated at the radius corresponding to the elastic-plastic interface by using the two (12.43) and (12.44), which hold for the plastic range and the elastic range respectively, must be equal. Given the continuity of stress function ϕ and relations (12.28), (12.26) and (12.25), continuity of stresses and displacements is also guaranteed at the elastic-plastic interface.

Boundary conditions for only rotating annular disks are as follows:

$$\begin{cases} \sigma_r = 0 & \text{for } r = r_i \\ \sigma_r = 0 & \text{for } r = r_e. \end{cases} \tag{12.45}$$

For only rotating solid disks, the boundary conditions are:

$$\begin{cases} u = 0 & \text{for } r = r_i \\ \sigma_r = 0 & \text{for } r = r_e. \end{cases} \tag{12.46}$$

These conditions are valid for any disks with no crown ring, be they solid or annular. For disks with crown ring, the second relations in (12.45) and (12.46) must be replaced with condition$\sigma_r = \sigma_{re}$ for $r = r_e$, where σ_{re} is the radial stress distribution at outer radius r_e, due to the centrifugal load of the crown ring and blading.

Differential equation (12.43) and (12.44), along with boundary conditions (12.45) and (12.46), are the fundamental relations that enable us to analyse the stress and strain states in the plastic region and in the elastic region of rotating disks consisting of homogeneous isotropic material, showing a curve $\sigma_e = \sigma_e(\varepsilon_e^p)$ which can be expressed by (12.35), with an arbitrary profile $h = h(r)$, having variable density along the radius according to a generic function $\gamma = \gamma(r)$ and subject to a thermal load featuring a generic temperature distribution along the radius $T = T(r)$.

Elastic-plastic analysis is structurally relevant only during disk design in cases where the disk is to be prestressed by overspeeding so as to generate a favourable state of compressive residual stress next to the inner radius of annular disks (solid disks are rarely prestressed, as they do not have a sufficient plastic reserve) or in the area where the stress state shows high equivalent stress from stress concentration effects.

However, when the disk is prestressed by overspeeding at the factory, neither the variability of density along the radius nor the temperature gradient are factors of interest. The disk to be prestressed is rotated at the factory at speeds higher than that which will be used in service, and which are sufficient to cause yielding in the inner portion of the disk. Overspeeding is performed at a constant temperature (usually room temperature) and without any blading on the crown ring and on the disks' lateral surfaces; from the standpoint of structural analysis, the presence of blades would correspond to a density variation along the radius. In this case, when determining the elastic-plastic interface and performing elastic-plastic analysis, (12.43) and (12.44) are (slightly) simplified, as terms in γ' and T' disappear.

From a design point of view, analysing a disk prestressed by overspeeding in the elastic range is extremely important, as it serves the threefold purpose of determining:

(1) the stress state in the elastic region of the prestressed disk; (2) the residual stress state on the disk, which is obtained by subtracting the purely elastic stress state calculated at the same speed and the same conditions used for prestressing, and therefore corresponding to the descending straight-line *AB* parallel to *OP* in Fig. 12.3a, from the elastic-plastic stress state as found above; (3) the actual stress state of the disk during service, resulting from superposing the above residual stress state on the purely elastic stress state, still corresponding to the ascending line segment *BA* in Fig. 12.3a, although this time calculated by means of (12.44) complete with terms in γ' and T', in order to allow for any likely temperature gradients as well as any density variations from blading on lateral surfaces or from slots provided in the crown ring as seats for blade attachments. As regards determining stress states due to temperature gradients ($T \neq 0$) and virtual density variations along the radius ($\gamma \neq 0$), see Sects. 7.3 and 7.4 of Chap. 7.

To integrate differential equation (12.43) and (12.44), functions $h = h(r)$, $\gamma = \gamma(r)$ and $T = T(r)$ must first be made explicit, and the grade and coefficients of the polynomial (12.35) with which the experimental curve $\sigma_e = \sigma_e(\varepsilon_e^p)$ is interpolated must be defined in accordance with the disk material. As regards function $h = h(r)$, the disks analysed here are axisymmetric and symmetric with respect to the mid–plane, and have non-linearly variable thickness according to relation (7.1). With regard to functions $\gamma = \gamma(r)$ and $T = T(r)$, relations (3.44) and (3.2) are used, because, as they are polynomial, they make it possible to interpolate any distribution of density and temperature along the radius.

In this case, the following must be substituted in (12.43) and (12.44), in addition to functions $h = h(r)$ and $\gamma = \gamma(r)$,

$$\frac{h'}{h} = -\frac{mk}{1 - mr} \tag{12.47}$$

$$\gamma' = \sum_{i=1}^{n} i\gamma_i r^{i-1} \tag{12.48}$$

$$T' = \sum_{i=1}^{n} ik_i r^{i-1} \tag{12.49}$$

$$h\omega^2 r[(3 + \nu)\gamma + \gamma' r] = h\omega^2 r \sum_{i=1}^{n} \left[(3 + i + \nu)\gamma_i r^i\right] \tag{12.50}$$

These expressions are derived from relations (7.1), (3.44) and (3.2).

In keeping with our purpose – to develop an analytical model for elastic-plastic analysis of rotating disks having a profile defined by function (7.1) and subjected to prestressing by overspeeding – process conditions in the factory are met ($\gamma = const$ and $T = const$) when the plasticization speed is reached in the prestressing equipment ($\omega = const$). Here, we will consider steel disks ($\nu = 0.3$, $\sigma_y = 480$ MPa; $E = 204$ GPa; $\gamma = 7{,}800$ kg/m^3), a material for which the Eulerian $\sigma_e = \sigma_e(\varepsilon_e)$ curve has been determined experimentally by means of a uniaxial tensile test on a

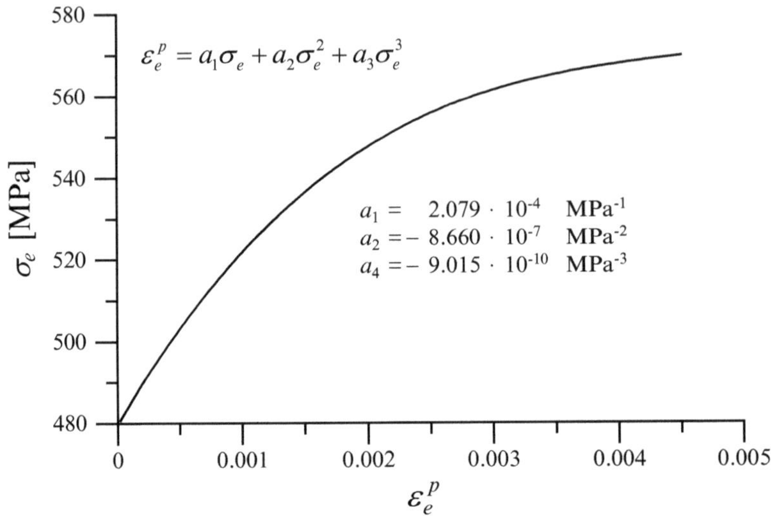

Fig. 12.4 Curve $\sigma_e = \sigma_e(\varepsilon_e^p)$ (equivalent stress/equivalent plastic strain) for turbine disk steel

standard specimen. The procedure illustrated in Fig. 12.3 can be used to obtain the $\sigma_e = \sigma_e(\varepsilon_e^p)$ curve shown in Fig. 12.4 from the experimental curve.

In accordance with (12.35), this curve has been interpolated with the following third-degree polynomial: $\varepsilon_e^p = a_1\sigma_e + a_2\sigma_e^2 + a_3\sigma_e^3$, with $a_1 = 2.079 \cdot 10^{-4}$ MPa^{-1}, $a_2 = -8.660 \cdot 10^{-7}$ MPa^{-2} and $a_3 = -9.015 \cdot 10^{-10}$ MPa^{-3}.

12.4 Elastic Analysis

The relations presented in Sects. 7.2.1, 7.2.2 and 7.2.3 of Chap. 7 can be used for elastic analysis of the disk considered here. These relations are based on the first approach illustrated in Sect. 1.3 of Chap. 1, which enables us to write a second order non-homogeneous differential equation in radial displacement u.

For demonstration purposes, it is useful to repeat the integration procedure for (12.44), which is formally similar to (7.3) and its specializations (7.26) and (7.40), applying respectively to a non-rotating disk subject to a temperature gradient along the radius given by (3.2), and to a rotating disk featuring density variation along its radius given by (3.44), not simply because it involves writing the particular integral in another form, but chiefly because a new procedure will be introduced for determining the second independent integral of the homogeneous equation associated with it.

Accordingly, under the conditions occurring during the prestressing by overspeeding, at an angular velocity ω higher than the rotational speed when yielding just begins (yielding onset or start of plasticization), but lower than the rotational speed bringing the elastic-plastic interface up to the outer radius (fully plastic state), the stress state in the elastic region of the disk is given by (12.44) without the terms in γ' and T' (for determining stress and strain variation with density and temperature variation along the radius, see Sects. 7.3 and 7.4 of Chap. 7), after replacing (12.47) in (12.44), i.e., by the relation:

$$\phi'' + \left(\frac{1}{r} + \frac{mk}{1 - mr}\right)\phi' - \left[\frac{1}{r^2} + \frac{vmk}{r(1 - mr)}\right]\phi = -(3 + v)\gamma\omega^2 h_0 r(1 - mr)^k. \tag{12.51}$$

The general solution of this equation is obtained by adding the solution of the associated homogeneous equation to the particular solution.

12.4.1 Solution of Homogeneous Differential Equation

The associated homogeneous equation of the differential equation (12.51) is as follows:

$$\frac{d^2\phi}{dr^2} + \left(\frac{1}{r} + \frac{mk}{1 - mr}\right)\frac{d\phi}{dr} - \left[\frac{1}{r^2} + \frac{vmk}{r(1 - mr)}\right]\phi = 0. \tag{12.52}$$

To find the solution of the above equation, it is best to introduce the dimensionless variable $\rho = r/r_e$, which will be used in the examples presented below. Taking (7.1), written in terms of ρ, into account and considering that $d/dr = (1/r_e)\, d/d\rho$ and $d^2/dr^2 = (1/r_e^2)\, d^2/d\rho^2$, (12.52) can be written in the following form:

$$\frac{d^2\phi}{d\rho^2} + \left(\frac{1}{\rho} + \frac{nk}{1 - n\rho}\right)\frac{d\phi}{d\rho} - \left[\frac{1}{\rho^2} + \frac{vnk}{\rho(1 - n\rho)}\right]\phi = 0. \tag{12.53}$$

Then, introducing the variable $t = n\rho$ and taking into account that $d\rho = dt/n$ and $d\rho^2 = dt^2/n^2$, (12.53) becomes:

$$\frac{d^2\phi}{dt^2} + \left(\frac{1}{t} - \frac{k}{t - 1}\right)\frac{d\phi}{dt} - \left[\frac{1}{t} + (vk - 1)\right]\frac{\phi}{t(t - 1)} = 0 \tag{12.54}$$

This equation is the Gauss hypergeometric differential equation, showing three singular points ($t = 0$; $t = 1$; $t = \infty$) and corresponding to the general Papperitz form given by (6.16), where (α, α'), (β, β') and (γ, γ') are the three pairs of roots of the determinant equations for the three singular points, with:

$$\alpha + \alpha' + \beta + \beta' + \gamma + \gamma' = 1 \tag{12.55}$$

As discussed in Sect. 6.2 of Chap. 6, this is because the sum of the roots of all determinant equations of a totally Fuchsian equation with $(i+1)$ singular points must be $(i-1)$. Thus, by comparing (6.16) with (12.54) and taking (12.55) into account, we can deduce the following relations:

$$\begin{aligned} &1-\alpha-\alpha'=1\,;\ 1-\gamma-\gamma'=-k\,;\ \beta\beta'=\nu k-1\,;\ \alpha\alpha'=-1\,;\\ &\gamma\gamma'=0\,;\ \beta+\beta'=-k \end{aligned} \tag{12.56}$$

Here, we obtain:

$$\begin{aligned} &\alpha=-1\,;\ \alpha'=1\,;\ \gamma=1+k\,;\ \gamma'=0\,;\ \beta=-\frac{k}{2}+\xi\,;\ \beta'=-\frac{k}{2}-\xi\,;\\ &\text{with } \xi=\frac{1}{2}\sqrt{k^2-4(\nu k-1)} \end{aligned} \tag{12.57}$$

According to the related general theory, (6.16) allows for two independent integrals, chosen from the four given by relations (6.20) as given here, and each corresponding to one of the exponents α and α' of the variable t:

$$\begin{aligned} \phi_{1a}&=t^{\alpha'}(1-t)^{\gamma}F(\alpha'+\beta+\gamma,\alpha'+\beta'+\gamma,1+\alpha'-\alpha,t)\\ \phi_{1b}&=t^{\alpha'}(1-t)^{\gamma'}F(\alpha'+\beta+\gamma',\alpha'+\beta'+\gamma',1+\alpha'-\alpha,t)\\ \phi_{1c}&=t^{\alpha}(1-t)^{\gamma}F(\alpha+\beta+\gamma,\alpha+\beta'+\gamma,1+\alpha-\alpha',t)\\ \phi_{1d}&=t^{\alpha}(1-t)^{\gamma'}F(\alpha+\beta+\gamma',\alpha+\beta'+\gamma',1+\alpha-\alpha',t). \end{aligned} \tag{12.58}$$

The first two of these expressions are true for exponent α' and the remaining two are true for exponent α; we can thus shift between one equivalent independent integral and the other, either of which correspond to the exponent α' or to the exponent α, simply by replacing γ with γ' and vice versa. In the case presented here, by taking into account relations (12.57), (12.58) change into:

$$\begin{aligned} \phi_{1a}&=t(1-t)^{1+k}F\left(2+\frac{k}{2}+\xi,2+\frac{k}{2}-\xi,3,t\right)\\ \phi_{1b}&=tF\left(1-\frac{k}{2}+\xi,1-\frac{k}{2}-\xi,3,t\right)\\ \phi_{1c}&=t^{-1}(1-t)^{1+k}F\left(\frac{k}{2}+\xi,\frac{k}{2}-\xi,-1,t\right)\\ \phi_{1d}&=t^{-1}F\left(-1-\frac{k}{2}+\xi,-1-\frac{k}{2}-\xi,-1,t\right). \end{aligned} \tag{12.59}$$

For convergence, the same considerations made for relations (6.21) in Sect. 6.2.2 also apply here.

In this context, it is necessary to select one of the first two relations (12.59) as the integral of (12.54). However, despite the apparent formal differences, the first two relations (12.59) actually coincide for any pair of values k and n (to demonstrate this, Fig. 12.5 shows the integrals ϕ_{1a} and ϕ_{1b} as functions of t for three different disk profiles, the first convex, the second concave and the third conical), i.e., $\phi_{1a} = \phi_{1b}$, it is possible to select as the integral of (12.54) the one deriving from the second relation (12.59), which is here designated ϕ_1 and transcribed as

$$\phi_1 = t^{\alpha'}(1-t)^{\gamma'}F(a,b,c,t) = tF\left(1-\frac{k}{2}+\xi, 1-\frac{k}{2}-\xi, 3, t\right) \tag{12.60}$$

where, for the sake of brevity, the following notation is used:

$$\begin{aligned} &a = \alpha' + \beta + \gamma' = 1 - \frac{k}{2} + \xi\,;\; b = \alpha' + \beta' + \gamma' = 1 - \frac{k}{2} - \xi\,; \\ &c = 1 + \alpha' - \alpha = 3 \end{aligned} \tag{12.61}$$

whereas $F(a, b, c, t)$ indicates the infinite hypergeometric series of the kind

$$\begin{aligned} F(a,b,c,t) = \; & 1 + \frac{a \cdot b}{1! \cdot c} \cdot t + \frac{a \cdot (a+1) \cdot b \cdot (b+1)}{2! \cdot c \cdot (c+1)} \cdot t^2 + \\ & + \ldots + \frac{a \cdot (a+1) \cdot \ldots \cdot (a+n-1) \cdot b \cdot (b+1) \cdot \ldots \cdot (b+n-1)}{n! \cdot c \cdot (c+1) \cdot \ldots \cdot (c+n-1)} \cdot t^n + \ldots \end{aligned} \tag{12.62}$$

This series slowly converges for $t = |n\rho| < 1$. This implies that $0 < n < 1$, i.e., that the parameter n is always less than 1, because the actual disk profile is defined, at most, in the domain $0 \le \rho \le 1$. More specifically, the hypergeometric series $F(a, b, c, t)$ is divergent if the real part $\Re(c - a - b) \le -1$, absolutely convergent if $\Re(c - a - b) > 0$ and, excluding the singularity point $t = n{\cdot}\rho = 1$, conditionally convergent if $-1 < \Re(c - a - b) \le 0$. Here, given that a, b and c are real coefficients and $(c - a - b) = 1 + k$, for tapered disks we have divergence for $k \le -2$, absolute convergence for $k > -1$ and conditional convergence for $-2 < k \le -1$.

For convergent disks, the range of the parameters n and k that appear in the relation (7.1) is therefore limited, as it must be $0 < n < 1$, with $k > -1$ for absolute convergence and $2 < k \le -1$ for conditional convergence. Finally, it should be noted that ϕ_1 vanishes for $t = 0$ and tends to a finite value for $t \to 1$.

Given that the first derivative of the stress function $\phi = \phi\,(r)$ must be calculated to determine the stress state (in this regard, see (12.30)), it should be borne in mind that the first derivative with respect to variable t in series (12.62) is as follows:

$$\frac{d}{dt}F(a,b,c,t) = \frac{ab}{c}F(a+1,\ \mathrm{b}+1, c+1, t) \tag{12.63}$$

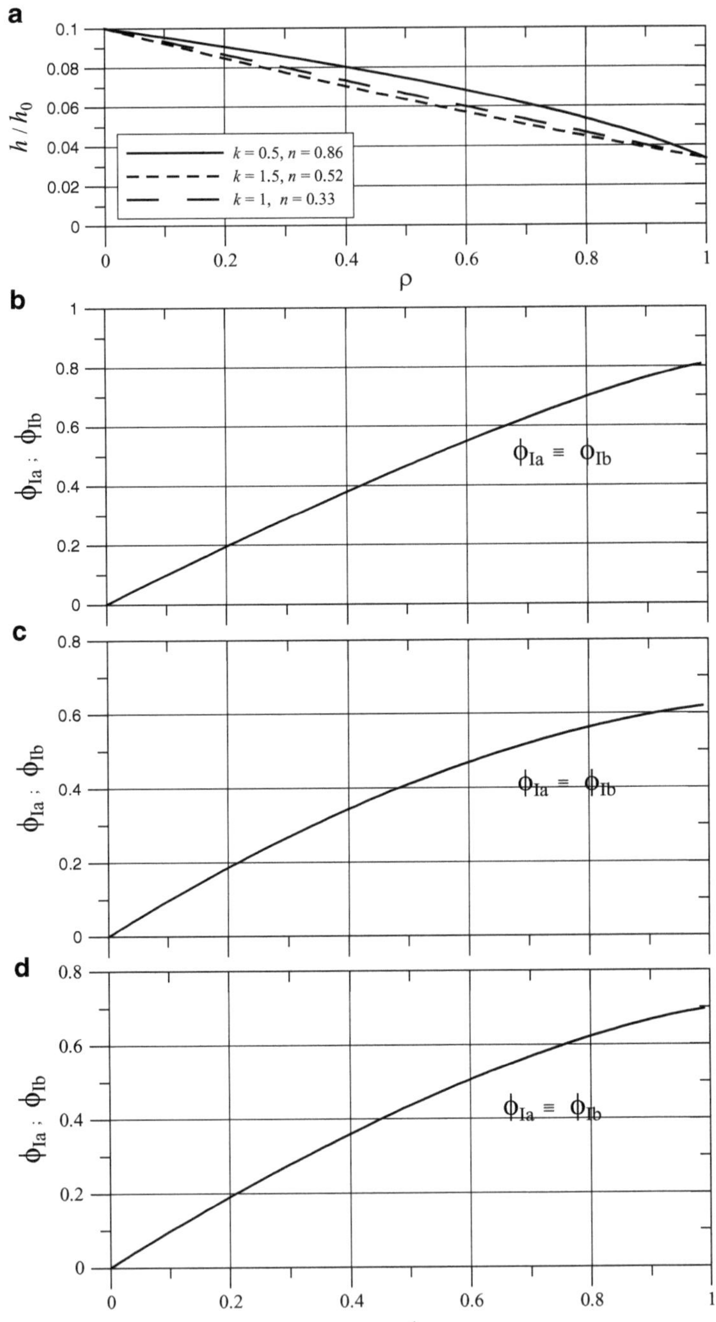

Fig. 12.5 (**a**) Convergent disks having variable profile according to (7.1), convex ($k = 0.5$; $n = 0.89$), concave ($k = 1.5$; $n = 0.52$) and conical ($k = 1; n = 0.33$), with $h_0/h_e = 3$; ϕ_{1a} and ϕ_{1b} versus t, for $0 \leq t \leq 1$, for steel disks ($\nu = 0.3$) having: (**b**) convex profile ($k = 0.5$; $n = 0.89$); (**c**) concave profile ($k = 1.5$; $n = 0.52$); (**d**) conical profile ($k = 1$; $n = 0.33$)

It is now necessary to find a second integral of homogeneous equation (12.54) which is independent of the first solution ϕ_1 given by relation (12.60). Without resorting to overly complex solutions (see, for example, those proposed by Klein [39] and used in Sects. 6.2.2 and 7.2.2), here it is possible to find other solutions of this homogenous equation based on two different independent integrals expressed by means of the power series in the variable $x = 1 - t$. By introducing this new variable, the homogeneous equation (12.54) becomes:

$$\frac{d^2\phi}{dx^2} + \left(\frac{1}{x-1} - \frac{k}{x}\right)\frac{d\phi}{dx} + \left[\frac{1}{x-1} - (vk-1)\right]\frac{\phi}{x(x-1)} = 0 \tag{12.64}$$

This is still a hypergeometric differential equation, featuring three singularity points ($x = 0$; $x = 1$; $x = \infty$) and corresponding to the general Papperitz form expressed by (6.16), after changing variable t to variable x. By applying the same procedure illustrated above, we can conclude that the integrals of (12.64) are the same as those obtained from relation (12.58), substituting roots α and α' with γ and γ' respectively, while β and β' are unchanged because they are correlated to the singular point at infinity. Therefore, the change of variable $x = 1 - t$ implies that the three singular points in the variable t ($t = 0$; $t = 1$; $t = \infty$) correspond, in order, to as many singular points in the variable x ($x = 1$; $x = 0$; $x = \infty$). In other words, with Riemann's notation, we can write

$$P\begin{pmatrix} 0 & \infty & 1 & \\ \alpha & \beta & \gamma & t \\ \alpha' & \beta' & \gamma' & \end{pmatrix} = P\begin{pmatrix} 1 & \infty & 0 & \\ \alpha & \beta & \gamma & x \\ \alpha' & \beta' & \gamma' & \end{pmatrix} = P\begin{pmatrix} 0 & \infty & 1 & \\ \gamma & \beta & \alpha & x \\ \gamma' & \beta' & \alpha' & \end{pmatrix}. \tag{12.65}$$

Consequently, we can write the relations (12.58) in the variable $x = 1 - t$, which thus become:

$$\begin{aligned}
\phi^*_{Ia} &= x^{\alpha'}(1-x)^{\gamma}F(\alpha' + \beta + \gamma, \alpha' + \beta' + \gamma, 1 + \alpha' - \alpha, x) \\
\phi^*_{Ib} &= x^{\alpha'}(1-x)^{\gamma'}F(\alpha' + \beta + \gamma', \alpha' + \beta' + \gamma', 1 + \alpha' - \alpha, x) \\
\phi^*_{IIa} &= x^{\alpha}(1-x)^{\gamma}F(\alpha + \beta + \gamma, \alpha + \beta' + \gamma, 1 + \alpha - \alpha', x) \\
\phi^*_{IIb} &= x^{\alpha}(1-x)^{\gamma'}F(\alpha + \beta + \gamma', \alpha + \beta' + \gamma', 1 + \alpha - \alpha', x).
\end{aligned} \tag{12.66}$$

Subsequently, by calculating three new pairs of roots (α, α'), (β, β') and (γ, γ') using the procedure outlined above and returning to the variable t, we obtain the following four integrals of the homogeneous equation (12.54):

$$\begin{aligned}
\phi_{Ia}^{*} &= t^{-1}F\left(-1-\frac{k}{2}+\xi,\,-1-\frac{k}{2}-\xi,\,-k,1-t\right)\\
\phi_{Ib}^{*} &= tF\left(1-\frac{k}{2}+\xi,1-\frac{k}{2}-\xi,\,-k,1-t\right)\\
\phi_{IIa}^{*} &= t^{-1}(1-t)^{1+k}F\left(\frac{k}{2}+\xi,\frac{k}{2}-\xi,2+k,1-t\right)\\
\phi_{IIb}^{*} &= t(1-t)^{1+k}F\left(2+\frac{k}{2}+\xi,2+\frac{k}{2}-\xi,2+k,1-t\right).
\end{aligned} \tag{12.67}$$

Figure 12.6 shows curves for the four integrals (12.67) for the same three pairs of values k and n considered in Fig. 12.5, each of which identifies one of the three disk profiles (convex, concave and conical) to be examined. It should first be noted that $\phi_{Ia}^{*}=\phi_{Ib}^{*}$ and $\phi_{IIa}^{*}=\phi_{IIb}^{*}$ for any pair of values k and n, provided that $k\neq 1$. In other words, only two of these four integrals are separate and independent of each other. For $k=1$ (converging and diverging conical disks), we still have $\phi_{IIa}^{*}=\phi_{IIb}^{*}$, while ϕ_{Ia}^{*} and ϕ_{Ib}^{*} are not defined in the real field. Consequently, to deal with any disk of actual design interest having one of the three types of profile represented in Fig. 7.2a, c, d, the second independent integral in homogeneous equation (12.54) must necessarily be that defined by $\phi_{IIa}^{*}=\phi_{IIb}^{*}$. As Fig. 12.6 shows, this integral is zero for $t=1$ and tends to infinity for $t\to 0$. It should also be noted that the four integrals (12.67) are not defined in the real field for $k=-1$ and $n=-1$ and, more generally, for any pair of negative values of the two parameters k and n. Consequently, the profiles represented in Fig. 7.2b cannot be treated with the method outlined here. Nevertheless, this does not limit the method's generality, as it can be seen from Fig. 7.2c that concave profiles can be obtained with $k>1$ and $-0.6<n<1$.

To solve the problems of interest to us here, we choose the integral given by (12.60) and that given by the fourth relation (12.67). The latter integral, despite its apparent formal differences, coincides with that expressed by the third relation (12.67). The two selected integrals, the first of which was designated above as ϕ_1 and the second which will here be designated as ϕ_2, with $\phi_2=\phi_{IIa}^{*}=\phi_{IIb}^{*}$, are thus:

$$\begin{aligned}
\phi_1 &= tF\left(1-\frac{k}{2}+\xi,1-\frac{k}{2}-\xi,3,t\right)\\
\phi_2 &= t(1-t)^{1+k}F\left(2+\frac{k}{2}+\xi,2+\frac{k}{2}-\xi,2+k,1-t\right).
\end{aligned} \tag{12.68}$$

Regarding the influence of parameters n and k on calculating the two integrals ϕ_1 and ϕ_2, it should be borne in mind, as the third factor c of the second hypergeometric series (12.68) must be positive, function ϕ_2 is subject to a constraint $k>-2$. By contrast, function ϕ_1 is not subject to constraints for parameter k.

For parameter n, there are no further constraints except those indicated in Sect. 7.1 of Chap. 7, which derive from the fact that once disk thickness $h=h_e$ at the

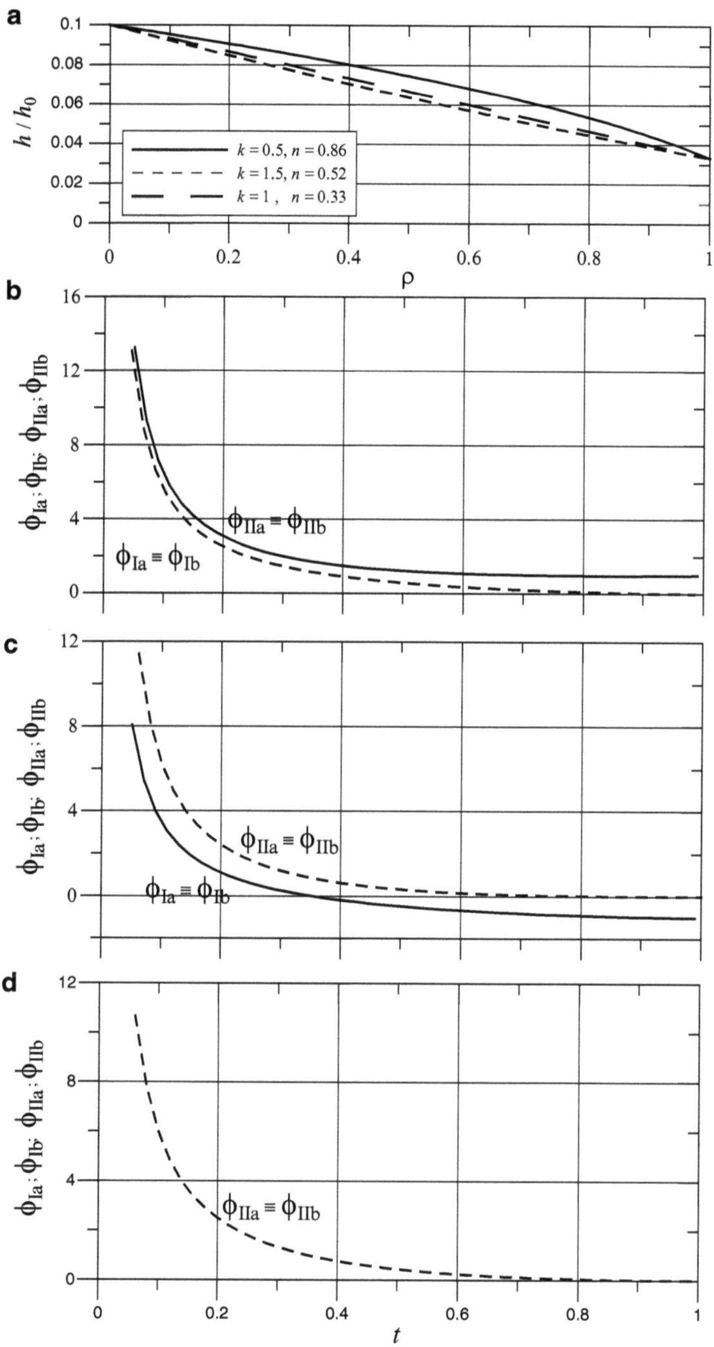

Fig. 12.6 (**a**) Convergent disks having variable profile, convex ($k = 0.5$; $n = 0.89$), concave ($k = 1.5$; $n = 0.52$) and conical ($k = 1$; $n = 0.33$), with $h_0/h_e = 3$; ϕ_{Ia}, ϕ_{Ib}, ϕ^*_{IIa} and ϕ^*_{IIb} versus t for $0 \leq t \leq 1$, for steel disks ($\nu = 0.3$) having: (**b**) convex profile ($k = 0.5$; $n = 0.89$); (**c**) concave profile ($k = 1.5$; $n = 0.52$); (**d**) conical profile ($k = 1$; $n = 0.33$)

outer radius, where $\rho = 1$, has been established as a design datum, we must have $n < 1 - (h_e/h_0)^{1/k}$ for the profiles to have physical meaning.

12.4.2 Particular Integral

To obtain the particular integral of (12.51) which, written in terms of the variable t, becomes:

$$\frac{d^2\phi}{dt^2} + \left(\frac{1}{t} + \frac{k}{1-t}\right)\frac{d\phi}{dt} - \left[\frac{1}{t^2} + \frac{\nu k}{t(1-t)}\right]\phi = At(1-t)^k, \tag{12.69}$$

with $A = -(3+\nu)\gamma\omega^2 h_0 (r_e/n)^3$, we put:

$$\phi_p = (1-t)^{k+1}\left(c_1 t + c_2 t^2\right), \tag{12.70}$$

where c_1 and c_2 are constants that can be determined by substituting (12.70), along with its first and second derivatives with respect to variable t, in (12.69) and by equalling to zero the coefficients of various powers in the variable r. We thus obtain:

$$c_1 = -\frac{3A}{[3+k(\nu+2)]\,[8+k(\nu+3)]}\,;\; c_2 = -\frac{A}{8+k(\nu+3)} \tag{12.71}$$

Then, by introducing relations (12.71) and notations A in (12.70), we obtain:

$$\phi_p = (3+\nu)\gamma\omega^2 h_0\left(\frac{r_e}{n}\right)^3 \frac{3+t[3+k(2+\nu)]}{[8+k(3+\nu)][3+k(2+\nu)]}\,t(1-t)^{k+1}, \tag{12.72}$$

which, written in terms of the variable ρ, becomes:

$$\phi_p = (3+\nu)\gamma\omega^2 h_0 \frac{r_e^3}{n^2} \frac{3+n\rho[3+k(2+\nu)]}{[8+k(3+\nu)][3+k(2+\nu)]}\,\rho(1-n\rho)^{k+1}. \tag{12.73}$$

Figure 12.7 shows distribution curves for particular integral ϕ_p as a function of ρ for three different pairs of parameters k and n, viz.: $k = 0.5$ and $n = 0.89$ (convergent disk having convex profile); $k = 1.5$ and $n = 0.52$ (convergent disk having concave profile); $k = 1$ and $n = 0.80$ (convergent disk having conical profile).

12.4.3 General Solution and Corresponding Stress and Strain State

The general solution of non-homogeneous differential equation (12.51) written as a function of t, and therefore of (12.69) – both governing the stress function ϕ and, by

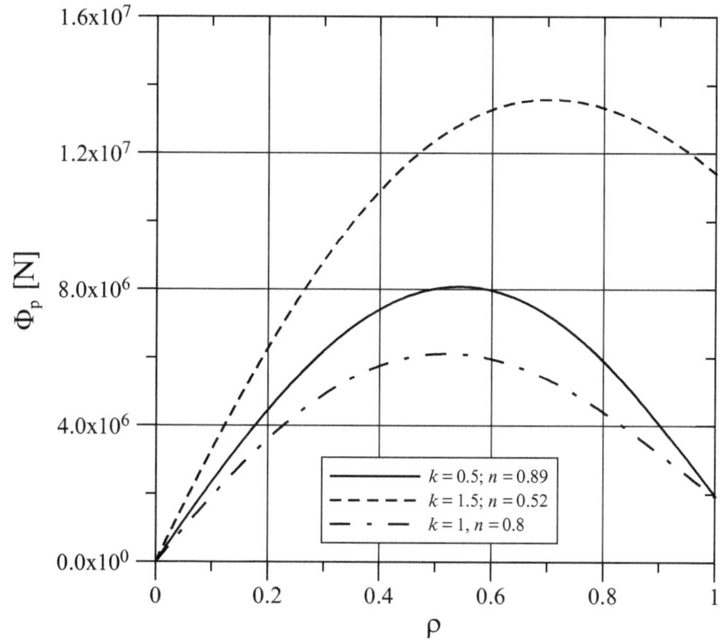

Fig. 12.7 Distribution curves of integrals $\phi_p = \phi_p(\rho)$ for three different convergent disk profiles

means of (12.30), the radial stress on disk – is obtained by adding the solution of the homogeneous equation (12.54), given by a linear combination of the two selected independent integrals ϕ_1 and ϕ_2, to the particular integral ϕ_p obtained earlier and given by relations (12.73). Thus, stress function ϕ as a function of ρ is given by:

$$\phi = C_1\phi_1 + C_2\phi_2 + \phi_p, \tag{12.74}$$

where C_1 and C_2 are integration constants that can be evaluated from the boundary conditions.

We can then find the stress state on the disk by substituting (12.74), along with its first derivative with respect to ρ, in (12.30), here given as a function of ρ.

$$\begin{aligned} \sigma_r &= \frac{\phi}{h r_e \rho} \\ \sigma_t &= \frac{1}{h r_e}\frac{d\phi}{d\rho} + \sigma_0 \rho^2, \end{aligned} \tag{12.75}$$

The strain state is then found by substituting (12.75) in (12.27) without temperature terms, and radial displacement $u = u(\rho)$ is obtained from the second geometrical relation (12.25) written for the elastic field.

12.4.4 Boundary Conditions and Integration Constants

12.4.4.1 Annular Disk

As it was indicated earlier, this is the most important disk type when prestressing is accomplished by overspeeding. In this case, the boundary conditions to be imposed for determining the constants appearing in (12.74) are:

$$\begin{cases} \sigma_r = 0 & \text{for } \rho = \beta \\ \sigma_r = 0 & \text{for } \rho = 1. \end{cases} \tag{12.76}$$

In other words, radial stress at the inner and outer radii must be zero, given that the disk is assumed to be only rotating and not subjected to surface forces on either the inner or outer radius.

For annular disks having no pointed tip at the outer radius r_e (cases of design interest), the three functions ϕ_1, ϕ_2 and ϕ_p show finite values both at the inner radius ($\rho = \beta = r_i/r_e$) and at the outer radius ($\rho = 1$). By imposing conditions (12.76), we obtain the following values of the two integration constants from the system of two equations with two unknown quantities obtained from (12.74) written for $\rho = \beta$ and for $\rho = 1$:

$$C_1 = \frac{(\phi_2)_1(\phi_p)_\beta - (\phi_p)_1(\phi_2)_\beta}{(\phi_1)_1(\phi_2)_\beta - (\phi_2)_1(\phi_1)_\beta} \text{ and } C_2 = -\frac{1}{(\phi_2)_\beta}\left[C_1(\phi_1)_\beta + (\phi_P)_\beta\right] \tag{12.77}$$

where the subscripts β and 1 refer to radius $\rho = \beta$ and $\rho = 1$.

12.4.4.2 Solid Disk

Although of less interest when prestressing by overspeeding is used, solid disks will also be analysed in order to demonstrate the generality of the method developed and proposed here. For this disk type, the boundary conditions to be imposed for determining the constants appearing in (12.74) are:

$$\begin{cases} \sigma_r = 0 & \text{for } \rho = 1 \\ u = 0 & \text{for } \rho = 0. \end{cases} \tag{12.78}$$

In this case, radial stress at outer radius must be zero and, given the assumption of axisymmetry as discussed in Sect. 2.1 of Chap. 2, radial displacement u on the rotational axis must also be zero.

By considering the second geometric relation (12.25) written for the elastic field and the second of Hooke's equation (12.27) without temperature terms, and imposing the second boundary condition (12.78), we obtain

$$\frac{1}{r_e h}\left\{C_1\left(\frac{d\phi_1}{d\rho}\right)_0+C_2\left(\frac{d\phi_2}{d\rho}\right)_0+\left(\frac{d\phi_p}{d\rho}\right)_0+\left(\sigma_0\rho^2\right)_0-(\nu h r_e\rho)_0\left[C_1(\phi_1)_0+C_2(\phi_2)_0+\left(\phi_p\right)_0\right]\right\}=0 \tag{12.79}$$

where the subscript 0 refers to the disk axis, i.e., for $\rho = 0$, and considering

$$\begin{gathered}(\phi_1)_0=0;\ (\phi_2)_0=\infty;\ \left(\phi_p\right)_0=0;\ \left(\frac{d\phi_1}{d\rho}\right)_0=\textit{finite value};\\ \left(\frac{d\phi_2}{d\rho}\right)_0=\infty;\ \left(\frac{d\phi_p}{d\rho}\right)_0=\textit{finite value}.\end{gathered} \tag{12.80}$$

From (12.79), we then find that we must necessarily have:

$$C_2 = 0. \tag{12.81}$$

Subsequently, imposing the first boundary condition (12.78) and considering (12.81) yields:

$$C_1 = -\left(\phi_p\right)_1/(\phi_1)_1. \tag{12.82}$$

12.4.5 Examples

For purposes of comparison, we will consider the same examples discussed in Sects. 6.6.3, 6.6.4, 7.6.3 and 7.6.4 for AISI 1060 steel disks ($\nu = 0.3$, $\sigma_y = 480$ MPa; $E = 204$ GPa; $\gamma = 7{,}800$ kg/m^3), having variable thickness according to relation (7.1) and rotating at constant angular velocity $\omega = 314$ rad/s. Six case studies will be presented: the first three for rotating annular disks and the others for the equivalent solid disks (equivalent in the sense that all these disks have the same profile geometry, except for the axial hole in the first three).

Once again, the results for the examples considered here will be compared both with those found using the hypergeometric functions introduced and described in Sect. 6.2.2 and 7.2.2, and with those obtained using finite element models with 4-node axially symmetric elements and three dofs per node. In general, the comparison indicates a perfect match between the FEM results and the analytical results obtained with the two different procedures whereby general equations governing the problem can be written in terms of radial displacement u and of stress function ϕ respectively and with the different hypergeometric functions introduced with these procedures. To avoid over-complication, the results from finite element analysis are not shown in the figures.

12.4.5.1 Rotating Annular Disks with $r_e < R$ and Having Constant Density

As has been stated several times, the majority of disks used in current machinery design feature a central hole of radius $r_i > 0$ and an outer radius $r_e < R$. For such disks, the range of variation for ρ is defined by: $\beta \leq \rho \leq 1$. Here, we will analyse three disks having converging profile and no load at the inner radius ($\sigma_{ri} = 0$) and outer radius ($\sigma_{re} = 0$), the first two with a ratio $h_e/h_0 = 1/3$, $r_i = 0.1$ m and $r_e = 0.8$ m. The first of these disks is convex ($k = 0.5$ and $n = 0.89$), the second concave ($k = 1.5$ and $n = 0.52$), while the third is conical with a ratio $h_e/h_0 = 1/5$, $r_i = 0.1$ m and $r_e = 0.8$ m. The first two disks thus coincide with those represented in Fig. 7.11.

Figure 12.8 a, b and c, which summarise the analytical results obtained with the procedure based on the stress function using the new hypergeometric functions described above, show curves for radial stress σ_r, hoop stress σ_t and radial displacement u versus ρ for these disk profiles. As a comparison of these results with those given in Fig. 7.11 indicates no differences, the two routes used to solve the differential equations governing the problem are perfectly equivalent.

12.4.5.2 Rotating Solid Disks with $r_e < R$ and Having Constant Density

Though they are not suitable for prestressing by overspeeding, solid disks are also widely used in current machinery design. Disks that are actually used feature an outer radius $r_e < R$; for such disks, the range for ρ is defined as: $0 \leq \rho \leq 1$. Here, the same three profiles studied previously are examined, but with no hole and no load at outer radius r_e ($\sigma_{re} = 0$), the first two with a ratio $h_e/h_0 = 1/3$ and $r_e = 1$ m. The first of these disks is convex ($k = 0.5$ and $n = 0.89$), the second concave ($k = 1.5$ and $n = 0.52$), while the third is conical with a ratio $h_e/h_0 = 1/5$ and $r_e = 0.8$ m. Here again, the first two disks thus coincide with those represented in Fig. 7.10.

Figure 12.8 d, e and f, which summarise the analytical results obtained with the procedure based on the stress function using the new hypergeometric functions described above, show curves for radial stress σ_r, hoop stress σ_t and radial displacement u versus ρ for these disk profiles. Once again, as a comparison of these results with those given in Fig. 7.10 indicates no differences, the two routes used to solve the differential equations governing the problem are perfectly equivalent.

12.5 Elastic-Plastic Analysis

We will now focus our attention on annular disks, given that – as has been mentioned several times – they show a higher plastic reserve than solid disks, so that prestressing them by overspeeding is undoubtedly of design interest.

Under the conditions occurring in the prestressing equipment by overspinning, at an angular velocity ω higher than the speed causing yielding onset at the inner radius (start of plasticization), but lower than the speed bringing the elastic-plastic

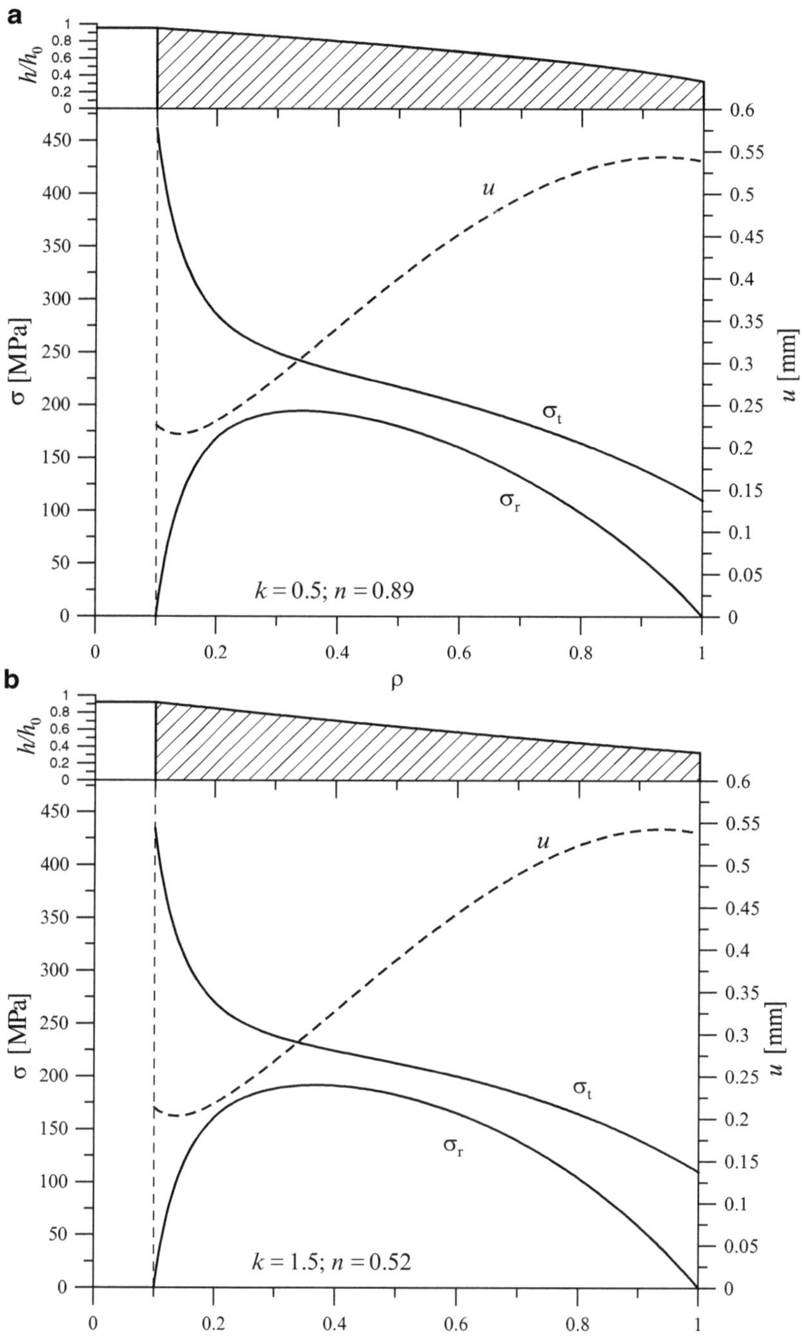

Fig. 12.8 (continued)

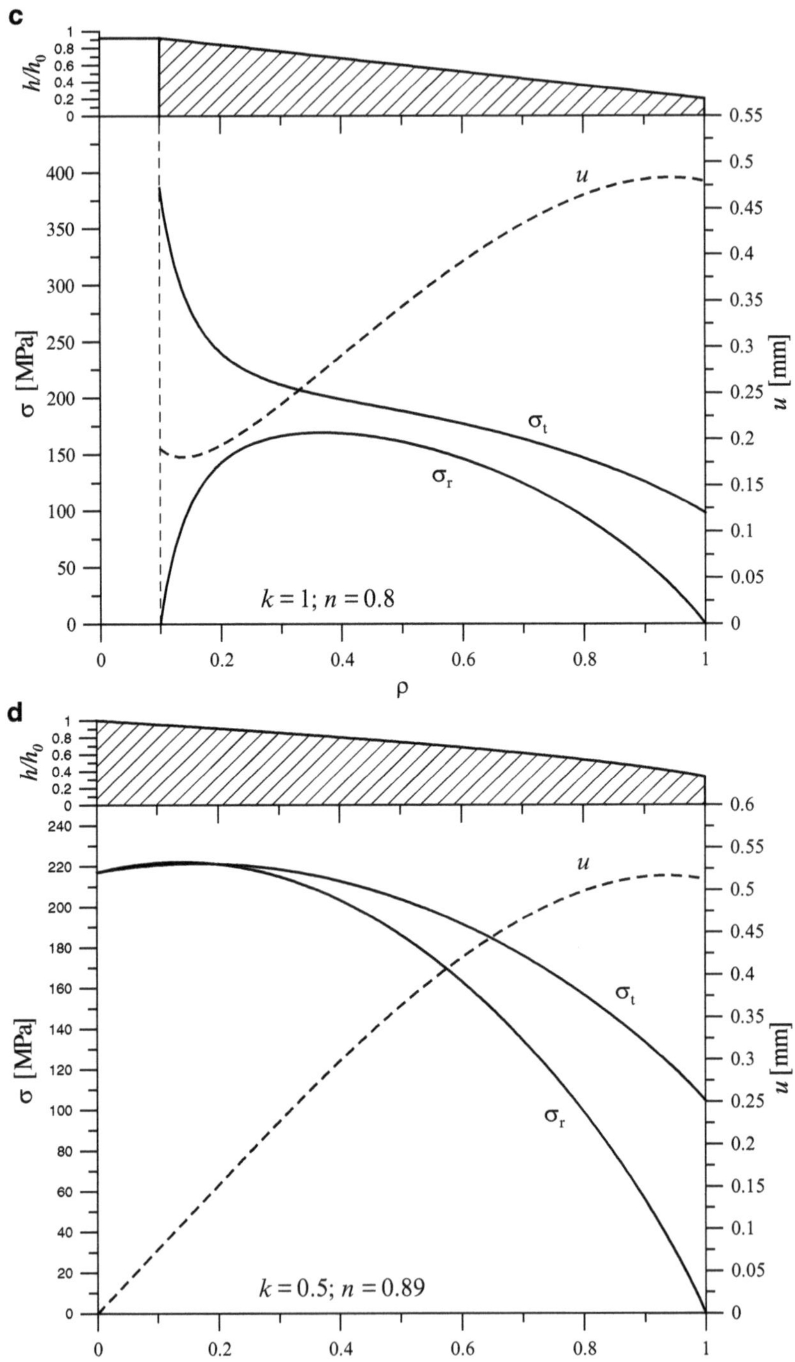

Fig. 12.8 (continued)

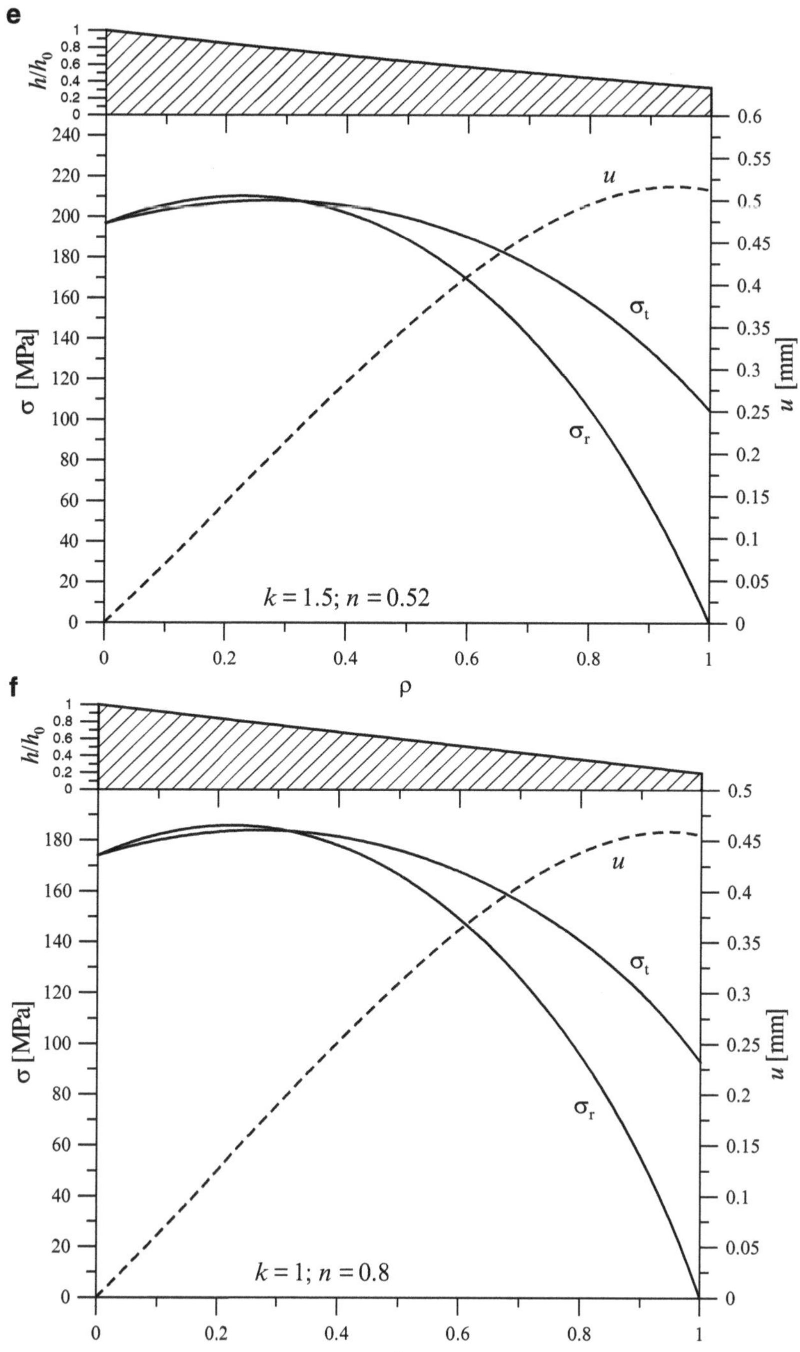

Fig. 12.8 Stress-distribution curves of σ_r, σ_t and displacement-distribution curve u versus ρ in converging disks having constant density and loaded in the elasticity field: (**a**) convex annular disk ($k = 0.5$ and $n = 0.89$); (**b**) concave annular disk ($k = 1.5$ and $n = 0.52$); (**c**) conical annular disk ($k = 1$ and $n = 0.8$); (**d**) convex solid disk ($k = 0.5$ and $n = 0.89$); (**e**) concave solid disk ($k = 1.5$ and $n = 0.52$); (**f**) conical solid disk ($k = 1$ and $n = 0.8$)

interface up to the outer radius (complete plasticization), the stress state in the plastic region of the non-linearly variable thickness disk – i.e., the region stressed beyond yielding – is given by (12.43), without γ' and T' terms, taking into account ratio h'/h given by equation (12.47) and polynomial (12.35) interpolating the experimental curve $\sigma_e = \sigma_e(\varepsilon_e^p)$, stopping at the third-degree term, with coefficients a_1, a_2 and a_3 equal to those shown in Fig. 12.4.

Beside the expected nonlinearities due to the behaviour of the material which is, as shown in Fig. 12.4, strain hardening (strain hardening is assumed to be isotropic), the differential equation (12.43) thus obtained (which is not given here for the sake of brevity) cannot be integrated in closed form, not even by resorting to series development. It can only be integrated numerically. Because this is an ordinary differential equation, it must meet specific continuity conditions (at the elastic-plastic interface between the inner plastic region and the outer elastic region, continuity must be assured for stress function ϕ, and therefore for radial stress σ_r, as well as for radial displacement u) in addition to specific boundary conditions (for annular disks, only rotating, radial stress must be zero at both the inner and outer radii). It can be integrated using, as an example, the MATLAB *bvp4c* solver, which employs a finite difference algorithm.

To reflect the design-related choices currently made for annular turbine disks prestressed by overspeeding, a predefined plastic radius is imposed in the following examples, and the program can be set to determine the corresponding angular velocity bringing the elastic-plastic interface up to this radius. The following continuity conditions are set at the elastic-plastic interface:

$$\begin{aligned} (\phi_{pl})_{r=r_p} &= (\phi_{el})_{r=r_p} \\ (\phi'_{pl})_{r=r_p} &= (\phi'_{el})_{r=r_p}, \end{aligned} \tag{12.83}$$

where subscripts *pl* and *el* refer to plastic and elastic fields governed by (12.43) and (12.44). The former relation entails continuity of radial stress σ_r, whereas the latter, combined with the former, entails continuity of hoop stress σ_t. The simultaneous continuity of σ_r and σ_t entails continuity of radial displacement u.

Once the plastic radius is defined, an iterative algorithm is set starting from an angular velocity range between ω_{min} and ω_{max}, specified at will, to identify the correct angular velocity correlated to the effective elastic-plastic junction radius. The iteration starts from a value ω_{min}, calculates equivalent stress σ_e at the imposed elastic-plastic junction radius, compares it to yield stress σ_y and if $\sigma_e < \sigma_y$ is obtained, the calculation is then reiterated starting from an angular velocity equal to $(\omega_{min} + \omega_{max})/2$. The iteration step is repeated using this new angular velocity value, and the new value of σ_e is calculated and compared to σ_y; if it is still $\sigma_e < \sigma_y$, then the angular velocity is changed again, this time as the arithmetic average of the last value used and ω_{max}. On the other hand, if the last calculation step yields $\sigma_e > \sigma_y$, then the angular velocity is changed to the arithmetic average of the last value used and ω_{min}. The calculation is reiterated to convergence, within a set variability range for σ_e between $(1 - \varepsilon')\sigma_y$ and $(1 + \varepsilon'')\sigma_y$, with ε' and ε'' arbitrary small values (as an example $\varepsilon' = 10^{-4}$ and $\varepsilon'' = 10^{-3}$); the choice of this range, clearly asymmetrical with respect to the yield stress σ_y, is justified by the need to avoid numerical errors at the zero value of curve $\sigma_e = \sigma_e(\varepsilon_e^p)$ in Fig. 12.4.

12.5.1 *Theoretical–Numerical Solution Case Studies and FEM Tests*

Because prestressing by overspeeding is only relevant in terms of design applications for annular disks – given that solid disks, although intrinsically stronger, do not have enough plastic reserve – the following calculation examples refer only to annular disks. To ensure non-restrictive comparison, the same three annular disks discussed above and subjected to elastic load are considered. Elastic results for these disks are shown in Fig. 12.8a, b and c.

We will thus consider three disks having converging profile and no load at the inner radius ($\sigma_{ri} = 0$) and outer radius ($\sigma_{re} = 0$), the first two with a ratio $h_e/h_0 = 1/3$, $r_i = 0.1$ m, $r_e = 1$ m and $h_0 = 0.1$ m. The first of these disks is convex ($k = 0.5$ and $n = 0.89$), the second concave ($k = 1.5$ and $n = 0.52$), while the third is conical with a ratio $h_e/h_0 = 1/5$, $r_i = 0.1$ m, $r_e = 0.8$ m and $h_0 = 0.1$ m. As a non-limiting example, the elastic-plastic interface will be located at radius $r = r_p = 0.2$ m ($\beta = 0.2$) for all three disks, which have a hole radius of $r_i = 0.1$ m.

Here again, all results pertaining to the three examples are compared to those obtained by means of FEA. The FE models also use 4-node axially symmetric elements with two dofs per node, but in this case the material is stressed beyond yielding. The figures for each example show both results from theoretical/numerical analysis and those obtained by means of FEA; the comparison indicates a perfect match between analytical-numerical results and from finite element analysis. It should be noted that several calculations were made using different elastic-plastic interfaces, and that analytical-numerical results were always confirmed by FE models, demonstrating the validity and complete reliability of the theoretical approach introduced and developed here. The basics of FEM analysis in the plastic field are outlined in Sect. 13.3.

The theoretical model with numerical integration is to be preferred because of the correctness and validity of the results it provides. Indeed, the theoretical-numerical model is of generally validity, and has clear advantages over the FE model, even when the latter is used in parametric form for increased flexibility, as it only requires that a differential equation be solved.

Figure 12.9a, b and c show stress-distribution curves for σ_r, σ_t, σ_e and displacement-distribution curves u as functions of ρ for the disk profiles described above. It should be noted that the analytical solution at the axis does not present the known stress field anomaly shown in the solution obtained by means of FE discretization. In these figures, numerical-analytical and FEA results are shown together.

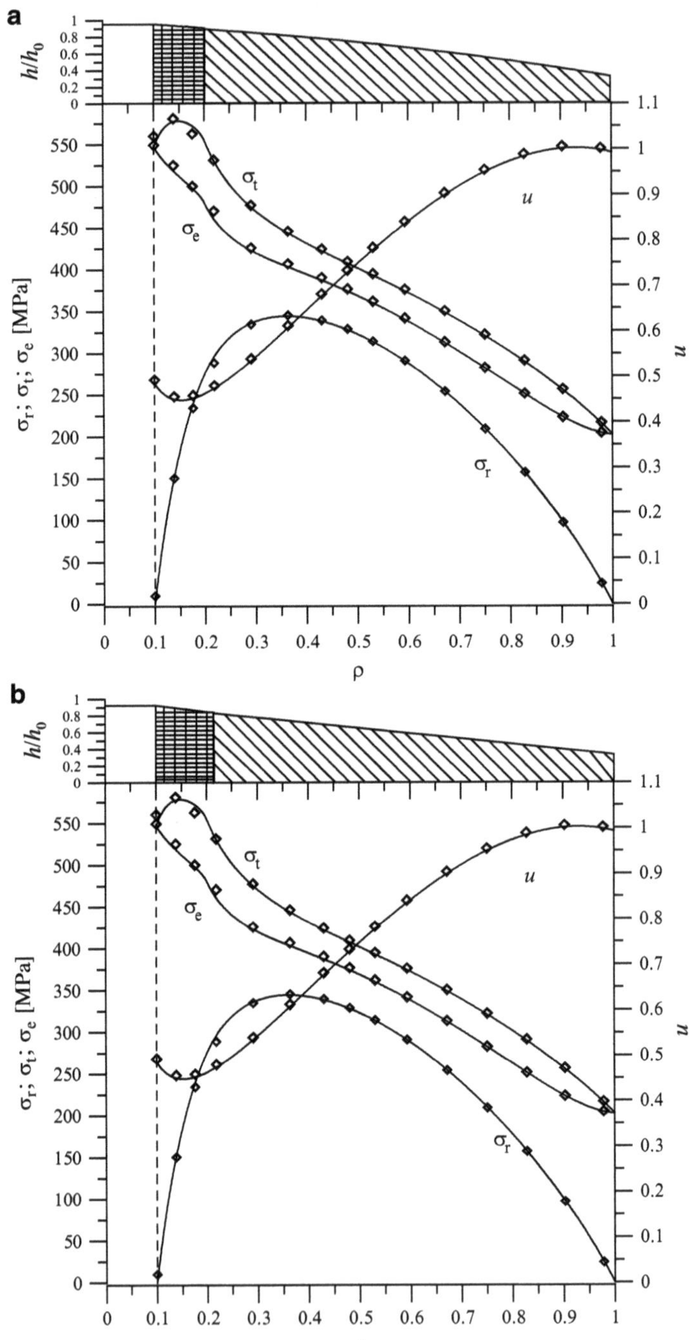

Fig. 12.9 (continued)

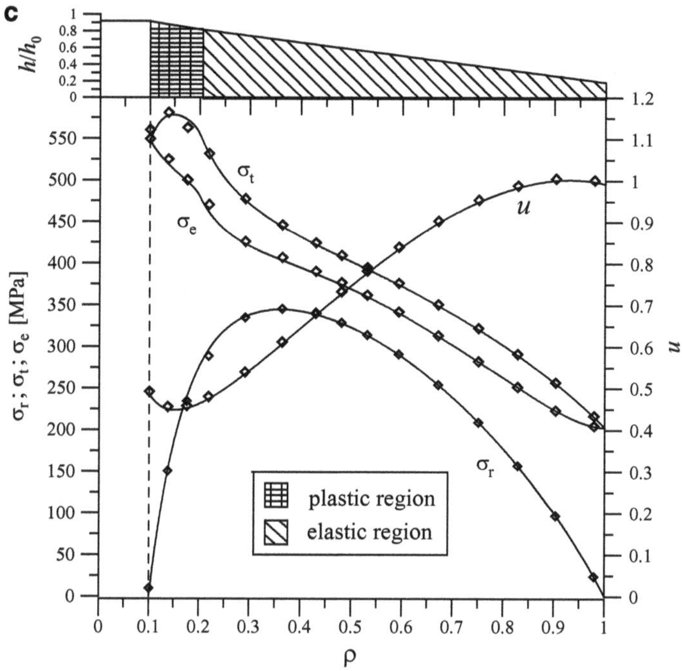

Fig. 12.9 Stress-distribution curves for σ_r, σ_t, σ_e and distribution curve u versus ρ in converging steel disks having constant density loaded beyond yielding at an angular velocity bringing the elastic-plastic interface up to radius $\rho = \rho_p = 0.2$: (**a**) convex annular disk ($k = 0.5$ and $n = 0.89$); (**b**) concave annular disk ($k = 1.5$ and $n = 0.52$); (**c**) conical annular disk ($k = 1$ and $n = 0.8$). Symbols refer to FEA results

12.5.2 Residual and Service Stresses

As was mentioned in the foregoing sections, the compressive stress state generated near the hole of an annular disk, prestressed beyond yielding when overspinning equipment speed is reduced to zero, is of considerable design interest. This precompression can be used to advantage at the design stage in order to reduce the maximum stresses, which are generally tensile, that result from service centrifugal loading. Assuming that when a disk that has been prestressed by overspeeding is decelerated, the material's behaviour is linear, or in other words, with reference to Fig. 12.3a, follows descending straight-line *AB* of the $\sigma_e = \sigma_e(\varepsilon_e)$ curve parallel to segment *OP*, and that angular acceleration is sufficiently low that its effects can be neglected, the residual stress state in the stationary disk can be calculated by subtracting the stresses due to centrifugal loading calculated with elastic relation (12.44) from the stresses in the plastic zone and the elastic zone expressed by (12.43) and (12.44), at the angular velocity that brought the elastic-plastic interface to the predetermined radius $r = r_p$.

By way of example, here we will consider only the convex annular disk whose elastic-plastic results are given in Fig. 12.9a. For this disk, the iterative calculation process indicates that the angular velocity ω to which the disk must be subjected during overspeeding in order to bring the elastic-plastic interface to the specified radius $r = r_p = 0.2$ m ($\rho = \rho_p = 0.2$) is approximately 423.1 rad/s.

Consequently, the radial, hoop and equivalent residual stress values as shown in Fig. 12.10 can be obtained by subtracting the results given by (12.44) without terms γ' and T' at the above angular velocity ($\omega = 423.1$ rad/s) from those shown in Fig. 12.9a. Stress distributions indicate that radial residual stress is compressive at all points, with a negative peak of slightly less than 50 MPa, located at an intermediate radius between the inner radius and the elastic-plastic interface, whereas hoop and equivalent residual stress values reach a negative peak of about 300 MPa at the inner radius.

As detailed in Sect. 12.6 below, such a compressive residual stress state may be used to reduce peak stress due to rotation in actual disk operating conditions. To demonstrate the effectiveness of this compressive prestressing, again by way of example, we will consider the same disk rotating in service at velocity $\omega = 400$ rad/s. As shown in Fig. 12.3a, a prestressed disk at a set angular velocity, when load is dropped to zero in order to bring about a compressive residual stress state, and then rotated at an angular velocity ω lower than the prestressing speed $\omega(\rho_p)$, will behave elastically following the ascending straight-line segment *BA* of the $\sigma = \sigma(\varepsilon)$ curve. The stress state on this disk under actual operating conditions will thus be the sum of residual stress and elastic stress at said operating speed.

Results for this specific case study are shown in Fig. 12.11, where dashed lines indicate radial, hoop and equivalent virtual elastic stress distributions at angular velocity $\omega = 400$ rad/s, whereas solid lines indicate the same stress values calculated as the superposition of residual stresses shown in Fig. 12.11 and the elastic stresses. The figure shows that the peak equivalent stress occurs next to the inner radius, and stays below yielding despite of virtual elastic stresses that locally exceed 750 MPa.

It is thus demonstrated that, for the prestressed disk in question (an annular disk with converging convex profile having: $r_e = 1$ m, $r_i = 0.1$ m and $k = 0.5$ and $n = 0.89$, prestressed at angular velocity $\omega = 423.12$ rad/s which brought the elastic-plastic interface up to radius $r = r_p = 0.2$ m), given the favourable residual stress state generated when the disk is decelerated from prestressing speed to a standstill as shown by the curves in Fig. 12.11, it is possible to reach an operating rotational speed $\omega = 400$ rad/s, i.e., over 25 % higher than the speed ($\omega \cong 320$ rad/s) when yielding just begins at the inner radius, without the maximum stress ever exceeding the yield stress value ($\sigma_y = 480$ MPa).

The curves shown in Fig. 12.11 are clearly continuous and provide a better interpretation of experimental findings than those obtained on the basis of the more restrictive simplifying assumptions used in Chap. 11. A number of researchers starting from Steele [68], who, however, worked with cylindrical solids under pressure, demonstrated experimentally that the two zone of the structure in the partially plastic state are not bounded by a well-defined cylindrical surface such as

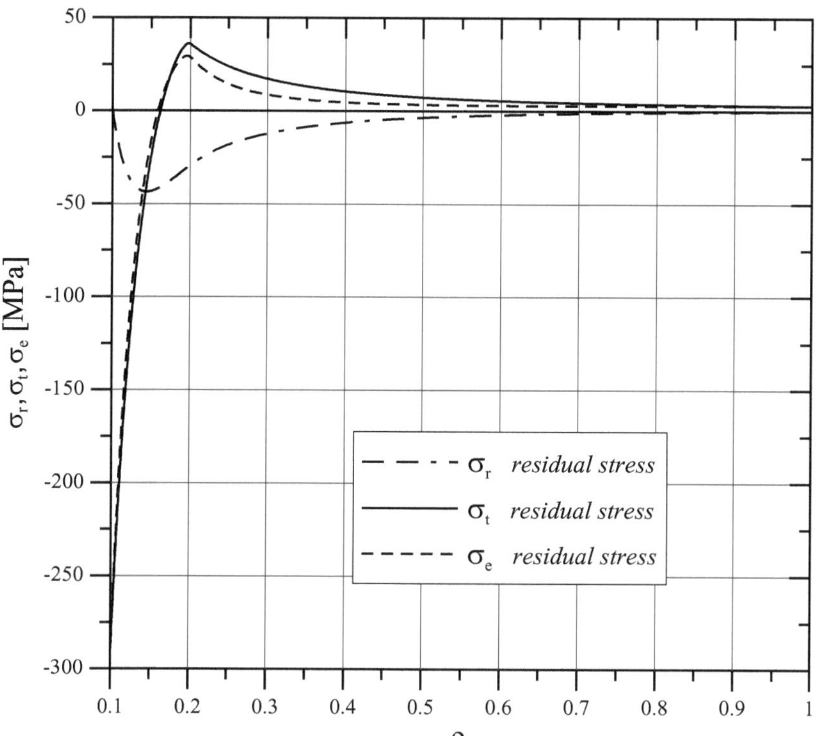

Fig. 12.10 Distribution curves of radial, hoop and equivalent residual stress σ_r, σ_t and σ_e in the convex profile disk shown in Fig. 12.9a, prestressed at velocity $\omega = 423.12$ rad/s, bringing the elastic-plastic interface up to radius $\rho = \rho_p = 0.2$

that at the elastic/plastic interface identified in Chap. 11, but by a transitional zone of annular cross section, where highly deformed elastic material is mixed with spiral or wedge-shaped layers of plastic material.

12.6 Design-Related Use of Plasticization and Limit-Design Factor

As was mentioned earlier, the stress state generated in the prestressed rotor is generally compressive precisely in those areas that will be subjected to tensile stresses in service. This favourable residual stress state can be used to advantage in order to reduce the maximum stresses due to rotation in actual service conditions. Precompression is generally produced by bringing the disk to an angular velocity which is higher than the velocity at the onset of plastic deformation. This load cycle must be included in the rotor manufacturing process, and must generally be followed by heat treatment at a temperature such that the stress state remains, and by the last finish machining operations. This type of prestressing is particularly appropriate

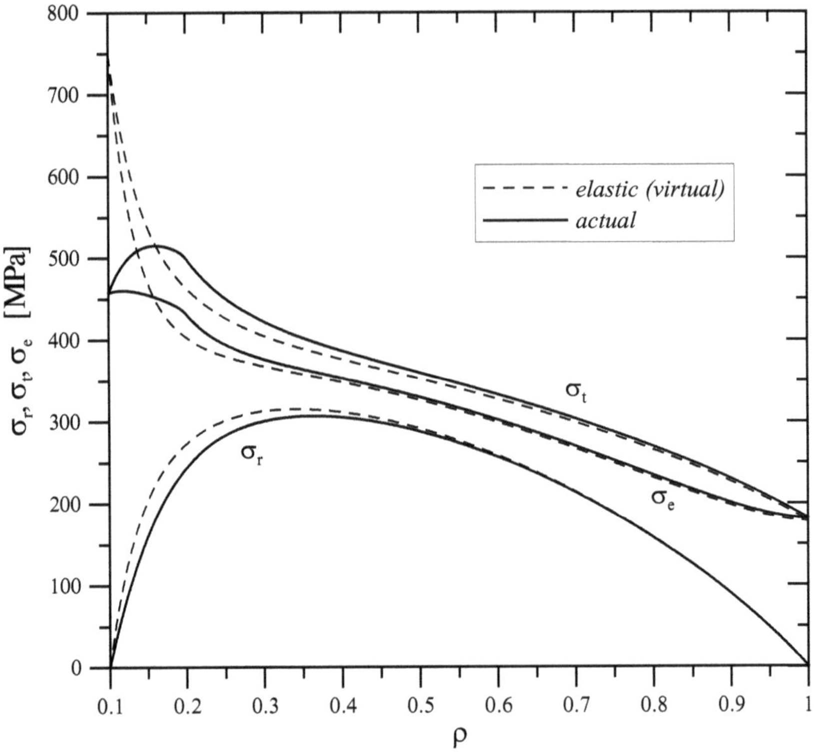

Fig. 12.11 Distribution curves of radial, hoop and equivalent stresses σ_r, σ_t and σ_e in a convex annular disk operating at velocity $\omega = 400$ rad/s, and prestressed at velocity $\omega = 423.12$ rad/s; actual stresses (*solid lines*) and virtual elastic stresses (*dashed lines*)

where high stress peaks are involved, as is the case of annular disks. Here, it is sufficient to carry out prestressing at a speed slightly above the elastic limit angular velocity, so that only a small zone around the hole enters the plastic state.

Figure 12.12 shows the qualitative behaviour of hoop stress σ_t under actual service conditions (curve *E*) in a constant-thickness annular disk that was brought to a speed above the elastic limit angular velocity at one stage of its manufacturing process. Curve *A* represents stress distribution σ_t at velocity $\omega(\rho_p)$ if the action were completely elastic; $\omega(\rho_p)$ is the velocity that brought the elastic-plastic interface to the desired radius ρ_p. Curve *B* represents stress distribution σ_t at the end of prestressing process by overspeeding at the same velocity $\omega(\rho_p)$. Curve *C* represents residual stress distribution σ_t corresponding to velocity ω (ρ_p); residual stresses are compressive at the inner radius and tensile at the outer radius, and are determined after the rotational velocity that brought the elastic-plastic interface to the desired radius ρ_p has dropped to zero by subtracting *A* from *B* at every value of the radius. Curve *D* represents elastic stress distribution σ_t corresponding to the actual operating speed $\omega > \omega_y$. Lastly, curve *E* represents the actual distribution of

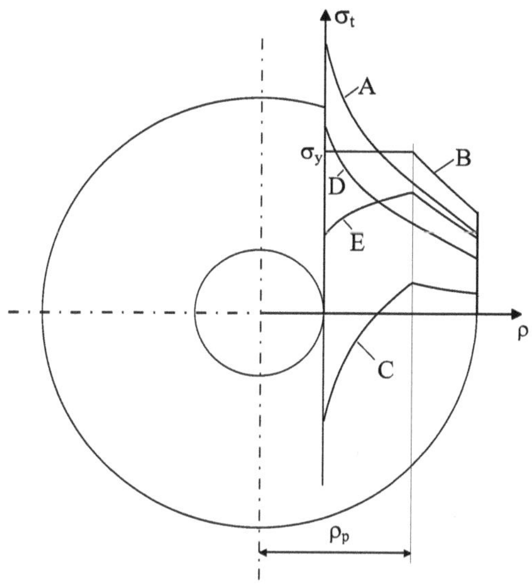

Fig. 12.12 Qualitative distributions of hoop stress in a constant-thickness annular disk overstressed by overspeeding at angular velocity ω (ρ_p): (*A*) stress σ_t of elastic range at velocity ω (ρ_p), (*B*) stress σ_t of elastic-plastic range at the same velocity, (*C*) residual stress σ_t obtained by subtracting (*A*) from (*B*), (*D*) stress σ_t of elastic range at actual operating speed $\omega > \omega_y$, (*E*) stress σ_t in actual operating conditions, as the sum (*C*) + (*D*)

hoop stress σ_t in service conditions, and is obtained by adding the residual stress state (curve *C*) and the elastic stress state at operating speed (curve *D*).

In solid disks, which as we have seen are much stronger than annular disks for any given conditions, prestressing is not advisable because the advantages thus achieved are too modest to outweigh the drawbacks of a complex overstressing process that adds significantly to the rotor's manufacturing process, and hence to its cost.

From the standpoint of the structural designer, it is advisable to consider the *limit-design factor L*, defined as the ratio of the loading for a completely plastic condition to the loading for a completely elastic condition with $\sigma_{max} = \sigma_y$. Here, the word loading is to be understood in the broadest sense, given that it may consist of a bending moment in bending of a beam, torque in twist of a circular bar, pressure in a vessel or circular cylindrical body, or rotational velocity in a rotor.

The general expression of the limit-design factor is:

$$L = \frac{\Lambda_p}{\Lambda_e} \tag{12.84}$$

where Λ_p and Λ_e are respectively the loading corresponding to the completely plastic condition and the loading corresponding to a completely elastic condition which, given the assumed $\sigma = \sigma(\varepsilon)$ curve, corresponds to the onset of yielding. In general, L assumes values between approximately 1.1 and 1.5 for rotors.

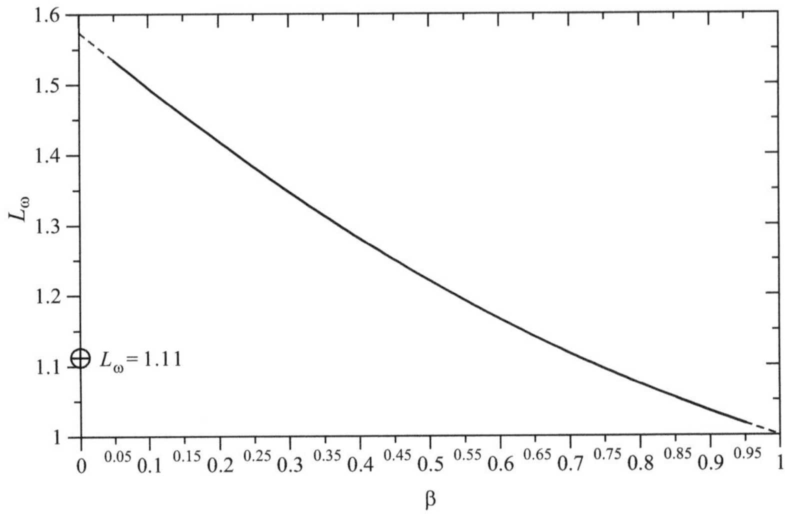

Fig. 12.13 Limit-design factor for a rotating annular disk with different values of ratio β, and for a solid rotating disk

For the annular disk consisting of linear elastic-perfectly plastic material discussed in Sect. 11.3, and bearing (11.6) and relation (11.19) in mind, we have:

$$L_\omega = \frac{\omega_p}{\omega_y} = \sqrt{\frac{3}{4} \cdot \frac{\left[(3+\nu) + \beta^2 \cdot (1-\nu)\right]}{\left(1+\beta+\beta^2\right)}} \tag{12.85}$$

It is clear that L_ω is independent of the material's yield point σ_y, but not of Poisson's ratio ν. In addition, it is heavily influenced by the value of β. Figure 12.13 shows the variation in the limit-design factor L_ω in a rotating annular steel disk ($\nu = 0.3$). As can be seen, the plastic reserve increases as β is reduced, and tends to zero for β tending unity. Values for L_ω are not very high: for example, β =1/6 corresponds to $L_\omega = 1.44$.

For the solid disk consisting of linear elastic-perfectly plastic material discussed in Sect. 11.4, the limit-design factor L_ω cannot be derived directly from (12.85) for $\beta = 0$ because of the singularity mentioned earlier. Starting from (11.23) and (11.29), we have the following for this solid disk:

$$L_\omega = \frac{\omega_p}{\omega_y} = \sqrt{\frac{3}{8} \cdot (3+\nu)} \tag{12.86}$$

The limit-design factor for a solid disk depends only on Poisson's ratio ν, and for a given material thus assumes a specific value (for steels, L_ω =1.11, as is also shown in Fig. 12.13).

Comparing a solid disk and an annular disk with a very small β value, both consisting of a linear elastic-perfectly plastic material, on the basis of the limit-design factor curve shown in Fig. 12.13, we could draw the mistaken conclusion that it is much more advantageous to use an annular disk because of its markedly higher plastic reserve. To make sounder design choices, it should first be borne in mind that, for any given outer radius (see Fig. 11.3), the elastic limit angular velocity for a solid disk is much higher than that for an annular disk (with $r_e = 0.2$ m, for example, a solid disk's elastic limit angular velocity will be around 40–50 % higher than that for an annular disk, and is a function of β). The fact that the effectiveness of prestressing differs for solid and annular disks should also be considered.

Nevertheless, as can be clearly seen from Fig. 12.13, the material's plasticity does not provide much margin of safety against uncontrolled increases in velocity. With prestressing, appreciable increases in velocity can be permitted only for annular disks with small β values. As is equally clear from Fig. 11.5, significant prestressing can be achieved by exceeding the elastic limit angular velocity by a small amount.

Chapter 13
Appendices

13.1 Appendix A: Rotating Bars, Paddles and Blades

Many practical applications involve bars or blades rotating around an axis, in which the body forces resulting from centrifugal loading generate internal stresses that must be kept under control. Take, for example, the solid bar shown in Fig. A.1, which has a circular cross section and rotates at angular velocity ω; the bar has the free edge, i.e. without end mass.

The elementary centrifugal force generated by a bar element of mass dm located between two cross sections of abscissa r and $r + dr$ relative to the rotational axis respectively is given by the relation:

$$dF = dm \cdot \omega^2 \cdot r = \gamma \cdot dV \cdot \omega^2 \cdot r = \gamma \cdot A \cdot \omega^2 \cdot r \cdot dr \tag{A.1}$$

where γ is the density of the bar material, dV is the elementary volume of the element in question, and A is the surface area of the cross section.

The resultant centrifugal force on a generic cross section at a distance r from the rotational axis and deriving from the body forces acting from r to L will thus be given by the relation:

$$F = \int_r^L \gamma \cdot A \cdot \omega^2 \cdot r \cdot dr = \gamma \cdot A \cdot \omega^2 \cdot \frac{(L^2 - r^2)}{2}. \tag{A.2}$$

It follows that the stress due to centrifugal load acting at radius r is given by:

$$\sigma_r = \frac{F}{A} = \gamma \cdot \omega^2 \cdot \frac{(L^2 - r^2)}{2}. \tag{A.3}$$

The maximum value of stress σ_r occurs for $r = 0$ and is $(\sigma_r)_{max} = \gamma \cdot \omega^2 \cdot L^2/2$.

V. Vullo and F. Vivio, *Rotors: Stress Analysis and Design*,
Mechanical Engineering Series, DOI 10.1007/978-88-470-2562-2_13,
© Springer-Verlag Italia 2013

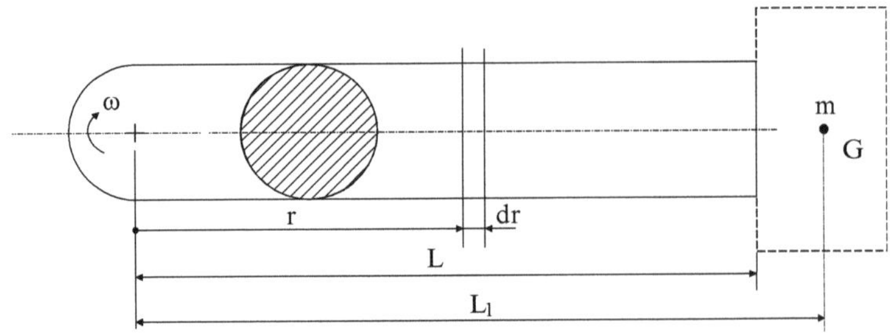

Fig. A.1 Solid cylindrical bar with circular cross section only rotating, with and without end mass

As we are dealing here with a uniaxial stress state, the total elongation of the portion of the rotating bar between generic axis r and the free end, where $r = L$, is given by the relation:

$$\ell = \int_r^L d\ell = \int_r^L \frac{\sigma_r}{E} \cdot dr = \int_r^L \gamma \cdot \omega^2 \cdot \frac{(L^2 - r^2)}{2E} \cdot dr. \tag{A.4}$$

Integrating this relation gives:

$$\ell = \frac{\gamma \cdot \omega^2}{2E} \cdot \left(\frac{2}{3} \cdot L^3 - L^2 \cdot r + \frac{r^3}{3} \right). \tag{A.5}$$

It follows from this relation that the total elongation of the bar occurs for $r = 0$ and is given by the relation:

$$\ell_t = \frac{\gamma \cdot \omega^2 \cdot L^3}{3 \cdot E}. \tag{A.6}$$

If a mass m whose center of gravity G is located at a distance L_1 from the rotational axis is connected to the end of the bar, as also shown in Fig. A.1, its effect must be taken into account. Consequently, relation (A.3) becomes:

$$\sigma_r = \gamma \cdot \omega^2 \cdot \frac{(L^2 - r^2)}{2} + \frac{m \cdot L_1 \cdot \omega^2}{A}. \tag{A.7}$$

We will now suppose that the solid bar has a conical profile and that its cross section is circular (Fig. A.2). Let A and A_0 be the surface areas of the cross sections at the generic distance r from the apex and at the rotational axis respectively. As these surface areas are clearly linked by the relation $A = A_0 \cdot r^2/L^2$, the elementary centrifugal force generated by a bar element of mass dm located between two cross sections of abscissa r and $r + dr$ relative to the apex respectively is given by the relation:

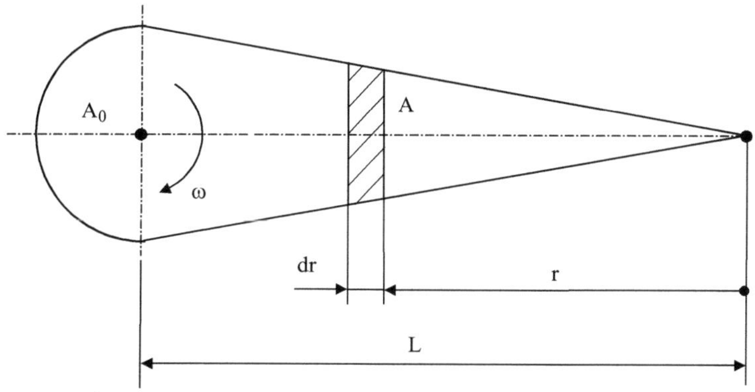

Fig. A.2 Solid conical bar with circular cross section only rotating

$$
\begin{aligned}
dF &= dm \cdot \omega^2 \cdot (L - r) = \gamma \cdot dV \cdot \omega^2 \cdot (L - r) \\
&= \gamma \cdot \omega^2 \cdot \left(A_0/L^2\right) \cdot (L - r) \cdot r^2 \cdot dr.
\end{aligned} \tag{A.8}
$$

The resultant centrifugal force on a generic section at a distance $(L - r)$ from the rotational axis and deriving from the body forces acting from the apex $(r = 0)$ to the cross section of abscissa r will thus be given by the relation:

$$
F = \frac{\gamma \cdot A_0 \cdot \omega^2}{L^2} \cdot \int_0^r (L - r) \cdot r^2 \cdot dr = \frac{\gamma \cdot A_0 \cdot \omega^2}{L^2} \cdot \left(\frac{L \cdot r^3}{3} - \frac{r^4}{4}\right). \tag{A.9}
$$

It follows that the stress due to centrifugal load acting at the generic abscissa r is given by:

$$
\sigma_r = \frac{F}{A} = \gamma \cdot \omega^2 \cdot \left(\frac{L \cdot r}{3} - \frac{r^2}{4}\right). \tag{A.10}
$$

This stress assumes its maximum value for $r = L$, where it is $(\sigma_r)_{max} = \gamma \cdot \omega^2 \cdot L^2/12$.

The total elongation of the portion of the rotating bar between generic abscissa r and the apex is given by the relation:

$$
\ell = \int_0^r d\ell = \int_0^r \frac{\sigma_r}{E} \cdot dr = \int_0^r \frac{\gamma \cdot \omega^2}{E} \cdot \left(\frac{L \cdot r}{3} - \frac{r^2}{4}\right) \cdot dr = \frac{\gamma \cdot \omega^2}{6 \cdot E} \left(L \cdot r^2 - \frac{r^3}{3}\right). \tag{A.11}
$$

It follows from this relation that the total elongation of the bar is with $r = L$ and is:

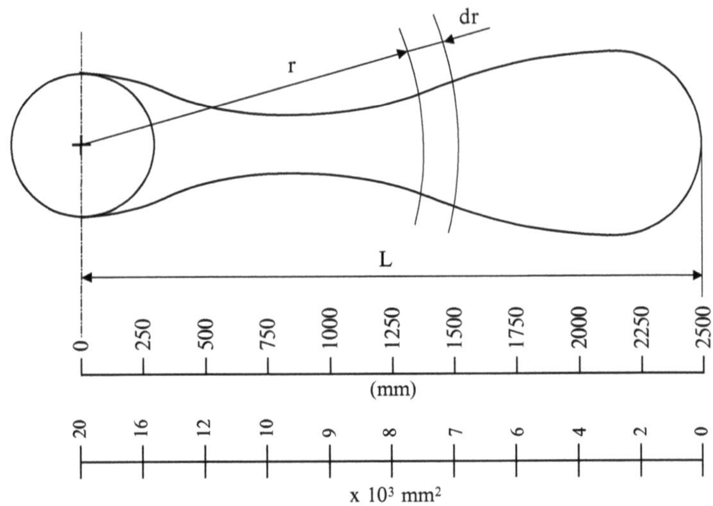

Fig. A.3 Rotating blade of given geometry

$$\ell_t = \frac{\gamma \cdot \omega^2 \cdot L^3}{12 \cdot E}. \tag{A.12}$$

If the rotating bar's profile cannot be expressed by a simple equation, it will not be possible to proceed analytically, as numerical methods or graphic procedures will be needed. The following example illustrates a procedure belonging to this latter family. We will consider the rotating blade shown in Fig. A.3, whose shape and the surface areas of the cross sections at predetermined distances r from the rotational axis are known. It will be assumed that the blade consists of aluminum and rotates at a velocity of 1,500 rpm.

The elementary centrifugal force generated by a blade element of mass dm located between two cross sections of abscissa r and $r + dr$ relative to the rotational axis respectively is given by the relation:

$$dF = \gamma \cdot A \cdot \omega^2 \cdot r \cdot dr. \tag{A.13}$$

where γ is the density of the material (for aluminum, $\gamma = 2{,}700\ \text{kg/m}^3$) and A is the surface area of the cross section at the generic distance r from the rotational axis. The resultant centrifugal force on a generic cross section of abscissa r deriving from the body forces acting from r to L will thus be given by the relation:

$$F = \gamma \cdot \omega^2 \cdot \int_r^L A \cdot r \cdot dr. \tag{A.14}$$

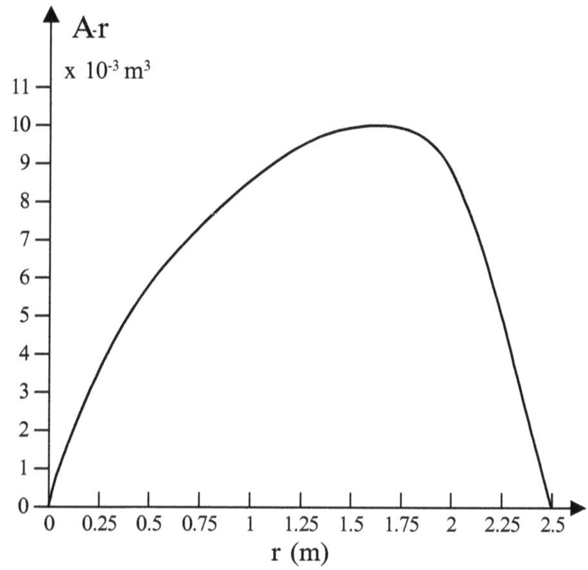

Fig. A.4 Product A·r versus radius r

When all magnitudes are expressed in consistent SI units (L and r in m, A in m^2, $\omega = 2\pi n/60 = 157$ rad/s), the foregoing relation becomes:

$$F = 6.655 \cdot 10^7 \cdot \int_r^{2.5} A \cdot r \cdot dr. \tag{A.15}$$

This relation is integrated graphically as follows. The product $A{\cdot}r$ is first plotted versus r, as shown in Fig. A.4. The centrifugal force acting on the generic cross section at distance r from the rotational axis is given by the area subtended by the portion of the curve between abscissa r and the end of the blade ($r = L$), multiplied by the factor $6.655{\cdot}10^7$. The stress σ_r due to centrifugal force in the generic cross section of area A is then obtained from the simple relation $\sigma_r = F/A$. Table A.1 summarizes the values of the magnitudes needed to calculated this stress.

The elongation of the portion of the blade between the generic cross section of abscissa r and the free end occurring as a result of centrifugal force is given by the relation:

$$\ell = \int_r^L \frac{\sigma_r}{E} \cdot dr = \frac{1}{E} \cdot \int_r^L \sigma_r \cdot dr. \tag{A.16}$$

The first step is to plot stress σ_r as a function of r as shown in Fig. A.5, after which this function is integrated graphically. Detailed calculation will be left to the

Table A.1 Numerical values of the magnitudes used to calculate radial stress σ_r

r (m)	$A \cdot r$ (10^{-3} m^3)	$\int_r^{2.5} A \cdot r \cdot dr.$ (10^{-3} m^4)	$F = 6.655 \cdot 10^7 \cdot \int_r^{2.5} A \cdot r \cdot dr.$ (MN)	$\sigma_r = \frac{F}{A}$ (MPa)
0	0	17.47	1.163	58
0.25	4.0	16.97	1.129	71
0.50	6.0	15.72	1.046	87
0.75	7.5	14.03	0.934	94
1.00	9.0	11.97	0.797	89
1.25	10.0	9.72	0.647	81
1.50	10.5	7.16	0.475	68
1.75	10.5	4.53	0.302	50
2.00	8.0	2.22	0.148	37
2.25	4.5	0.66	0.044	22
2.50	0	0	0	0

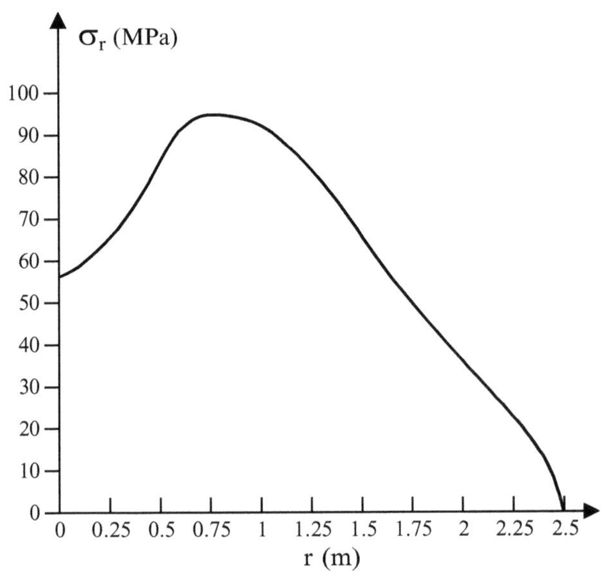

Fig. A.5 Stress σ_r versus radius r

reader. Noting that we have $E = 70$ GPa for aluminum, the resulting total elongation is thus:

$$\ell = \frac{1}{E} \cdot \int_0^{2.5} \sigma_r \cdot dr = 2.24 \cdot 10^{-3}\text{m}.$$

13.2 Appendix B: In-Depth Analysis of the Solution of the Hypergeometric Differential Equation

Both Honegger and Giovannozzi note that: (1) the independent integrals u_1 and u_2 are convergent within the interval $0 < t < 1$; (2) convergence decreases as t increases and, starting around $t = 1/2$, several terms of the series expansion are needed to calculate the function value with suitable accuracy; (3) with even higher t values, calculation becomes more intricate, especially in view of the calculation tools available at the time these authors wrote. Accordingly, both split the interval $0 < t < 1$ into two contiguous parts $0 < t \leq 1/2$ and $1/2 < t < 1$ bounded by $t = 1/2$, and use the integral u_1 and a linear combination of the two independent integrals u_1 and u_2 in the interval $0 < t \leq 1/2$, whereas in the interval $1/2 < t < 1$ they make use of other solutions of associated homogeneous equation (6.6) based on two different independent integrals expressed by means of power series in the variable $x = (1 - t)$.

By introducing this new variable, the differential equation (6.6) becomes:

$$\frac{d^2u}{dx^2} + \left(\frac{1}{x} + \frac{1}{x-1}\right) \cdot \frac{du}{dx} + \left[-(1-v) - \frac{1}{x-1}\right] \cdot \frac{u}{x \cdot (x-1)} = 0 \qquad \text{(B.1)}$$

This is still a hypergeometric differential equation, again featuring three singularity points ($x = 0$; $x = 1$; $x = \infty$), and corresponding to the general form:

$$\frac{d^2u}{dx^2} + \left(\frac{1-\alpha-\alpha'}{x} + \frac{1-\gamma-\gamma'}{x-1}\right) \cdot \frac{du}{dx} + \left[-\frac{\alpha \cdot \alpha'}{x} + \frac{\gamma \cdot \gamma'}{x-1} + \beta \cdot \beta'\right] \cdot \frac{u}{x \cdot (x-1)} = 0 \quad \text{(B.2)}$$

As this has the same structure as (6.16), applying the procedure described earlier to it allows us to conclude that its integrals are the same as those given by relations (6.20), after substituting factors α and α' with γ and γ'respectively, while β and β' remain unchanged. As a result of the variable change $x = (1 - t)$, the three singularity points of variable t ($t = 0$; $t = 1$; $t = \infty$) correspond respectively to as many singularity points of variable x ($x = 0$; $x = 1$; $x = \infty$). In other words, using Riemann's notation, the following can be specified:

$$P\begin{pmatrix} 0 & \infty & 1 & \\ \alpha & \beta & \gamma & t \\ \alpha' & \beta' & \gamma' & \end{pmatrix} = P\begin{pmatrix} 1 & \infty & 0 & \\ \alpha & \beta & \gamma & x \\ \alpha' & \beta' & \gamma' & \end{pmatrix} = P\begin{pmatrix} 0 & \infty & 1 & \\ \gamma & \beta & \alpha & x \\ \gamma' & \beta' & \alpha' & \end{pmatrix}. \qquad \text{(B.3)}$$

The first independent integral $\bar{u}_1$ in (6.6) will thus be expressed in the following form:

$$\bar{u}_1=(1-x)\cdot F(\gamma'+\beta+\alpha,\gamma'+\beta'+\alpha,1-\gamma'-\gamma,x)=(1-x)\cdot F(a,b,c,x)=$$
$$=(1-x)\cdot F\left(\frac{3}{2}+\xi,\frac{3}{2}-\xi,1,x\right)=$$
$$=(1-x)\cdot\left[1+\frac{\left(\frac{3}{2}+\xi\right)\cdot\left(\frac{3}{2}-\xi\right)}{1!\cdot 1}\cdot x+\frac{\left(\frac{3}{2}+\xi\right)\cdot\left(\frac{3}{2}+\xi+1\right)\cdot\left(\frac{3}{2}-\xi\right)\cdot\left(\frac{3}{2}-\xi+1\right)}{2!\cdot 1\cdot 2}\cdot x^2+\ldots\right] \tag{B.4}$$

which, reintroducing variable t, becomes:

$$\bar{u}_1= \quad t\cdot F\left(\frac{3}{2}+\xi,\frac{3}{2}-\xi,1,[1-t]\right) \tag{B.5}$$

Here again, the second independent integral $\bar{u}_2$ must be calculated with a domain change of the independent variable, as we are again dealing with factors in the hypergeometric series that assume infinite value. With a procedure similar to that used in Sect. 6.2.2 to calculate u_2, it can be demonstrated that the following expression can be used for $\bar{u}_2$:

$$\bar{u}_2=\bar{u}_1\cdot \ln x+(1-x)\cdot\sum_{i=1}^{\infty}\bar{C}_i\cdot x^i=\bar{u}_1\cdot\ln(1-t)+t\cdot\sum_{i=1}^{\infty}\bar{C}_i\cdot(1-t)^i \tag{B.6}$$

where

$$\bar{C}_i=\frac{\prod_{m=1}^{i}\left(\frac{3}{2}+\xi+m-1\right)\cdot\left(\frac{3}{2}-\xi+m-1\right)}{i!\cdot i!}\cdot\sum_{m=0}^{i-1}\left(\frac{1}{\frac{3}{2}+\xi+m}+\frac{1}{\frac{3}{2}-\xi+m}-\frac{2}{1+m}\right). \tag{B.7}$$

With the procedure used by Honegger and by Giovannozzi, two pairs of independent integrals are found for the homogeneous differential equation 6.6, the first (u_1 and u_2) being valid within the interval $0\le t\le 1/2$ and the second ($\bar{u}_1$ and $\bar{u}_2$) valid within the interval $1/2<t\le 1$. However, it can readily be seen that at point $t=1/2$ separating the two contiguous parts of the interval, the domain of the internal integrals u_1 and u_2 does not match the domain of the two external integrals $\bar{u}_1$ and $\bar{u}_2$. Continuity of integral u_1 beyond $t=1/2$ up to $t=1$ can be obtained by means of a linear combination of $\bar{u}_1$and $\bar{u}_2$, given that any linear combination of two independent integrals is sufficient solution to satisfy the differential equation in question.

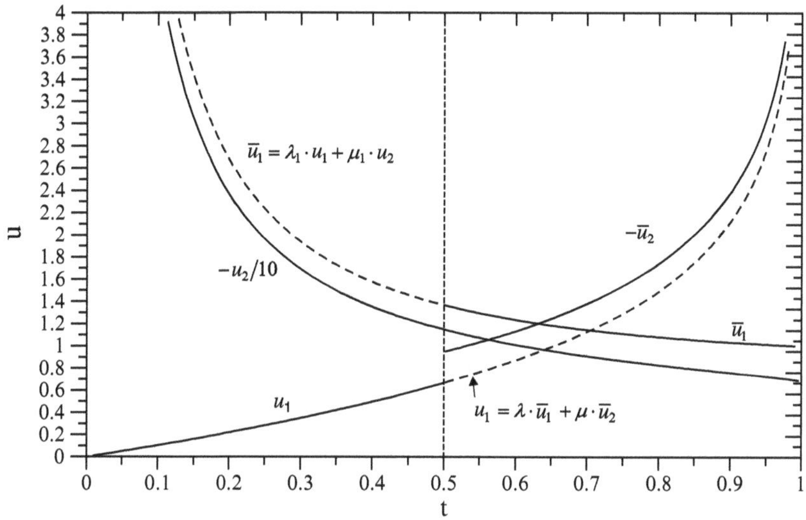

Fig. B.1 Functions u_1 and $\bar{u}_1$ used by Honegger and Giovannozzi for conical disks, and functions u_2 and $\bar{u}_2$

In this connection, Honnegger and Giovannozzi define two constants λ and μ, so that, for $t = 1/2$, the continuity of the function and of its first derivative are satisfied, in order to have:

$$
\begin{aligned}
u_1 &= \lambda \cdot \bar{u}_1 + \mu \cdot \bar{u}_2 \\
\frac{du_1}{dt} &= \lambda \cdot \frac{d\bar{u}_1}{dt} + \mu \cdot \frac{d\bar{u}_2}{dt}.
\end{aligned}
\tag{B.8}
$$

A relation is thus obtained – the first of relations (B.8) – whereby u_1 can be calculated beyond $t = 1/2$, i.e., within the interval of independent variable $1/2 < t \leq 1$. Similarly, and again following the procedure outlined by Honegger and Giovannozzi, a function $\bar{u}_1 = \lambda_1 \cdot u_1 + \mu_1 \cdot u_2$ expressed as a linear combination of u_1 and u_2 can be found by using two new constants λ_1 and μ_1. This function is valid within the interval $0 \leq t \leq 1/2$.

To summarize, both Honegger and Giovannozzi split the interval $0 \leq t \leq 1$ into two component partial intervals $0 \leq t \leq 1/2$ and $1/2 < t \leq 1$ and, as shown in Fig. B.1, use integral u_1 obtained from relation (6.22) within the interval $0 \leq t \leq 1/2$, and linear combination $u_1 = \lambda \cdot \bar{u}_1 + \mu \cdot \bar{u}_2$ obtained from the first relation (B.8) within the interval $1/2 < t \leq 1$. In addition, they use integral $\bar{u}_1$ obtained from relation (B.5) within the interval $1/2 \leq t \leq 1$, and linear combination $\bar{u}_1 = \lambda_1 \cdot u_1 + \mu_1 \cdot u_2$(equivalent to the first relation (B.8)), with λ_1 and μ_1 as constants to be found by imposing the continuity of the function and its first derivative at point $t = 1/2$, within the interval $0 \leq t < 1/2$. Both researchers express the solution of the associated homogeneous equation (6.6) as a linear combination of the two independent integrals, viz.:

$$u = C_1 \cdot u_1 + C_2 \cdot \bar{u}_1, \tag{B.9}$$

where C_1 and C_2 are integration constants to be determined by imposing boundary conditions, whereas $u_1$and $\bar{u}_1$feature different expressions depending on whether the interval $0 \leq t \leq 1/2$ or the interval $1/2 < t \leq 1$ applies.

Lastly, it should be noted that, as shown in Fig. 6.1b, e, f, g the function $h = h_0 \cdot (1 - t)$ can also describe the geometry of diverging conical disks. In the general case where the apex V does not converge on the axis (Fig. 6.1e, f, g) and $t > 1$, it is advisable to introduce the new variable $y = 1/t$, as does Giovannozzi, to solve (6.6) analytically. By replacing this variable in (6.6), a hypergeometric differential equation is found which is formally similar to (6.16) and (B.1). By following the same procedure used for these two equations, we again reach the conclusion that the integrals of the hypergeometric differential equation thus obtained are the same as those given by relations (6.20), after substituting factors α and $\alpha'$with β and β', while γ and γ' remain unchanged. As a result of the variable change $y = 1/t$, the three singularity points of variable t ($t = 0$; $t = 1$; $t = \infty$) correspond respectively to as many singularity points of variable y ($y = \infty$; $y = 1$; $y = 0$). In this case, using Riemann's notation, the following can be specified:

$$P\begin{pmatrix} 0 & \infty & 1 & \\ \alpha & \beta & \gamma & t \\ \alpha' & \beta' & \gamma' & \end{pmatrix} = P\begin{pmatrix} \infty & 0 & 1 & \\ \alpha & \beta & \gamma & y \\ \alpha' & \beta' & \gamma' & \end{pmatrix} = P\begin{pmatrix} 0 & \infty & 1 & \\ \beta & \alpha & \gamma & y \\ \beta' & \alpha' & \gamma' & \end{pmatrix}. \tag{B.10}$$

We will not discuss the further developments (which are entirely similar to those that have been described) or the procedures involved in introducing new variables related to variable t in order to improve calculation of hypergeometrical series, as does Giovannozzi, in specific sub-intervals of interval $1 < t < \infty$.

Where the conical disc has lateral faces converging on the axis of rotation (see Fig. 4.1a), it is in any case preferable to calculate stress and strain states by means of a closed form formulation based on Stodola's hyperbolic profile disk.

13.3 Appendix C: The Finite Element Method for Elastic-Plastic Problems

Many structural problems involve nonlinearities, which may be both geometric and in the material's behavior. The latter include cases where, as a result of applied loads, stress and strain states are generated in the material that exceed the non-linear elastic limits, so that the material is stressed beyond yielding. With the finite element method, all nonlinear problems, regardless of the nature of the nonlinearities, are generally solved by reducing them to a sequence of linear steps.

The procedure thus involves writing the equilibrium equations $\{Q\} = [K]\{\delta\}$ in the following incremental form:

$$\{\Delta Q\} = [K]\{\Delta\delta\} \tag{C.1}$$

where $\{\Delta Q\}$ is the total or equivalent nodal force increment matrix, and the stiffness matrix $[K]$ is a function of displacements $\{\delta\}$, as the problem is nonlinear. In the following steps, the current vector $\{\delta\}$ is the sum of the previous vectors $\{\Delta\delta\}$, while the current matrix $[K]$, called the *tangent stiffness matrix*, is used to calculate the next step $\{\Delta\delta\}$. The vector $\{\delta\}$ and the matrix $[K]$ are then updated, and we are ready to perform another step. With this process, the true stress-true strain curve is approximated by means of a succession of straight segments.

As we know from *continuum mechanics*, if the stress-strain relationships are linear or nonlinear but still elastic, there is a one-to-one correspondence between stress and strain. By contrast, if there are plastic deformations, this correspondence will no longer be one-to-one, as any given stress state can be produced by many different strain paths. As the focus here is on plasticity, we will ignore the effects of other nonlinearities such as the geometric nonlinearities resulting from large deformations. The algorithms used to solve these problems are in any case entirely general, and apply independently of the nature of the nonlinearities exhibited.

Solving plastic problems with the finite element method is based on the Levy-Mises plasticity theory, which is an *incremental theory* or *plastic flow theory*, as it relates stress increments to strain increments. According to this theory, the increment of equivalent or effective plastic strain $d\varepsilon_e^p$ is defined by the contributions of the single, separate increments of plastic strain and is expressed with reference to a three-dimensional rectangular system O(x, y, z) of non-principal axes in the form:

$$d\varepsilon_e^p = \frac{\sqrt{2}}{3}\left\{\left(d\varepsilon_x^p - d\varepsilon_y^p\right)^2 + \left(d\varepsilon_y^p - d\varepsilon_z^p\right)^2 + \left(d\varepsilon_z^p - d\varepsilon_x^p\right)^2 + \frac{3}{2}\left[\left(d\gamma_{xy}^p\right)^2 + \left(d\gamma_{yz}^p\right)^2 + \left(d\gamma_{zx}^p\right)^2\right]\right\}^{1/2}, \tag{C.2}$$

where index p, in addition to the other subscripts of known meaning, denotes the plastic contribution. In accordance with the Von Mises criterion for triaxial stress, and again with reference to the system of non-principal axes, yielding begins when the equivalent or effective stress σ_e reaches and exceeds a specific limit value; σ_e is given by the relation:

$$\sigma_e = \frac{\sqrt{2}}{2}\left[\left(\sigma_x - \sigma_y\right)^2 + \left(\sigma_y - \sigma_z\right)^2 + \left(\sigma_z - \sigma_x\right)^2 + 6\left[\tau_{xy}^2 + \tau_{yz}^2 + \tau_{zx}^2\right]\right]^{1/2}. \tag{C.3}$$

With reference to the uniaxial tensile test, beyond yielding, or in other words in the plastic range where Poisson's ratio v is 0.5, we have:

$$\sigma_e = \sigma_x \text{ and } d\varepsilon_e^p = d\varepsilon_x^p \tag{C.4}$$

In the $\sigma_e = \sigma_e(\varepsilon_e)$ curve shown in Fig. 12.3a, P represents the onset of yielding ($\sigma = \sigma_s$); for $\sigma = \sigma_A$, with $\sigma_A > \sigma_s$, the total equivalent strain ε_e is the sum of elastic strain ε_e^e, which can be completely recovered when load is removed, and plastic strain ε_e^p, i.e., $\varepsilon_e = \varepsilon_e^e + \varepsilon_e^p$. The material hardens as a result of plastic strain ε_e^p. If the specimen is loaded again, further yielding will not take place until σ_e exceeds σ_A. It is assumed that yielding will occur for $\sigma_e > \sigma_A$ regardless of the type of stress considered (i.e., tensile, compressive or multiaxial), i.e., that the material anisotropy effects generated by plastic deformation – including the Bauschinger effect described earlier – can be neglected. In other words, the isotropic strain hardening criterion is considered to apply.

As we saw in Sect. 12.3, the slope H' at any point of the $\sigma_e = \sigma_e(\varepsilon_e^p)$ curve (Fig. 12.3b) is given by relation (12.33). The increments of equivalent stresses, of total strains and of plastic strains are given by the following relations, which all involve 6×1 vectors:

$$\begin{aligned}\{d\sigma_e\} &= \{d\sigma_x, d\sigma_y, \ldots, d\tau_{zx}\}\\ \{d\varepsilon_e\} &= \{d\varepsilon_x, d\varepsilon_y, \ldots, d\gamma_{zx}\}\\ \{d\varepsilon_e^p\} &= \left\{d\varepsilon_x^p, d\varepsilon_y^p, \ldots, d\gamma_z^p\right\}.\end{aligned} \tag{C.5}$$

The deviatoric stresses s_x, s_y, . . ., are defined by the relations:

$$\begin{aligned}&s_x = \sigma_x - \sigma_m; \qquad &&s_y = \sigma_y - \sigma_m; \qquad &&s_z = \sigma_z - \sigma_m;\\ &s_{xy} = \tau_{xy}; &&s_{yz} = \tau_{yz}; &&s_{zx} = \tau_{zx};\end{aligned} \tag{C.6}$$

where $\sigma_m = (\sigma_x + \sigma_y + \sigma_z)/3$ is the average stress. By differentiating relation (C.3), we then obtain:

$$d\sigma_e = \{R\}^T\{d\sigma\} \quad \text{where} \quad \{R\} = \frac{3}{\sigma_e}\left\{\frac{s_x}{2} \quad \frac{s_y}{2} \quad \frac{s_z}{2} \quad s_{xy} \quad s_{yz} \quad s_{zx}\right\} \tag{C.7}$$

It should be noted here that $\sigma_e^2 = 9\tau_{oct}^2/2 = 3J_2$, where τ_{oct} is the octahedral shear stress and J_2 is the second invariant of the deviatoric stresses. The theory used here is also called the J_2 *flow theory*. It should also be borne in mind that the flow rule associated with the Von Mises yield criterion is represented by the Prandtl-Reuss relation, which states that:

$$d\varepsilon^p = \{R\}d\varepsilon_e^p. \tag{C.8}$$

This means that there will be increments of plastic strain $\{d\varepsilon^p\}$ when an increment of equivalent plastic strain $d\varepsilon_e^p$ occurs as the result of a stress state $\{R\}$. The corresponding stress increment $\{d\sigma\}$ can be written as a function of the increments of elastic strain $\{d\varepsilon^e\}$ in the following shape:

$$\{d\sigma\} = [E]\{d\varepsilon^e\} \text{ or } \{d\sigma\} = [E]\Big(\{d\varepsilon^*\} - \{d\varepsilon^p\}\Big) \tag{C.9}$$

with

$$\begin{aligned}\{d\varepsilon^e\} &= \{d\varepsilon\} - \{d\varepsilon^p\} - \{d\varepsilon^T\} - \{d\varepsilon^C\} \text{ and}\\ \{d\varepsilon^*\} &= \{d\varepsilon\} - \{d\varepsilon^T\} - \{d\varepsilon^C\}\end{aligned} \tag{C.10}$$

where $[E]$ is the conventional matrix of elastic constants, while the five strain vectors appearing in the first relation (C.10) represent the increments of elastic, total, plastic, thermal (or initial) and creep strains respectively. Obviously, if there is no contribution from thermal loading or creep, we must put $\{d\varepsilon^T\} = 0$ and $\{d\varepsilon^C\} = 0$ in this relation.

With the finite element method, we first determine the vector $\{d\varepsilon\}$. Relations (C.8) and (C.9) can be used to find the vector $\{d\sigma\}$ only after $d\varepsilon_e^p$ has been calculated. To find $d\varepsilon_e^p$, relation (C.8) is substituted in (C.9) and both members of the relation thus found are premultiplied by $\{R\}^T$. We then substitute $\{R\}^T\{d\sigma\} = H'\, d\varepsilon_e^p$, as obtained from relations (C.4), (12.34) and (C.7), and obtain:

$$d\varepsilon_e^p = [W]\{d\varepsilon^*\} \text{ with } [W] = \frac{\{R\}^T[E]}{H' + \{R\}^T[E]\{R\}} \tag{C.11}$$

Substituting relation (C.11) in (C.8) and the result thus found in relation (C.9) gives the following incremental stress-strain relation, similar to the elastic relation $\{\sigma\} = [E]\,\{\varepsilon\}$, but applying to elasto-plastic behavior:

$$\{d\sigma\} = [E^*]\Big(\{d\varepsilon\} - \{d\varepsilon^T\} - \{d\varepsilon^C\}\Big) = [E^*]\{d\varepsilon^*\} \tag{C.12}$$

where

$$[E^*] = [E] - [E]\{R\}[W] \tag{C.13}$$

Matrix $[E^*]$ is symmetric and also applies for elastic-perfectly plastic materials, for which $E_T = H' = 0$ (Fig. 12.3).

In linear elastic problems, stresses and strains depend on load, but not on how the load state is reached: in other words, the sequence of loads is not important. In elastic-plastic problems, on the other hand, the results depend on the loading sequence. Loading is considered proportional when the stresses at any point of the continuum in question maintain the same ratio to one another throughout loading.

The assumption of isotropic strain hardening holds true for proportional loads. For non-proportional loads, as for loads that change sign, this assumption is less valid, though it is still a sufficiently satisfactory approximation for many problems involved in practical applications. Other strain hardening criteria lead to different matrixes $[E^*]$; in this connection, it should be noted that a number of models have attempted to unify and generalize the strain hardening criteria, even in cases where softening phenomena are also involved.

It should also be borne in mind that plasticity is independent of time. The equations involved are often written in terms of rate, with the sole purpose of highlighting the nature of plastic flow. Thus, for example, (C.8) becomes $\{\dot{\varepsilon}^p\} = \{R\}\{\dot{\varepsilon}^p_e\}$. This, however, is a topic that goes beyond our scope here.

References

1. Abdul-Mihsein, M.J., Bakr, A.A., Parker, A.P.: Stresses in axisymmetric rotating bodies determined by the boundary integral equation method. J. Strain. Anal. Eng. **20**(2), 79–86 (1985)
2. Abramowitz, M., Stegun, I.H.: Handbook of Mathematical Functions. Dover Publications, Mineola (1972)
3. Belloni, G., Lo Conte, A.: Costruzione di Macchine: Resistenza dei Materiali e Sicurezza. Hoepli, Milano (2002)
4. Banerjee, P.K., Butterfield, R.: Boundary Element Methods in Engineering Science. McGraw-Hill, London (1981)
5. Biezeno, C.B., Grammel, R.: Technische Dynamik. Springer, Berlin (1953)
6. Botto, P.: Calcolo delle Sollecitazioni Centrifughe in Dischi Conici e a Spessore Costante con Peso Specifico Variabile Lungo il Raggio. L'Aerotecnica **33**(5), 159–169 (1953)
7. Burr, A.H.: Mechanical Analysis and Design. Elsevier, New York (1982)
8. Capetti, A.: Motori termici. UTET, Torino (1964)
9. Coates, W.M.: The state of stress in thin-walled pressure vessels. Trans. Am. Soc. Mech. Eng. **51**, 829 (1929)
10. Coates, W.M.: The state of stress in full heads of pressure vessels. Trans. Am. Soc. Mech. Eng. **52**: APM 52–12, 117–131 (1930)
11. Den Hartog, J.P.: Advanced Strength of Materials, 1st edn. McGraw-Hill, New York (1952)
12. Donath, M.: Die Berechnung rotierender Scheiben und Ringe. Springer, Berlin (1912)
13. Dubbel, H.: Taschenbuch für den Maschinenbau 15. Springer, Berlin/Heidelberg (1983)
14. Eraslan, A.N., Argeşo, H.: Limit angular velocities of variable thickness rotating disks. Int. J. Solids. Struct. **39**(12), 3109–3130 (2002)
15. Eraslan, A.N., Orçan, Y.: On the rotating elastic–plastic solid disks of variable thickness having concave profiles. Int. J. Mech. Sci. **44**, 1445–1466 (2002)
16. Faupel, J.H.: Engineering Design: A Synthesis of Stress Analysis and Materials Engineering. Wiley, New York (1964)
17. Fenner, D.N.: Engineering Stress Analysis: A Finite Element Approach with FORTRAN 77 Software. Ellis Horwood, Chichester (1987)
18. Fenner, R.T.: Finite Element Methods for Engineers. The Macmillan Press, London (1975)
19. Fenner, R.T.: Engineering Elasticity: Application of Numerical and Analytical Techniques. Ellis Horwood, Chichester (1986)
20. Ford, H., Alexander, J.M.: Advanced Mechanics of Materials, 2nd edn. Ellis Horwood, Chichester (1977)
21. Forsyth, A.R.: A Treatise on Differential Equations, 6th edn. Dover Publications, Mineola (1996)

V. Vullo and F. Vivio, *Rotors: Stress Analysis and Design*,
Mechanical Engineering Series, DOI 10.1007/978-88-470-2562-2,

© Springer-Verlag Italia 2013

22. Gamer, U.: Elastic–plastic deformation of the rotating solid disk. Arch.Appl. Mech. **54**(5), 345–354 (1984)
23. Gamer, U., Mack, W.: Thermal stress in an elastic–plastic disk exposed to a circular heat source. J. Appl. Math. Phys **36**, 568–580 (1985)
24. Genta, G.: Calcolo di Resistenza degli Organi Rotanti e dei Recipienti Cilindrici. Levrotto & Bella, Torino (1996)
25. Gibbs, C.W. (ed.): Compressed Air and Gas Data, 2nd edn. Ingersoll-Rand, Woodcliff Lake (1971)
26. Giovannozzi, R.: Studio delle Sollecitazioni Termiche in Dischi Conici, in Dischi a Spessore Costante e in Dischi a Profilo Arbitrario Mediante Scomposizione in Tronchi Parziali. L'Aerotecnica **30**(6), 308–315 (1950)
27. Giovannozzi, R.: Calculation, tabulation and uses of some functions occurring in the theory of the conical disc subjected to centrifugal and thermic stresses. In: Proceedings of II Congresso Aeronautico Europeo, pp. 30.1–30.12, Scheveningen (1956)
28. Giovannozzi, R.: Tavole per il Calcolo di Dischi Conici a Spessore sia Decrescente, sia Crescente verso l'Esterno, con Peso Specifico e Temperatura Variabili Lungo il Raggio. L'Aerotecnica **36**(5), 333–355 (1956)
29. Giovannozzi, R.: Costruzione di Macchine, 5th edn. Pàtron, Bologna (1980)
30. Gradshteyn, I.S., Ryzhik, I.M.: Tables of Integrals, Series and Products. Academic, San Diego (2000)
31. Grammel, R.: Ein neues Verfahren zur Berechnung Rotierender Scheiben. Dinglers Polytechnical Journal **338**, 217–219 (1923)
32. Güven, U.: Elastic plastic stresses in a rotating annular disk of variable thickness and variable density. Int. J. Mech. Sci. 34, 133–138 (1992)
33. Haddad, Y.M.: Mechanical Behaviour of Engineering Materials. Static and Quasi-Static Loading, vol. 1. Kluwer, Dordrecht (2000)
34. Heyman, J.: Rotating disks. Insensitivity of design. In: Proceedings of the Third US National Congress of Applied Mechanics, ASME, New York (1958)
35. Honegger von, E.: Festigkeitsberechnung von Kegelschalen mit linear veränderlicher Wandstärke. Dissertation, Zurich (1919)
36. von Honegger, E.: Festigkeitsberechnung von Rotierenden Konischen Scheiben. Zeitschrift für Angewandte Mathematik und Mechanik **7**(2), 120–128 (1927)
37. Juvinall, R.C.: Engineering Consideration of Stress, Strain and Strength. McGraw-Hill, New York (1967)
38. Kirsch, G.: Die Theorie der Elastizität und die Bedürfnisse der Festigkeitslehre. Zeitshrift des Vereines deutscher Ingenieure **42**, 797–807 (1898)
39. Klein, F.: Vorlesungen über die hypergeometrische Funktion. Springer, Berlin (1933)
40. Lamé, G., Clapeyron, B.P.E.: Mémoire sur l'équilibre intérieur des solides homogènes. Mémoires de l'Académie des Sciences **7**, 465–562 (1833)
41. Lamé, G.: Leçons sur la théorie mathématique de l'élasticité des corps solides. Bachelier, Parigi (1852)
42. László, F.: Geschleuderte umdrehungskörper im gebiet bleibender deformation. Zietschrift für Angewandte Mathematik and Mechanik ZAMM **5**, 281–293 (1925)
43. Lenard, J., Haddow, J.B.: Plastic collapse speeds for rotating cylinders. Int. J. Mech. Sci. **14**(5), 285–292 (1972)
44. Love, A.E.H.: A Treatise on the Mathematical Theory of Elasticity, 4th edn. Dover Publication, Mineola (1944)
45. Ludwik, P.: Elemente der Technologischen Mechanik. Springer, Berlin (1909)
46. Ma, G., Hao, H., Miyamoto, Y.: Limit angular velocity of rotating disc with unified yield criterion. Int. J. Mech. Sci. **43**(5), 1137–1153 (2001)
47. Manna, F.: Rotating discs of unconventional profile. Meccanica **3**(4), 1–9 (1968)
48. Manson, S.S.: The determination of elastic stresses in gas-turbine disks. NACA Report #871, 241–251 (1947)

49. Martin, H.M.: Rotating disks of conical profile. Engineering **115**, 1–3 (1923)
50. Mase, G.E.: Continuum Mechanics. McGraw-Hill, New York (1970)
51. Massa, E., Bonfigli, L.: Costruzione di Macchine. Voll. I e II, Masson Italia, Milano (1981)
52. Massonnet, C., Save, M.: Calcolo a Rottura delle Strutture: Strutture Spaziali, vol. II. Zanichelli, Bologna (1968)
53. Meissner, E.: Das Elastizitätsproblem für dünne Schalen von Ringflächen-Kugel-oder Kegelform. Physikalische Zeitschrift **14**, 343–349 (1913)
54. Millenson, M.B., Manson, S.S.: Determination of stresses in gas turbine disks subjected to plastic flow and creep. NACA Report #906 (1948)
55. Nadai, A.: Theory of Flow and Fracture of Solids. McGraw-Hill, New York (1950)
56. Odqvist, F.K.G.: Theory of creep under action of combined stress with application to high temperature machinery. In: Proceedings of the Royal Swedish Institute for Engineering Research, vol.141, p. 31. Stockholm (1936)
57. Prager, W., Hodge, P.G.: Theory of Perfectly Plastic Solids. Wiley, New York (1951)
58. Prosciutto, A.: Elementi di Costruzione di Macchine. Pàtron, Bologna (1952)
59. Rees, D.W.A.: Elastic–plastic stresses in rotating disks by von Mises and Tresca. Zeitschrift fur Angewandte Mathematik und Mechanik **79**(4), 281–288 (1999)
60. Reid, S.R.: On the influence of acceleration stresses on the yielding of disks of uniform thickness. Int. J. Mech. Sci. **14**(11), 755–763 (1972)
61. Ramberg, W., Osgood, W.R.: Description of stress–strain curves by three parameters. NACA Technical Note #902 (1943)
62. Saada, A.S.: Elasticity: Theory and Applications, 2nd edn. Krieger Publishing Company, Malabar (1993)
63. Saravanamuttoo, H.I.H., Rogers, G.F.C., Cohen, H.: Gas Turbine Theory, 5th edn. Pearson Prentice Hall, Harlow (2001)
64. Shames, I.H., Cozzarelli, F.A.: Elastic and Inelastic Stress Analysis. Taylor & Francis Ltd., Philadelphia (1997)
65. Skidmore, W.E.: Bursting test of rotating disks typical of small gas turbine design. Proc. Soc. Exp. Stress Analysis **8**(2), 29–48 (1951)
66. Smirnov, V.: Cours de Mathematiques Superieures, Tome III. Editions MIR, Moscow (1972)
67. Stanley, T.A., Lester, P.A.: Stresses in rotating disks at high temperatures. J. Appl. Mech. **13** (1), A45–A52 (1946)
68. Steele, M.C.: Partially plastic thick-walled cylinder theory. J. Appl. Mech. **19**, 133–140 (1952)
69. Sterner, S.C., Saigal, S., Kistler, W., Dietrich, D.E.: A unified numerical approach for the analysis of rotating disks including turbine rotors. Int. J. Solids. Struct. **31**(2), 269–277 (1994)
70. Stodola, A.: Dampf und Gas-Turbinen, 6th edn. Julius Springer, Berlin (1924)
71. Swift, H.W.: Plastic instability under plane stress. J. Mech. Phys. Solids **1**, 1–18 (1952)
72. Thompson, A.S.: Stresses in rotating discs at high temperature. J. Appl. Mech. **13**, A45–A52 (1946)
73. Timoshenko, S.P.: Strength of Materials. Part I and Part II, 3rd edn. D. Van Nostrand Co, Princeton (1956)
74. Timoshenko, S.P., Goodier, J.N.: Theory of Elasticity, 3rd edn. McGraw-Hill, New York (1970)
75. Tricomi, F.G.: Equazioni differenziali, 4th edn. Boringhieri, Torino (1967)
76. Ugural, S.C., Fenster, S.K.: Advanced Strength and Applied Elasticity, 4th edn. Prentice Hall, Upper Saddle River (2003)
77. Vivio, F., Vullo, V.: Elastic stress analysis of rotating converging conical disks subjected to thermal load and having variable density along the radius. Int. J. Solids Struct. **44**(24), 7767–7784 (2007)
78. Vullo, V., Vivio, F.: Elastic stress analysis of non-linear variable thickness rotating disks subjected to thermal load and having variable density along the radius. Int. J. Solids. Struct. **45** (20), 5337–5355 (2008)
79. Vivio, F., Vullo, L.: Elastic–plastic analysis of rotating disks having non-linearly variable thickness: residual stresses by overspeeding and service stress state reduction. Ann. Solid. Struct. Mech. **1**(2), 87–102 (2010)

80. Weiss, H.J., Prager, W.J.: Aero. Sci. **21**(3), 196–200 (1954)
81. Westergaard, H.M.: Theory of Elasticity and Plasticity. Wiley, New York (1952)
82. Wislicenus, G.F.: Fluid Mechanics of Turbomachinery. Dover Publications, Mineola (1965)
83. You, L.H., Long, S.Y., Zhang, J.J.: Perturbation solution of rotating solid disks with nonlinear strain-hardening. Mech. Res. Commun. **24**(6), 649–658 (1997)
84. You, L.H., Tang, Y.Y., Zhang, J.J., Zheng, C.Y.: Numerical analysis of elastic–plastic rotating disks with arbitrary variable thickness and density. Int. J. Solids. Struct. **37**(52), 7809–7820 (2000)
85. You, L.H., Zhang, J.J.: Elastic–plastic stresses in a rotating solid disk. Int. J. Mech. Sci. **41**(3), 269–282 (1999)
86. Zudans, Z., Yen, T.C., Steigelmann, W.H.: Thermal Stress Tecniques in the Nuclear Industry. Franklin Institute Research Laboratories/Elsevier, New York (1965)
87. Brebbia, C.A.: The Boundary Element Method for Engineers. Pentech, Plymouth (1978)
88. Calderale, P.M., Vivio, F., Vullo, V.: Thermal stresses of rotating hyperbolic disks as particular case of non linearly variable thickness disks. J. Therm. Stresses. (2012, in press)
89. Cook, R.D.: Concepts and Applications of Finite Element Analysis, 2nd edn. Wiley, New York (1981)
90. Güven, U.: Elastic plastic stresses in a rotating annular disk of variable thickness and variable density. Int. J. Mech. Sci. **34**, 133–138 (1992)
91. Zienkiewicz, O.C.: The Finite Element Method, 3rd edn. McGraw Hill, London (1977)

Index

V. Vullo and F. Vivio, *Rotors: Stress Analysis and Design*, Mechanical Engineering Series, DOI 10.1007/978-88-470-2562-2,

© Springer-Verlag Italia 2013

S

T

U

V

MIX
Papier aus verantwortungsvollen Quellen
Paper from responsible sources
FSC® C105338

If you have any concerns about our products,
you can contact us on
ProductSafety@springernature.com

In case Publisher is established outside the EU,
the EU authorized representative is:
Springer Nature Customer Service Center GmbH
Europaplatz 3, 69115 Heidelberg, Germany

Printed by Libri Plureos GmbH
in Hamburg, Germany